GREAT PLATTE VALLEY ROUTE

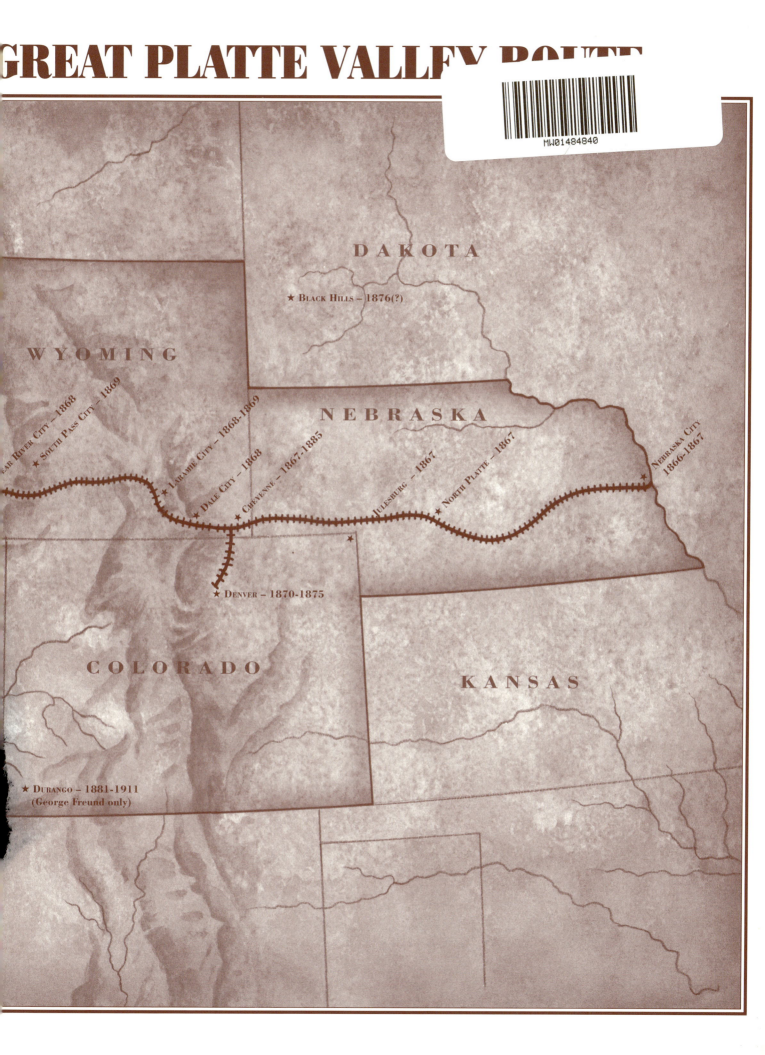

DAKOTA

★ BLACK HILLS – 1876(?)

WYOMING

NEBRASKA

BEAR RIVER CITY – 1868

★ SOUTH PASS CITY – 1869

★ LARAMIE CITY – 1868-1869

★ DALE CITY – 1868

★ CHEYENNE – 1867-1885

JULESBURG – 1867

★ NORTH PLATTE – 1867

NEBRASKA CITY 1866-1867

★ DENVER – 1870-1875

COLORADO

KANSAS

★ DURANGO – 1881-1911
(George Freund only)

MW01484840

FREUND & BROTHER

This work is dedicated to my wife,
"Red" Balentine,
my best friend, companion, and
typist for forty-nine years,

and to

the Sharps Collectors Association,
whose constant prodding and asking
"When is the book coming out?"
kept the author's nose to the grindstone.

FREUND & BRO.

PIONEER GUNMAKERS TO THE WEST

LONG RANGE RIFLES
FINE
GUNS MADE TO ORDR.

F. J. Pablo Balentine

The publication of the book was made possible, in part, by a grant from the Kinnucan Arms Chair of the Buffalo Bill Historical Center, Cody, Wyoming.

Library of Congress Card Number 95-75009
ISBN 1-882824-13-X trade hardcover,
     1-882824-14-8 deluxe limited edition

First Printing July, 1997
Manufactured in the United States of America

Another fine collector arms volume from
Graphic Publishers
Newport Beach, California 92660 USA

Acknowledgments

The production of a research work of this scope is impossible without the assistance of many fellow collectors and the dedicated individuals of public and private museums, libraries, and other institutions. They are the people still interested in the significant role played by firearms in the founding of our country and the development of the West. These individuals, from both the public and private sectors, are the lifeblood of research in the quest for the knowledge and history of firearms.

Research Sources

The American Rifleman
National Rifle Association
Washington, DC
E.G. Bell, Editor

Animas Museum
La Plata Co. Historical Society
Durango, Colorado
Robert McDaniel, Director

Autry Museum of Western Heritage
Los Angeles, California
James H. Nottage,
V.P. and Chief Curator

Bayerisches National Museum
Munich, Germany
Dr. Lorenz Seelig, Director

Butterfield & Butterfield, Auctioneers
San Francisco, California
Greg Martin

Buffalo Bill Historical Center
Cody Firearms Museum
Cody, Wyoming
Howard Michael Madaus, Curator

Denver Public Library
Denver, Colorado
Western History Department

Gun Digest
Articles by John Barsotti in the 1957 11th and 1958 12th Editions

Heritage Plantation of Sandwich Collection
Sandwich, Massachusetts
James Cervantes

Minnesota Historical Society
St. Paul, Minnesota
Jeffrey P. Tordoff, Cataloger

National Cowboy Hall of Fame
Oklahoma City, Oklahoma
Richard Rattenbury

Theodore Roosevelt National Park
Medora, North Dakota
Bruce M. Kaye, Chief of Interpretation

Smithsonian Institution
Washington, DC
Harry Hunter
Sarah Rittgers

Union Pacific Railroad
Omaha, Nebraska
Don Snoddy, Museum Director
Bill Kratville, Photographic Consultant

University of Wyoming
American Heritage Center
Cheyenne, Wyoming
Emmett D. Chisum, Researcher

Wyoming State Museum
Cheyenne, Wyoming
Jean Barnard, Senior Historian

Special mention is due the following individuals for their assistance and other valuable contributions:

John Barsotti, author and early researcher of the Freund Brothers
John Berthour, Frank W. Freund's grandson, who granted access to the Freund Family Collection
Robert Borcherdt, fellow collector and Freund Brothers researcher
Howard Michael Madaus, Curator of the Cody Firearms Museum

Robert L. Moore, Jr., M.D., friend, fellow collector, researcher, and owner of the original Sharps Rifle Company records
Frank Sellers, collector, researcher, and author of *Sharps Firearms*, whose assistance and constructive criticism guided me toward the completion of this work

Contributors

James O. Aplan	Fred Fellows	Robert Holter	Linton McKenzie
Larry Barrett	Gary Fellows	John Horning	Lance Peterson
Fred Borcherdt	Ken R. Freund	Doug Jahnke	Ron Peterson
Richard Bringer	Gary Gallup	Burke Johnson	Gary Roedl
Dennis Brooks	Bruce Graham	LeCleroq Jones	William B. Ruger
Robert T. Buttweiler	Charles Grimes	Gerald Kelver	Chris Schneider
David Carter	Tommy Haas	Richard (Dick) Labowski	David Tawney
Dennis Coberly	Richard (Dick) Hammer	William LaRue	Lloyd Tillett
Gerald Denning	Margaret Brock Hanson	Glen Marsh	Rex Thrower
Lloyd Dietrich	John Hartman	Greg Martin	Ed Webber
Bill Faust	Marion McMillan Huseas	Jerry Mayberry	R.L. Wilson
			Randy Wright

Photography Credits

Peter Beard and G. Allen Brown
Cecil's Photography
Susan Einstein

John Fox
Rudolph Hemminger
Greg Marrs
Jerry Mayberry

Richard Rattenbury
Charles Semmer
the Author

American Society of Arms Collectors

OFFICE OF THE PRESIDENT
WILLIAM A. GARY
P.O. BOX 1301
PRESCOTT, ARIZONA 86302 U.S.A.

October 16, 1996

To whom it may concern,

The American Society of Arms Collectors hereby endorses and takes great pleasure in recommending "FREUND & BRO., PIONEER GUNMAKERS TO THE WEST" by F.J. "Pablo" Balentine. The publications review board concurs that this treatise is well documented and highly informative. It will be the definitive work for historians and collectors interested in the Freund brothers and their work.

William A. Gary
President
The American Society of Arms Collectors

William H. Guthman, Chairman
Publications Review Board

April 30, 1997

The Cody Firearms Museum of the Buffalo Bill
Historical Center, through its Kinnucan Arms
Chair, has reviewed the initial manuscript of Mr.
F.J. "Pablo" Balentine's study *Freund & Bros., -
Pioneer Gunmakers to the West*, and found it to be
well researched and capably written. On the basis
of that review, Mr. Balentine was awarded a grant
for the completion of his project, which covered
the history of the Freund brothers during their
gunmaking and marketing endeavors from 1866 to
1910 in what would become the states of Colorado
and Wyoming.

It is my opinion, as curator of the Cody Firearms
Museum, that Mr. Balentine's research has resulted
in a monograph that is factual and informative.
His documentation indicates an extensive use of
national and local records, the latter including
the surviving papers of the Freund family.
Moreover, Mr. Balentine's writing style affords
the owner of his work enjoyable and easy reading.
Mr. Balentine's *Freund & Bros., Pioneer Gunmakers
to the West*, is a significant addition to the
scholarly works on firearms in the American West,
both for collectors and historians of the West.
It's definitive text will stand as the "bible" on
the subject for decades. I heartily recommend it
to all interested in our Western culture.

Sincerely yours,

Howard Michael Madaus
Robert W. Woodruff Curatorial Chair
Curator, Cody Firearms Museum

HMM:gps

**BUFFALO BILL
HISTORICAL
CENTER**

BUFFALO BILL MUSEUM
CODY FIREARMS MUSEUM
McCRACKEN LIBRARY
PLAINS INDIAN MUSEUM
WHITNEY GALLERY OF
WESTERN ART

P.O. BOX 1000 CODY, WYOMING 82414
(307) 587-4771

OFFICE OF THE CURATOR

Hartford, Conn. March 1, 1997 _____ 187_

TO WHOM IT MAY CONCERN:

The Sharps Rifle Company of Hartford and Bridgeport, Connecticut, manufactured one of the earliest commercially successful breechloading rifles in America. Following the Civil War, in which it passed a bloody baptism by fire, the Sharps made its way westward in the hands of frontier ranchers, as well as market- and buffalo hunters.

In the West, however, the Sharps rifle was found to have one serious shortcoming. After repeat firing, dirt or powder residue could build up in the chamber, preventing easy loading and ejection of cartridges. The Sharps Company apparently did not consider this a major defect.

But Frank and George Freund, of Cheyenne, Wyoming Territory, did, and with potential danger. With their patented camming-action breech-block, the brothers earned everlasting fame for their improvements to Sharps and other rifle actions.

As owner of the original Sharps Rifle Company records, I have delved in the Sharps-Freund relationship. F.J. Pablo Balentine's research adds much to the Sharps Company's story, and provides a fresh outlook into the Freund brother's firearms innovations and alterations.

It is my sincere belief that FREUND & BRO. PIONEER GUNMAKERS TO THE WEST is a valuable addition to the literature of single-shot rifles, and especially those made by Sharps.

R. L. Moore, Jr. M.D.
R.L. Moore, Jr. M.D.

OLD RELIABLE
(TRADE MARK)

R. L. MOORE, JR., M. D.
310 MAGNOLIA STREET
PHILADELPHIA, MISSISSIPPI 39350

ANTIQUE GUN COLLECTOR
AND OWNER OF THE ORIGINAL
Sharps Rifle Co. RECORDS

▲CFA.18   SEE CHAPTER TWELVE FOR DESCRIPTIONS   ▼CFA.14

▲CFA.07 (upper), CFB.19   ▼CFA.12

▲CFA.09 ▼DCB.04

▲CFB.24

▼CFA.16

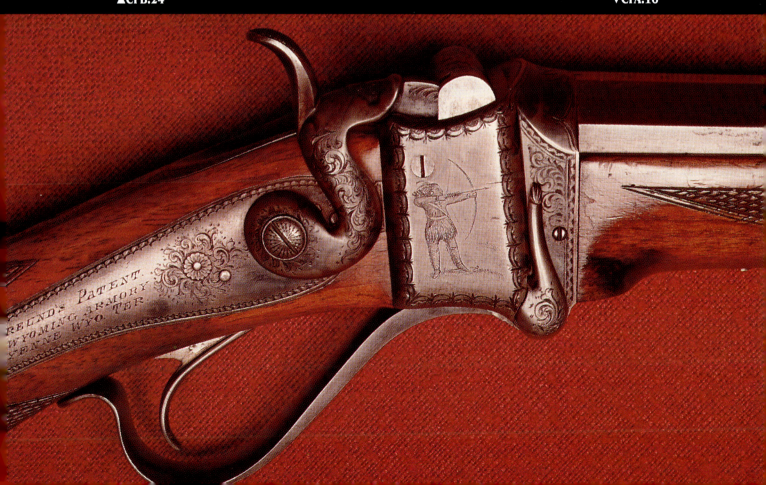

FREUND'S PATENT.
WYOMING ARMORY
CHEYENNE WYO. TER.

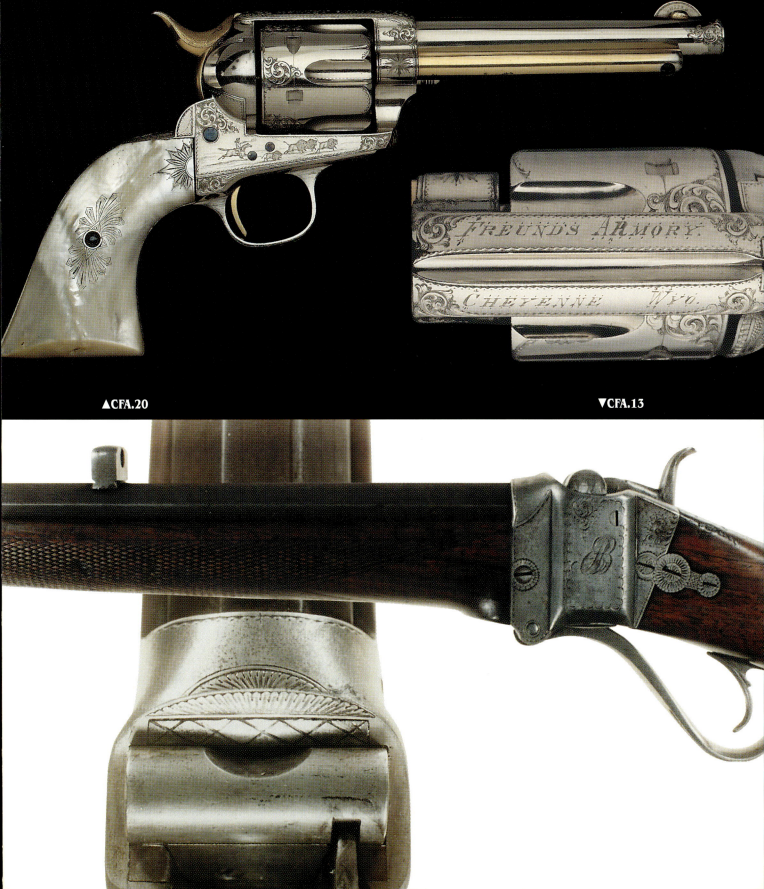

▲CFA.20

▼CFA.13

▲CFB.25

▼CFA.10

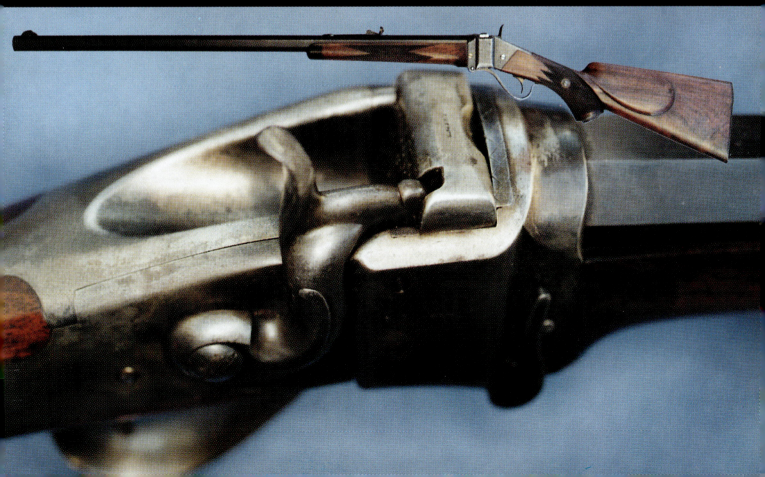

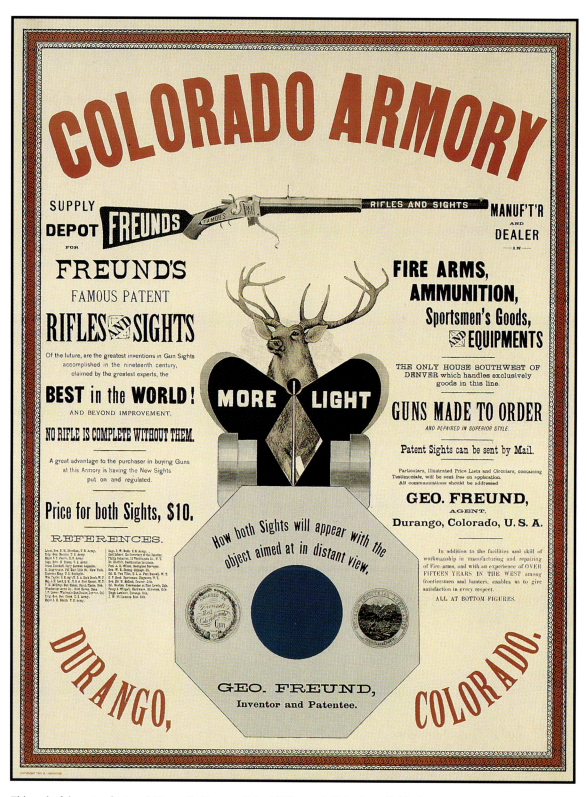

This colorful poster featured "Freund's Famous Patent Rifles and Sights", available from George Freund's Colorado Armory in Durango. (Courtesy Richard Labowski)

Table of Contents

CHAPTER ONE: The Early Years 1

CHAPTER TWO: The Railroad Years 9

CHAPTER THREE: The Denver Years 35

CHAPTER FOUR: The Cheyenne Years 51

CHAPTER FIVE: The New Jersey Years 101

CHAPTER SIX: The Durango Years 113

CHAPTER SEVEN: The Freund Patents 127

CHAPTER EIGHT: The Freund Rifles 205

CHAPTER NINE: The Freund Sights 225

CHAPTER TEN: The Court Cases 243

CHAPTER ELEVEN: The Freund Catalogs 251

CHAPTER TWELVE: The Surviving Specimens 309

APPENDIX I: Endorsements 357

APPENDIX II: Freund Family Tree 380

INDEX: 381

CHAPTER ONE

THE
EARLY
YEARS

*T*he opening of the great American West during the first three quarters of the nineteenth century saw a rapidly-developing need for hunting and defensive weapons, and especially firearms. Brothers Frank and George Freund of Colorado and Wyoming, along with the Brownings (father and son) of Utah, were prominent among the pioneer gunsmiths and gunmakers to the Western frontier.

The Freund brothers came west with the Union Pacific railroad, setting up gunsmith shops in tent cities that sprang up at the end of the tracks. Like thousands of other footloose men after the Civil War, they followed the shining steel rails in their march across Nebraska, Wyoming, and Utah.

One might well wonder how the Freunds came to have the talent to make and repair firearms, while other emigrants labored under the elements grading and laying track and rail, often under threat from marauding Indians. Perhaps a study of their background will give us a clue as to why they chose gunsmithing as their career, and how they became so accomplished in that chosen field.

The old German surname *Freundt* or *Freund* derives from the German word *freund*, and literally means "friend." The earliest recorded occurence of the surname Freund is found in the *Wurtembergishes Urkandenbuch*, and lists one Ulrich Freundt in the year 1281. (The final "T" frequently was added to names due to German regional dialects, and differences in pronunciations.)

On the following page is a list of gun makers named Freund who lived and operated in Germany from the advent of the flintlock mechanism.[1] (The dates shown are the approximate dates that they built guns.)

Carl Freund; Furstenau, *circa* 1730-1760.

Carl Freund; Furstenau, *circa* 1790-1810.

Christian Freund; Furstenau, Prussia, 1793-1811.

Christoph Wilhelm Freund; Furstenau, *circa* 1740-1780.[2]

Georg C. Freund; Furstenau, 1700-1730; maker of left-hand flintlocks in about 1700.

Jean Louis Freund; Strassburg, 1773-1793 (from Furstenau). Known to have worked in Strassburg in 1793.

John Christoph Wilhelm Freund; Furstenau, *circa* 1782-1800.

Johannes Freund; Suhl weapons manufactory, 1846.

Ludwig Freund; Bern, 1793; may be Jean Louis Freund.

In the Munich Museum there are at least twenty-three rifles made by

A silhouette drawing of the Freund brothers' parents, Wilhelm and Maria (Sitze) Freund, done in Heidelberg about 1830. Early engravers picked up extra money during their travels doing these portrait silhouettes. (Freund Family Collection)

the early Freund gunmakers (*see* examples on pages 3 and 4).[3]

Too, in the literature of early German gunmakers one finds the names of Freund and Furstenau (an early German city where the Freunds worked) mentioned together regularly.

Into this heritage the Freund brothers were born at or near Heidelberg,

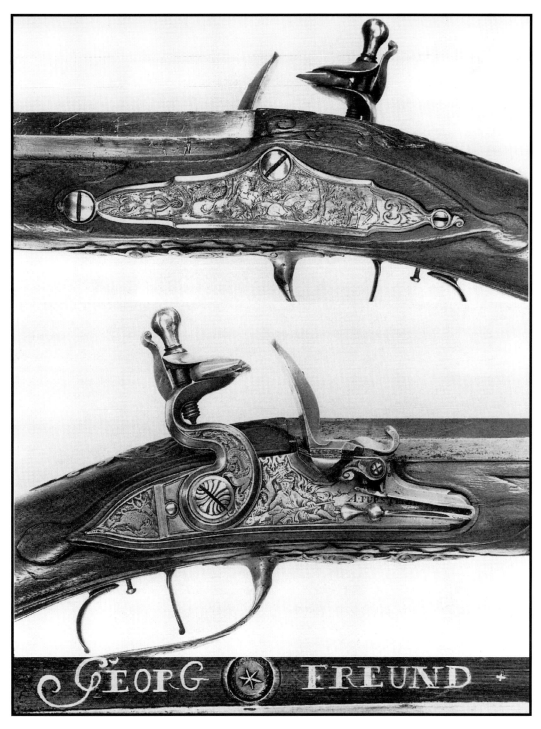

A carved fullstock flintlock rifle, circa 1730. Top of breech inlaid in brass letters, "GEORG * FREUND". Dolphin engraved on triggerguard; hunter with rifle reclining under tree, dog, rabbit, boar, "A FURSTENAU", on lockplate; reclining goddess Diana, dogs, and stag on left sideplate. Overall length is 43.7 inches; octagonal barrel is 28.75 inches long; caliber .28, rifled with 7 grooves. (Munich Bavarian Museum)

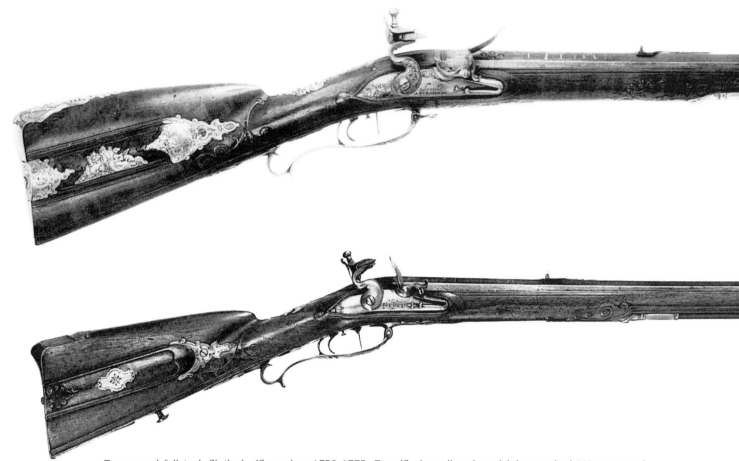

Two carved fullstock flintlock rifles, *circa* 1750-1775. Top rifle has silver barrel inlay marked "1"; engraved lockplate marked "C/Freund/A FURSTENAU". Overall length is 39.75 inches; octagonal barrel 24.8 inches long; caliber .55, rifled with 7 grooves. Bottom rifle has silver barrel inlay marked "2"; engraved lockplate marked "C/FREUND"; silver thumbplate on patchbox. Barrel top flats inlaid "C Wilhelm Freund A FURSTENAU". Gold plated brass mountings. (Munich Bavarian Museum)

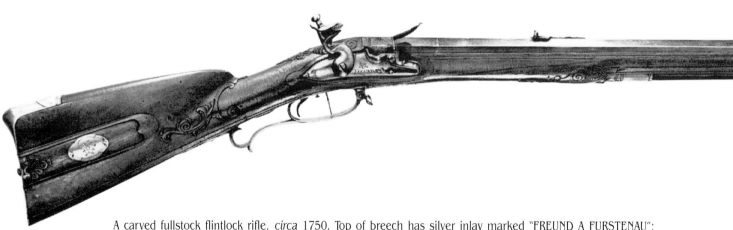

A carved fullstock flintlock rifle, *circa* 1750. Top of breech has silver inlay marked "FREUND A FURSTENAU"; engraved lockplate marked "C/FREUND". Cheekpiece inlaid with 8-ray star, of bone; brass mountings engraved with blossoms and leaf garlands. Overall length 42.72 inches; octagonal barrel 24.8 inches long; caliber .55, rifled with 7 grooves. (Munich Bavarian Museum)

Germany— Frank in 1837 and George in 1840— to Wilhelm and Maria (Sitze) Freund. There were six children in the family: Frank William, George, Elsie, William, Louise, and another sibling that this author has been unable to trace. Louise was married to Baron Esmond de Forester, Captain-Adjutant-Major of the 70th Territorial Regiment, who had been decorated with the military medal Chevelier of the Legion of Honor of the French Army. Louise died at Handfeuer Waffen, Deutches Kunstverlag, at the age of 39 years on September 15, 1885.[4]

Early craftsmen in Germany, as in other European countries, belonged to guilds, associations of merchants and artisans intended to promote the interests and welfare of their members. Craft guilds were formed along the lines of particular trades, such as gunsmiths, and were social as well as economic in their scope.

The craft guilds were composed of three levels of membership based on accomplishment: apprentices, journeymen, and master craftsmen. Usually a youth would begin his apprenticeship between the ages of fourteen and nineteen. Although the length of apprenticeship varied widely depending on the area, and the boy's ability, seven years was a common period to serve in that initial phase of one's career. Often, apprenticeships were limited (or more readily available) to boys with family ties to guild members.

These highly skilled artisans were understandably proud of their products, many of which were stamped or engraved with their makers' names and location. The master German gunsmiths preferred to use marks consisting of figures and initials, to which were added inspection or control marks of the guild, and often the year of manufacture. From this age-old custom Frank and George Freund inherited the habit of profusely marking their work.

At the time that Frank, the eldest son, came into the apprenticeship program the guilds were slowly dying out. But the system of apprenticeship still was the only way for a young man to gain entrance into a respected trade. Coming from a long line of gunmakers, Frank mastered the skills of metalworker, woodworker, and engraver as a young man. Like many others before him he journeyed to Austria, France, and England, to study under other masters of gunmaking, and there gained valuable experience.

After returning to his native Germany Frank followed another familiar pattern and joined the military, and around 1856 was a member of the 75th Infantry Regiment, Reserve Battalion, in Bremen. Later he was to tell his youngest daughter, Jeanette, that there always would be a place in the German Army for anyone named Freund.

Showing the restless spirit which was to mark his life thereafter, twenty-

Frank W. Freund in full dress uniform (including *Pickelhaube* helmet), while a member of the 75th Infantry Regiment, Reserve Battalion, in Bremen, Germany *circa* 1856. (Freund Family Collection)

year-old Frank William Freund emigrated to America, arriving in New York in 1857. Full of confidence and courage, for the next few years Frank Freund further developed his gunmaking skills while working for the Remington Arms Company at Ilion, New York.

Apparently Frank left Remington with the company owing him back wages. A letter still in the possession of the Freund family and written on July 31, 1910 by John V. Schmidt, an old family friend, states:

> *P.S. Perhaps you can find out the address of Eliphalet Remington from his brother. I remember that he still owes you some money. If it does not do any good, all you lose is the time you are going to spend. Among gentlemen there should not be outlawing of honest debts.*[5]

True to his military heritage, Frank W. Freund served in the Union Army during the last eighteen months of the Civil War, as a member of the First New York Engineers. At some later time he was joined in America by his younger brother, George. Unfortunately, information about the Freund brothers' early years together in the New World is scarce, as very little of it was recorded during that period of their lives.

Frank W. Freund in the uniform of a Federal soldier, as he appeared during the American Civil War *circa* 1864-65. (Freund Family Collection)

1. *Der Neue Stockel*, Volume I. Eugene Heer, Herausgeber: Journal-Verlag Schwend GMBH Schwabisch Hall, 1978.
2. The attribution of "C. Freund" signed weapons is very difficult. It appears that Carl, Christoph Wilhelm, and Georg were apprenticed in the same workshop.
3. Handfeuer Waffen, *Kataloge des Bayerischen Nationalmuseum,* Band XIX, Deutches Kunstverlag, 1988.
4. Freund Family Collection.
5. *Ibid*. Eliphalet Remington II, whom Frank W. Freund had worked for, died in 1861. The company declared bankruptcy in 1886, and was reorganized under new ownership. Apparently John V. Schmidt was unaware of the above; it is unlikely that a claim for back wages would have been paid under the circumstances, especially after fifty-plus years had elapsed.

The Freund brothers' "Wyoming Armory," in Cheyenne, Wyoming Territory. From the guns pictured, this photograph would have to date from about 1874–1876. Frank W. Freund stands third from left; George Freund is fifth from left. (Freund Family Collection)

CHAPTER TWO

THE
RAILROAD
YEARS

*F*ollowing the end of the Civil War in 1865, the opening up of the great American West brought Manifest Destiny — America's mid-19th century policy of expansion between the Atlantic and the Pacific — once again to the forefront. With it came the railroads, and the resultant mining camps and ranches, as well as the Army. The chief function of the latter was to keep the Indians in check, and to allow the whites free movement throughout the land.

Frank W. Freund, like thousands of other veterans released from their wartime bonds, turned his face westward. Unlike many of his contemporaries, however, especially those from the South, Frank was a skilled craftsman with a little money to invest in an inventory.

During 1866 the Union Pacific railroad was off to a grand start, building a line across the vast prairies of Nebraska, Wyoming, and Utah, eventually to meet up with the Central Pacific, which was building eastward. They would join on May 10th, 1869, at a lonely spot in the Utah desert called Promontory Point.

By the sixteenth of June 1866, Frank W. Freund had established a gun shop at Nebraska City, on the Missouri River in far-eastern Nebraska Territory. It marked the first recorded appearance of a Freund on the Western scene.

On the following pages are chronological listings published in newspapers of the end-of-track towns, as the Freund brothers followed the railroad westward.

Nebraska City, Nebraska Territory

Freund, F.W., Nebraska City gunmaker. (Nebraska *Statesman*, Nebraska City, N.T., 16 June 1866, page 2).

F.W. Freund, Nebraska City, manufacturer and dealer in arms and ammunition. (Nebraska City *News*, 4 January 1867, page 4).

F.W. Freund and Brother, Nebraska City, gunsmiths and dealers in arms and ammunition. (Nebraska City *News*, 19 January 1867, page 3).

Freund and Brother gunsmiths, guns made to order. (Nebraska City *News*, 28 January 1867, page 3).

Freund and Brother, Nebraska City, gunsmiths have a branch shop at North Platte. (Nebraska City *News*, 15 May 1867, page 3).

F.W. Freund, Nebraska City, of Freund and Co. left for Julesburg. (Nebraska City *News*, 22 July 1867).

F.W. Freund, Nebraska City, of Freund and Brother, is leaving for Julesburg. Mr. Picard will attend to the business in his absence. (Nebraska City *News*, 27 July 1867, page 1).

F.W. Freund returned from Cheyenne to settle up his old business. (Nebraska City *News*, 30 December 1867, page 3).

The following is an example of one of the Freund & Brother advertisements that appeared in the frontier newspapers of the day, this one in the Nebraska City *News* of January 19th, 1867:[1]

Sportsmen's Depot

F.W. Freund and Bro., Gunsmiths
Sporting Apparatus, Ammunition, etc.
Eley's wire cartridge, Eley's wad, Waterproof caps
Agents for DuPont's celebrated powder
Repairing done in a workmanlike manner
Short notice

It is interesting to note that for the first time Frank included the words "and Bro." in the above ad. Both Frank and George Freund were fine workmen, skilled in their trade, but it was Frank's talent as an inventor that would make the Freund name so well-known throughout the pioneer West.

The older of the two brothers, Frank was the leader and superior craftsman, possessed of a keen eye and skilled at design, engraving, metalsmithing and woodworking. While Frank always was working on a new invention or patent idea, or off on a trip somewhere, George tended to remain more in the background. He seemingly was more content in the roles of shopkeeper and merchandiser, at least until the 1880s.[2]

At the time, Nebraska City was a lively town on the Missouri River.

Situated at the eastern end of the emigrant and wagon roads that stretched across the empty prairies, the town was headquarters for a number of freight outfits that included the famous Russell, Majors & Waddell. Here steamboats unloaded their cargoes, to be transported overland on plodding bull trains to the mining camps, ranches, and Army posts of the new land. But the same railroad that built Nebraska City also ended its brief glory, as the tracks followed the sun ever-westward.[3]

North Platte, Nebraska Territory

North Platte was first in a long line of infamous "Hell-on-wheels" towns, the wide-open, anything-goes, end-of-track camps that invariably sprang up to service the lusty needs of rail-layers and roustabouts — a notoriously hard lot.

These places attracted the rankest assortment of rough and rowdy camp followers ever assembled on the Western frontier, who went at it twenty-four hours a day, including Sundays. Card sharks, soiled doves, shivs, drunks, Bowery toughs, flim-flammers of all creeds and persuasions — the flotsam and jetsam of humanity plied their trades openly from within canvas tents or rude wooden false-fronts. When the rails moved on so did they, and they never looked back.

The end-of-track reached North Platte on October 24th, 1866. At the time, the town's population already stood at about 1,000 souls, but by the time the construction workers began to gather for the winter that number rose to around 5,000. Soon, too, the rough element converged to prey on those assembled to build the Union Pacific in its race to Promontory Point.

General Jack Casement and his brother, Dan, held the contract to lay the rails, and they had four rail cars specially built to serve their operations. The cars held the Casement brothers' office, as well as kitchens, dining halls, and bunkhouses for their crews. In addition, the cars carried a thousand rifles in racks located everywhere it was convenient. A glass-plate picture exists made by railroad photographer Andrew J. Russell, who had served as a captain in the U.S. infantry during the late Civil War. It pictures General John S. Casement and his outfit circa 1867-1868 along the U.P.R.R. tracks. On the roofs of the cars are tents and wooden sheds built by the men, apparently to escape the summer heat and the vermin. After having been detached to make photographs for the U.S. Military Railroad Construction Corps during the war, Russell now was working as a company man for Major General Grenville Dodge, chief engineer for the railroad.

Despite the Freund brothers' advertising themselves as "pioneer armorers to the Union Pacific Rail Road," a search of records fails to reveal any evidence that the Freunds ever were actually in the employ of the U.P.R.R. To set the record straight, herewith is presented a letter from Wm. G. Murphy, U.P.R.R. director of public relations, dated December 7, 1949:

> *The names of Frank W. and George Freund do not appear anywhere in any of these records (i.e., Chief Superintendent General S. Reed's, the Casement brothers' or another contractor's account books). We know that the Casement brothers' construction train was practically a self-contained mobile combination of office, sleeping quarters, commissary, workshop, and arsenal. Its equipment included a thousand rifles, principally U.S. Army Springfields; the men were trained and drilled in defense against attacks of the hostile Indians. One of the cars in the construction train was fitted up as a blacksmith shop. It seems likely that repairs of the weapons in the Casement equipment were made by their own skilled men. The small arms, revolvers, and pistols (private property of the individuals) were repaired by the tradesmen making that their regular business.*
>
> *In the absence of any positive evidence in our records to establish the fact, I believe I can say conclusively that the Freund brothers had no official or employee relationship with the Union Pacific company.*

However, what with all the guns being carried by the 5,000 to 10,000 tracklayers and other workers in these rough-and-tumble trains and towns, certainly there was an ample supply of gunsmithing work available to the Freunds, in addition to a considerable trade in ammunition and new and used firearms. Of interest is the fact that the brothers must have been so busy during this period that Frank did not have the time to build fine guns, or to apply for any patents, until the Freunds had settled in at Denver during the early 1870s.

The Freund's branch shop in North Platte was short-lived, as the town itself was literally dismantled and transported to the next end-of-track.

Julesburg, Colorado Territory

The Union Pacific reached Julesburg on June 25th, 1867, and before the end of July most of the town that had been North Platte was reassembled and in operation in the new territory of Colorado. This latest camp was even larger and more vice-ridden than the first had been.

Here in Julesburg a certain clique of gamblers and other shady characters decided to seize control of the town, and refused to pay the Union Pacific for the lots they occupied. So the U.P.R.R. wired General Jack Casement, and told him to take as large a force as he deemed necessary to clean up the situation. Casement hand-picked two hundred of his brawniest trackmen, armed them, loaded them onto a train, and steamed eastward. On confronting the gang leaders, and their refusal to move, Casement ordered his men to open fire and not to care who they hit. Before Casement left, the surviving rebels were begging him to let them pay for the quarters they occupied in Julesburg. Others of their number were allowed to stay on for free, forever occupying unmarked graves on the edge of town.

Photographer Russell managed to fix in time the real character of Jack Casement, who to his men seemed "seven feet tall and tough as nails." Along with his Cossack hat and bullwhip, the portrait captured the strength and psychological stature of the general, who in reality stood barely five feet tall.

By the middle of August 1867, the Union Pacific was advertising that it was ready to begin selling lots in yet another new town. The Freund's stay in Julesburg was of short duration, as they beat even the first trains to Cheyenne.

Cheyenne, Dakota Territory

The Union Pacific tracks reached Cheyenne on November 17, 1867, and the first train that reached the town was made up of flatcars piled high with the numbered sections of buildings that had been dismantled and moved from Julesburg. But Frank and George Freund already had been in Cheyenne two months by then, as witnessed by their ad that appeared in the first newspaper in town, N.A. Baker's Cheyenne *Daily Leader*, on September 19th:

Freund & Bro.
Manufacturers and Importers of
GUNS, PISTOLS, AND CUTLERY

East side of Eddy Street, Cheyenne, Dakota

Sporting apparatus and all kinds of Fixed
and Loose Ammunition. Double, Single Barrelled
Rifles and Shot Guns made to order.
Every kind of repairing done with
NEATNESS AND DISPATCH

Agents for
E.I. DuPont & deNemours & Co's.
celebrated
SPORTING & MINING POWDER!

Cheyenne was North Platte and Julesburg all over again, but even bigger, more vigorous, and more lawless; and, according to the Union Pacific, it was the gambling capital of the world. There were six bonafide theaters, and at least seventeen "variety halls," which usually meant a saloon, theater, and fancy bordello combined under one roof. The *Daily Leader* ran a regular column under the standing head "Last Night's Shootings," and it is said that a local magistrate named Colonel Luke Martin levied a ten-dollar fine on any man who drew a gun on another person within the Cheyenne city limits, regardless of "whether he hit or missed."

This was pioneer Dakota Territory, and its capital, Yankton, was far away. Law enforcement was practically non-existent.

Cheyenne's population soared to ten thousand following the arrival of the steel rails. Even legitimate firms were doing exceptionally well. Probably the Freund brothers could not have been in a better area for the success of their gun repair and sporting goods business.

The first known photograph (*see* page 8) of a Freund Brothers store is of the one in Cheyenne, the only railroad town on the U.P.R.R. main line where Frank had a shop (or "armory," as he chose to call it) continuously throughout the entire period he lived in the West. Freund's Wyoming Armory was located on Eddy Street (later the name was changed to Pioneer Street).

During those early times in Cheyenne, men lived and carried on their businesses in whatever they could find for shelter— be it dugout, tent, shanty, or canvas-and-frame shack. Several thousand such primitive structures were scattered over the townsite in the first months. Some of them must have been fairly large, in order to accommodate the dance halls, gambling halls, and saloons (and the attendant crowds of humanity that piled into them for entertainment on paydays). Like many boomtowns that mushroomed, in time Cheyenne's crude, slovenly look faded as more substantial and permanent buildings rose.

But whatever it may have lacked in appearance, early Cheyenne made up for in liveliness. As in all railroad terminus towns, a lawless and often violent way of life prevailed among a considerable segment of the population. Due to that problem, and the Indian danger out along the advancing rails, most men carried or at least owned a gun of some kind.

No doubt the Freund brothers did a good business selling arms and ammunition, and repairing firearms, even at this early date in Cheyenne. They were listed among the first nineteen dealers and jobbers of the newly-organized Winchester Repeating Arms Company of New Haven, Connecticut. In addition to the latest brass-frame Model 1866 Winchester, the brothers likely

The impressive Dale Creek Bridge in Wyoming, at Mile 550. Instead of a more economical and secure earthen fill, this unsafe bridge, 700 feet long and 126 feet above the creek bed, was built. It swayed dangerously in modest winds, and passing trains were slowed, despite guy wires and cables having been attached to diminish the sway. The Freund brothers had a gun shop at Dale City, on the rails between Cheyenne and Laramie, for a short period of time. (Photograph by William Henry Jackson; courtesy Union Pacific Railroad Museum Collection)

also handled many other types of longarms and handguns, and repaired everthing from muzzleloaders and shotguns to new breechloaders and repeaters like the Henry, Winchester, and Spencer. The most popular revolvers still were the big Colt and Remington percussion models.[3]

But the tracks continued moving ever-westward. By March of 1868 the winter weather had started to break, and the rails were snaking their way up and over the Black Hills (not the Black Hills of present-day South Dakota, but those to the west of Cheyenne). By the last week of April the Union Pacific had worked its way up to Sherman, at 8,242 feet the highest point on the U.P.R.R. Beyond Sherman, the railroad continued pushing westward across Wyoming.

Dale City, Dakota Territory

Just a few miles beyond Sherman the Union Pacific had built the Dale Creek bridge, 126 feet above almost-dry Dale Creek. The 700-foot-long structure was guyed with ropes and wires against the ever-present Wyoming winds and the shock of rolling trains (which of necessity were obliged to slow down when crossing). The bridge was completed on April 27th, 1868.

The sole mention this author has found of the Freund brothers' presence at Dale City, was an advertisement that ran in *The Sweetwater Mines* on May 22nd, 1868. The ad mentioned the Freund's store in Cheyenne, also a branch on Main Street, Dale City, Dakota Territory (*see* page 26). Certainly it was a short-lived location, as the head-of-track reached Laramie City on June 18th. Laramie was a division point on the U.P.R.R., and as such was a more substantial town that would continue to grow.

Laramie City, Dakota Territory

Laramie City was destined to live the high-life of a terminus town for all of three months. There, Freund & Bro. again set up shop, selling guns, pistols, and ammunition, and doing repairs. Other than their Cheyenne Armory, the Freund's shop in Laramie City would operate longer than any other they had along the Union Pacific route.

The Laramie shop was still in operation during the summer of 1869, when Freund & Bro. advertised their "Sweetwater Armory" there. Fortunately, a good photograph of that store exists today.

There have been reports that the Freund brothers also had gun shops in Benton, Green River, or perhaps Bryan City, Wyoming, but the author has been unable to find evidence to confirm this. Doubtless if the brothers had a presence in those towns, it was only in the form of an agent who received orders or guns for

The Freund's original gun store at Laramie City, W.T. When the brothers first arrived with the railroad at Laramie City in early 1868, it was in Dakota Territory. By July 25th of that year — and the time this photo was taken — it had become Wyoming Territory. (Photo by A.C. Hull, 1868; courtesy Union Pacific Railroad Museum Collection)

Frank W. Freund, circa 1866, when he was about thirty years old. (Freund Family Collection)

repair, who then shipped them on the railroad back to Laramie City or Cheyenne.

Eastern excursionists were then beginning to flock over the Union Pacific line, caught up in the exitement of "pioneering," and they found it great sport to shoot buffalo from the safety of a train. Many were equipped with shiny new Ballard or Sharps rifles, eager for a shot at the Western buffalo, antelope, mountain deer, and even the grizzly bear found along or nearby the railroad right-of-way. Finding themselves in the right spot at the right time, Frank and George Freund must have done a great business supplying the tourists with sporting ammunition and other necessities.

Still, the Casement brothers and their track-laying crews continued pushing westward across Wyoming. The next boomtown in which we find the Freund brothers is Bear River City, also known as "Beartown."

Bear River City, W.T. was one of the "Hell-on-wheels" towns along the Union Pacific. It was located on White Sulphur Creek, 965 miles west of Omaha, near the present Utah-Wyoming border. Note the Freund's large rifle-sign over the street, in the middle of the photo, above. (A.D. Russell photo, 1868)

Bear River City, Wyoming Territory

Before the track-layers reached Bear River City in October of 1868 it had been a tie hack and loggers' camp; afterward, it was considered by many as the worst Hell-on-wheels town of them all.

Soon Bear River City was home to 2,000 people, and 140 buildings were erected there seemingly overnight. It boasted more than its share of riffraff and hard cases, plus a newspaper, the *Frontier Index,* published by one Leigh Freeman, a militant enthusiast of vigilante justice.

Editor Freeman was early run out of town by the rough crowd, and his press was broken up by some two hundred track graders whose friends were incarcerated in the Beartown jail. The vigilantes retaliated, and their first burst of rifle fire killed seventeen or eighteen of the graders. The fight quickly ended with the graders retreating back to their camp, and "peace" of a sort returned to the town.

The end came soon for Bear River City, too, as the end-of-track moved past the town, emptying it of its roistering population. With all of the excitement and shooting, Bear River City of course had been home to a Freund & Bro. gun shop, with their trademark rifle-sign above one of its board-and-canvas buildings. But the brothers' stay there doubtless was a short one, as the railroad moved westward into Utah Territory. When the U.P.R.R. failed to put even a switch or siding at Bear River City, its people moved on, and within a year the town did not exist at all.

Ever following the tracks of the advancing railroad, the Freunds' next logical stop would have been Ogden, Utah Territory. But there was a major drawback in the town in that highly-regarded Morman gunsmith Jonathan Browning had settled in Ogden in 1852, and his gun shop was well established (with fourteen-year-old John Moses Browning working in his father's shop part-time) by the time the U.P.R.R. arrived in 1869. So the Freund brothers instead moved on to Salt Lake City, and to Corinne, a few miles to the north.

Salt Lake City, Utah Territory

The Union Pacific had hoped to run a line around the south end of the Great Salt Lake and through Salt Lake City, but had to give up on the idea as it would have added seventy-six miles of track and many steep grades at substantial extra cost. So Salt Lake City was bypassed to the north by the rails, just as Denver had been earlier, and the directors had the unpleasant task of breaking the news to Brigham Young that his city— his monument— would not be on the main line of the railroad.

Not surprisingly, the Mormon leader was furious, and threatened to throw his support to the competing Central Pacific if they would build their line along the southern route. But the C.P.R.R. agreed with the Union Pacific, and Young finally had to compromise for a branch line serving Salt Lake City. Thus was the Utah Central Rail Road built, connecting the capital city with Ogden on the main line. Construction of the spur was completed on the tenth day of January, 1870.

Freund & Bro. "Pioneer Gun Store" in Salt Lake City, Utah Territory. Frank W. Freund is holding the shotgun in the center of the photograph, which probably was taken in 1868 or 1869. (Freund Family Collection)

By then the Freund brothers were already established in Salt Lake City, advertising the following in the *Salt Lake Directory and Business Guide* for 1869:[4]

———◆———

We are Agents for the
WINCHESTER PATENT REPEATING RIFLES AND CARBINES.
For the whole west.

The rifles are capable of firing eighteen
times in succession without reloading, and capable of being fired twice in one
second. Also for the
LEE FIRE ARMS COMPANY,
E.I. DUPONT DE NEMOURS & CO'S.
CELEBRATED SPORTING AND MINING POWDER, etc.

———◆———

In the Freund Family Collection there is a copy of *Mackey's Masonic Ritualist or Monitorial Instructions*, inscribed:

With Fraternal Regards
to
F.W. Freund
by
Th. Schenk
Salt Lake City
December 25,
1868

so perhaps Frank Freund was already established in Salt lake City as early as Christmas of 1868. It is fortunate that a good photograph exists of the Freunds' Salt Lake City gun shop, which was located on East Temple (now Main) Street between First and Second Streets South. An advertising broadside from that time and place is reproduced here, too (*see* page 24).

Corinne, Utah Territory

Corinne was a town of an entirely different character than the Mormon capital. Situated northwest of Ogden and just twenty-eight miles from Promontory Point (where the east- and westbound rails joined), Corinne did not lack in saloons or other rowdy entertainment for the track crews. In a March 1869 issue of the local newspaper a reporter described the town as "built of canvas and board shanties. The place is fast becoming civilized, several men having been killed there already; the last one was found in the river with four bullets in him." The same reporter ventured that between Promontory Summit and Brigham City, thirty-six miles to the east, there were three hundred whiskey shops, all "developing the resources of the territory. There are many heavy contractors on the Promontory, but the heaviest firm I have heard of is named 'Red Jacket' [whiskey]. I notice nearly every wagon that passes has a great many boxes with this name."

The pioneer Western photographer, William Henry Jackson, also passed through Corinne in 1869, and left us with two pictures of the town's dusty main street. Prominent is the familiar wooden rifle-sign of the Freund brothers' business, set up in front of the tent shop of John Kupfer, watchmaker and jeweler.

Obviously, a number of the Freunds' shops during this period were railroad branches of their main Cheyenne Armory, and were operated only briefly at the head-of-track towns. Perhaps Frank and George Freund even stood among

Freund brothers' gun shop located at Corinne, Utah Territory. John Kupfer's jewelry store also housed a hardware store in addition to the gun shop. (Photo by William Henry Jackson)

the throng at Promontory Point on May 10th, 1869, witnesses to the excitement as steel rails finally joined the continent. As the golden spike was ceremoniously driven home by directors of the U.P.R.R. and the C.P.R.R., the word "Done!" was flashed eastward by telegraph, and a great cry rang through the assembled crowd.

With the completion of the railroad came the disappearance of many of the towns which had sprung up and flourished briefly during the construction days. Other towns settled down to orderly growth; some became cities. The track gangs followed new excitements, such as the gold rushes in Montana and Idaho Territories, and many businesses followed them.

Once again the Freund brothers were seeking yet another location for their gun shop, perhaps with the thought of a permanent place in a growing town.[5] This time, however, they turned their eyes eastward.

A view looking toward the other end of Corinne's Main Street. The familiar rifle-sign of Freund & Bro. can be seen over the street in the left background. (Photo by William Henry Jackson)

The Sweetwater Mines.

NATIONAL MINERAL LAND LAW.

Be it enacted by the Senate and House of Representatives of the United States of America, in Congress assembled:

SECTION 1. That the mineral lands of the public domain, both surveyed and unsurveyed, are hereby declared to be free and open to exploration and occupation by all citizens of the United States, and those who have declared their intention to become citizens, subject to such regulations as may be prescribed by law, and subject also to the local custom or rules of miners in the several mining districts so far as the same may not be in conflict with the laws of the United States.

SEC. 2. *And be it further enacted*, That whenever any person or association of persons claim a vein or lode of quartz or other rock in place, bearing gold, silver, cinnabar, or copper, having previously occupied and improved the same according to the local custom or rules of miners in the district where the same is situated, and having expended, in actual labor and improvements thereon an amount of not less than $1,000, and in regard to whose possession there is no controversy or opposing claim, it shall and may be lawful for said claimants or association of claimants, to file in the local land office a diagram of the same so extended laterly or otherwise as to conform to the local laws, customs, and rules of miners, and to enter such tract and receive a patent therefor, granting such mine, together with the right to follow said vein or lode with its dips, angles, and variations to any depth, although it may enter the land adjoining, which land adjoining shall be sold subject to this condition.

SEC. 3. *And be it further enacted*, That upon the filing of the diagram as provided in the 2d section of this Act, and posting the same in a conspicuous place on the claim, together with a notice of intention to apply for a patent, the Register of the Land Office shall publish a notice of the same in a newspaper published nearest to the location of said claim, and shall also post such notice in his office for the period of ninety days; and after the expiration of said period, if no adverse claim shall have been filed, it shall be the duty of the Surveyor General, upon application of the party, to survey the premises and make a plat thereof, endorsed with his approval, designating the number and description of the location; the value of the labor and improvements, and the character of the vein exposed; and upon the payment to the proper officer of five dollars per acre, together with the cost of such survey, plat and notice, and giving satisfactory evidence that said diagram and notice have been posted on the claim during said period of ninety days, the Register of the Land Office shall transmit to the General Land Office said plat, survey and description, and a patent shall issue for the same thereupon, but said plat, survey or description, shall in no case cover more than one vein or lode, and no patent shall issue for more than one vein or lode, which shall be expressed in the patent issued.

SEC. 4. *And be it further enacted*, That when such location and entry of a mine shall be upon unsurveyed lands, it shall and may be lawful, after the extension thereto of the public surveys, to adjust the surveys to the limits of the premises according to the location and possession and plat aforesaid, and the Surveyor General may, in extending the surveys, vary the same from a rectangular form to suit the circumstances of tons of miners: Provided, That no location hereafter made shall exceed two hundred feet in length along the vein for each locator, with an additional claim for discovery to the discoverer of the lode, with the right to follow such vein to any depth, with all its dips, variations and angles, together with a reasonable quantity of surface for the convenient working of the same, as fixed by local rules: And provided further, That no person may make more than one location on the same lode, and not more than three thousand feet shall be taken in any one claim by any association of persons.

SEC. 5. *And be it further enacted*, That as a further condition of sale, in the absence of necessary legislation by Congress, the local Legislature of any State or Territory may provide rules for working mines, involving easements, drainage and other necessary means to their complete development; and those conditions shall be fully expressed in the patent.

SEC. 6. *And be it further enacted*, That, whenever adverse claimants to any mine located and claimed as aforesaid, shall appear, before the approval of the survey, as provided in the third section of this Act, all proceedings shall be stayed until a final settlement and adjudication in the courts of competent jurisdiction of the rights of possession to such claim, when a patent may issue as in other cases.

SEC. 7. *And be it further enacted*, That the President of the United States be, and is hereby, authorized to establish additional land districts, and to appoint the necessary officers under existing laws whenever he may deem the same necessary for the public convenience in executing the provisions of this Act.

SEC. 8. *And be it further enacted*, That the right of way for the construction of highways over public lands, not reserved for public uses, is hereby granted.

SEC. 9. *And be it further enacted*, That, whenever, by priority of possession, rights to the use of water for mining, agricultural, manufacturing or other purposes, have vested and accrued, and the same are recognized and acknowledged by the local customs, laws, and the decisions of courts, the possessors and owners of such vested rights shall be maintained and protected in the same; and the right of way for the construction of ditches and canals for the purposes aforesaid is hereby acknowledged and confirmed: Provided, however, That whenever, after the passage of this Act, any person or persons shall, in the construction of any ditch or canal, injure or damage the possession of any settler on the public domain, the party committing such injury or damage shall be liable to the party injured for such injury or damage.

SEC. 10. *And be it further enacted*, That, wherever, prior to the passage of this Act, upon the lands heretofore designated as mineral lands which have been excluded from survey and sale, there have been homesteads made by citizens of the United States, or persons who have declared their intention to become citizens, which homesteads have been made, improved and used for agricultural purposes, and upon which there have been no valuable mines of gold, silver, cinnabar or copper discovered, and which are properly agricultural lands, the said settlers or owners of such homesteads shall have a right of pre-emption thereto, and shall be entitled to purchase the same at the price of one dollar and twenty-five cents per acre, and in quantity not to exceed one hundred and sixty acres, or said parties may avail themselves of the provisions of the Act of Congress approved May 20, 1862, entitled "An Act to secure homesteads to actual settlers on the public domain," and Acts amendatory thereof.

SEC. 11. *And be it further enacted*, That upon the survey of the lands aforesaid, the Secretary of the Interior may designate and set apart such portions of the said lands as are clearly agricultural lands, which lands shall thereafter be subject to pre-emption and sale as other public lands of the United States, and subject to all the laws and regulations applicable to the same.

Approved July 26th, 1866.

Miscellaneous Advertisements.

NOTICE TO TRAVELERS BOUND FOR SWEETWATER!

Idaho Bridge and Ferry Company!

THE

NEAREST, BEST and MOST DIRECT

Route from the State of Oregon, and the Territories of Washington, Idaho and Montana, to the

SWEETWATER MINES,

Is by the way of Soda Springs and the Crossing of Smith's Fork of Bear River, and Upper Crossing of Hams Fork on the old SUBLIT'S CUT-OFF, across Green River.

Travelers by taking this route will greatly shorten the distance traveled on other routes, and find plenty of feed for their animals.

E. D. PIERCE & CO., Proprietors.

mh25 3m

TRAVELERS FOR SWEETWATER AHOY!

MOSES BYRNE

Desires to call the attention of miners and others bound for the mines from the West, to the fact that he keeps constantly on hand and for sale at reasonable rates

HAY,
GRAIN,
FLOUR,
MULES,
HORSES,
WAGONS,
etc., etc., etc.

He is also prepared to furnish entertainment for man and beast at his Ranch, which is situated at the junction of the Overland and Muddy roads, at which point most of the travel for Sweetwater leaves the Overland road, it being considered much the shortest route from the West to the mines. mh25 3m

GILES & Co.'s

PIONEER PONY EXPRESS.

Through to Fort Bridger in 24 hours.

General Express Forwarders.

We are prepared to carry Bank Notes, Bullion, Gold and Silver Coin, Parcels and Express Freight, at greatly reduced rates.

Collections and Commissions promptly attended to.

ARRIVES:—Tuesday and Friday 8 P. M.
DEPARTS:—Monday and Thursday 6 A. M.

Office at Atlantic City at Decker & Hamblin's store.

For particulars apply at the office, South Pass Avenue, South Pass City.

my13tf SCHERMIER & MORRIS, Agts.

MERCHANTS EXCHANGE,

Second South Street, Salt Lake City.

WAGENER & ENGELBRECHT, Proprietors.

The choicest brands of liquors and cigars constantly on hand.

Also

Two fine Billiard Tables.

fe15-3m

BLACKSMITHING.

E. STEEL,

Corner of Price and Grant streets, South Pass City, D. T.,

Begs leave to inform his friends and the public generally, that he has established a large and commodious

BLACKSMITHSHOP

In South Pass City, and is prepared to do all kinds of work in his line with neatness and dispatch. mh25-3m

PRINTING.

THE PIONEER NEWSPAPER

—AND—

JOB PRINTING HOUSE

—OF—

WYOMING TERRITORY.

THE "SWEETWATER MINES"

IS PUBLISHED EVERY

WEDNESDAY AND SATURDAY

—BY—

WARREN & HAZARD,

And carefully mailed in strong wrappers to any address in the Union.

SUBSCRIBE FOR THE PIONEER

NEWSPAPER OF WYOMING

The MINES will be devoted to the development of the mineral and agricultural resources of our new Territory, and to keeping its readers informed of the current events of the day.

The "Sweetwater Mines" will only contain a limited number of advertisements, our purpose being to make our journal a reading sheet dependent upon its circulation for support; consequently those of our readers wishing to avail themselves of the advantages of our advertising space will see the necessity of promptly sending in their orders.

Cheyenne Advertisements.

Cornforth Brothers,

GROCERIES AND LIQUOR MERCHANTS,

—AND—

GENERAL OUTFITTERS,

Have always on hand

PICKS, SHOVELS, GOLD PANS, BLASTING POWDER, FUSE,

And everything the miners are in need of.

We have the most select stock to be found in the West, and guarantee Eastern prices, freight added.

Give us a call.

Corner of Eddy and Seventeenth streets,

CHEYENNE, D. T. pr4 6m

LIDDELL, ROBERTSON & BROWN,

Galbraith Buildings, Sixteenth street,

CHEYENNE, D. T.,

Offer for sale a large stock of

GROCERIES

of every description.

Also a large stock of

Whiskies, Wines and Brandies,

Tobaccoes and Fine Cigars.

Also all kinds of

Cased Fruits and Liquors,

PROVISIONS,

GRAIN, FLOUR and BACON.

Our old freighting friends, and parties outfit-

Freund & Bro.,

Manufacturers and Importers of

GUNS and PISTOLS, CUTLERY, &c.

East side of Eddy street, between 16th and 17th,

CHEYENNE, D. T.

Branch: Main street, Dale City, D. T.

Sporting Apparatus,

and all kinds of

Fixed and Loose Ammunition.

SOLE AGENTS

for the whole West, for the celebrated

Winchester's Patent

Repeating Rifle and Carbine,

called the

18 shooter.

Also Agents for

E. J. Dupont de Nemours & Co.'s

celebrated

Sporting and Mining Powder.

All orders promptly attended to. pr4-3m

CROCKERY, GLASSWARE, MIRRORS, CUTLERY, LAMPS, CHANDELIERS,

SILVER and BRITTANIA WARE, WINDOW GLASS,

Coal Oil,

FURNITURE and BEDDING,

AND

FURNISHING GOODS

of all descriptions.

We have a complete outfit of heavy ware for hotels and restaurants; also an elegant assortment of bar fixtures.

We keep the largest stock of goods in the West and will furnish the Mining trade and Freighters at New York prices, freight added.

N. B.—All orders promptly filled. Any information given that is desired. fe15-6m

MACHINERY.

GRIFFITH & WEDGE,

ZANESVILLE, OHIO,

Manufacturers of

Portable Engines

AND

Saw Mills.

For full particulars, Price Lists, or Printed circulars, write to

GRIFFITH & WEDGE,

ZANESVILLE, OHIO.

fe15-3m

EAGLE WORKS

Manufacturing Co.

FOUNDED

OVER A QUARTER OF A CENTURY

P. W. GATES, President.

MANUFACTURE PORTABLE AND STATIONARY STEAM ENGINES AND BOILERS, SUGAR CANE MILLS, EVAPORATORS, ROCK BREAKERS, STAMP MILLS, MINING MACHINERY, GARD'S PATENT BRICK MACHINES, LATHES, PLANERS, DRILLS, SCREW CUTTERS, HAY AND COTTON PRESSES, SAW MILLS, FLOUR MILLS, MILL FURNISHING, etc.

Office, 49 Canal Street, CHICAGO.

SEND FOR CIRCULARS.

Thos. Chalmers, Supt. Ralph Gates, Sec & Treas't. Asst. " W. L. Chase, Gen Agt N. Draughtsman. Mabbs & Bro., Agents, Lake Superior.

3m

EWELRY!

A large and well selected stock of

uine American and European

D AND SILVER WATCHES, LD AND SILVER CHAINS, AMONDS, PEARLS, RUBIES, ADIES CHATLINS, SILVER WARE, CLOCKS, GOLD, SILVER AND STEEL SPECTACLES. EYE, OPERA AND FIELD GLASSES of EVERY DESCRIPTION.

mplete assortment of California Jewelry e to suit the times,

ALL GOODS WARRANTED.

g engaged a competent and practical maker I am now prepared to warrant all r one year.

lry repairing promptly attended to.

I. WATTERS.

door South of the New York Store, E Street, Salt Lake City. m15tf

h: Cheyenne City, Dakota Ter.

THE

EVERE HOUSE,

H. A. CARTER, Proprietor.

SECOND SOUTH STREET,

Near Snow's corner, Salt Lake City, U. T.

HOURS FOR MEALS:

Breakfast	from 6	to	10:30	o'clock	A. M.
Lunch	" 12	"	2	"	P. M.
Dinner	" 4	"	7	"	"

The Proprietor begs leave to announce to the traveling public that no expense has been spared in refitting the above named house, which will be kept hereafter as a First Class Hotel, and that suitable accommodations will be furnished to all who will patronize the establishment.

fe15-3m

Historical Research Collection, Wyoming State Museum

South Pass City, Wyoming Territory

Gold had been reported in the South Pass region as early as 1842. There was sporadic mining in the area from then until June of 1867, when the "Carissa" mine was discovered, followed by the "Miners Delight" three months later. A number of additional mines soon were opened, and thus began the South Pass mining boom, which lasted from 1867 to 1869.

As the rush developed, towns were founded near the mines. The first to be laid out was South Pass City, on Willow Creek about a half-mile below the Carissa mine; others nearby were Atlantic City and Spring Gulch, also known as Hamilton City.

Located fifteen miles north of the Oregon Trail, South Pass City by 1869 had a population of 1,597 souls and that same year was named seat of Sweetwater County. By 1872 and the end of the mining boom, South Pass City's population had dropped to around 300. Today both South Pass and Atlantic City remain alive, as romantic monuments to their gold rush past. Mining has been revived at various times over the years, and in 1994 dewatering of the Carissa mine was begun. Perhaps yet another gold rush will occur at this historic old site.[6]

The Freund brothers came to the South Pass area in 1869 and first advertised in *The Sweetwater Mines* on May 27th, listing only their Cheyenne Armory in Dakota Territory and a branch location on Main Street in Dale City, D.T.[7] The first mention of a Freund & Bro. gun shop in South Pass City— the impressive-sounding "Sweetwater Armory"— appeared in the July 14, 1869 edition of *The Sweetwater Mines*.[8]

Another Freund & Bro. advertisement, this in the South Pass *News* of Wednesday October 27th, 1869, reads as follows:[9]

—Freund & Bro. would say to the citizens of South Pass and vicinity, that their establishment will be closed about the 6th of November next. Those wishing a supply of arms or ammunition for the winter, should call soon; and persons having arms at our shop for repairs, are requested to call, pay charges, and take them away: otherwise they will be stored away with our stock until next spring.
n20-3t   FREUND & BRO.

Freund & Bro.'s reconstructed Sweetwater Armory at South Pass City, Wyoming. The upstairs was— and still is— used as a meeting place by the Masonic Lodge. (Author's photo, 1994)

The South Pass *News* carried a small ad in its April 9, 1870 edition, announcing that as of the previous November 6th "Freund & Brother" had closed their store there for the winter, but that ammunition still could be had from Goodman & Franks' wholesale liquor establishment.

Records show, however, that by the middle of 1870 the Freunds had returned to South Pass City.

That summer, the Arapahoes had fled their camps, and were engaged in attacks on towns and travelers throughout the Sweetwater mining district. The authorities stepped in, providing 100 guns and ammunition to the residents of South Pass. But those guns were not deemed to offer adequate protection, and the citizens resorted to taking arms, which were to be returned to the brothers after the violence had passed, from the gunshop of Freund & Bro. They were to be paid $5.00 for each weapon damaged while in use, and later submitted a bill to the county commissioners for $656.47. It is unknown how the Freunds arrived at that figure.[12]

As an indication of the seriousness of their intentions toward permanence

The South Pass News.

Volume 1.   South Pass City, Wyoming Territory, Saturday, April 9, 1870.   Number 56.

THE SOUTH PASS NEWS.

PUBLISHED EVERY SATURDAY BY
S. W. RUSSELL.

THE INDIAN OUTRAGES.

The Remains of Gen. Thomas.

The Meeting on Tuesday.

THE INDIAN QUESTION.

Practical Views of Pioneers.

Resolutions Adopted—Company Organized.

PREAMBLE AND RESOLUTIONS

in South Pass City, the Freunds did purchase a business building there, a replica of which still stands. The original structure was damaged by fire at a later date, and reconstructed by Wyoming Lodge No. 2 of the Ancient Free and Accepted Masons. Today the building's lower floor is arranged consistent with what would have been found in most of the tent or board stores of the Freunds and other frontier merchants. The front room was used as a sales and display area, while the back room was devoted to gunmaking and repairs, and to storage. The upstairs was utilized as a meeting room by the Masons during the Freunds' time, and still is being used for that purpose today.

At about this same time, several "rumors" surfaced concerning the Freunds, which will now be addressed:

Frank W. Freund in the uniform of a Knight Templar, circa 1870. (Freund Family Collection)

Frank W. Freund's name appears in the 1869 records of the Virginia City, Montana Territory Masonic Lodge. However, nothing exists to indicate that the brothers ever had a shop there, and a search of the records in Virginia City does not reveal that the Freunds ever owned property in that town.

On April 24, 1869 Frank Freund received four certificates of entry into the Masonic Order in Virginia City. It is the contention of this author that Frank Freund took his degrees in Masonry there only in preparation for starting a chapter or council in South Pass City. He remained a life-long member of the Masons and always paid his dues to the council in Cheyenne.

Some sources state that young John M. Browning worked for the Freund brothers; or that Frank Freund worked for Jonathan Browning, or in the mines

One of four ornate society membership certificates presented to Frank W. Freund by the Masonic Order in Virginia City, Montana in 1869, prior to the formation of a chapter in South Pass City. (Freund Family Collection)

as a machinist. But it is established fact that fourteen-year-old John Browning was working for his father in 1869; Frank Freund probably spent very little time in South Pass City during 1869, as he would have been busy traveling between Virginia City, Salt Lake City, Corinne, and Cheyenne attending to the gun business; and Jonathan Browning would have been equally occupied operating his Ogden gun shop, tannery, and sawmill, as well as venturing into real estate and the manufacture of plows, mill irons, and cut nails.

Perhaps the above rumors exist due to the fact that the Freund brothers' South Pass City building originally belonged to a man named J.W. Browning, of Ogden in Utah Territory. That same Mr. Browning sold the property to J.O. Farmer on May 6, 1869. Frank W. Freund bought the building from Hugo Rohn on the following 14th of October, apparently just in time to leave South Pass City for the winter.[11]

Just where the Freund brothers spent the winter of 1869-70 is not recorded. Perhaps they returned to attend to business at their armory in Cheyenne, where the weather was milder, and from there began to study the business climate of a burgeoning town one hundred miles to the south— Denver.

1. Nebraska City *News*, 19 January 1867. Courtesy Nebraska State Historical Society, Donald F. Danker, archivist.
2. "Freund & Bro., Gunmakers on the Frontier" by John Barsotti, in *Gun Digest*, Eleventh Edition, 1957. Hereafter cited as "Barsotti," with Part and page numbers.
3. *Ibid.*
4. *Ibid.*
5. *Ibid.*
6. Author's visit to South Pass City, Wyoming in June, 1994.
7. *The Sweetwater Mines*, 27 May 1868, page 4. Courtesy American Heritage Center, University of Wyoming, Laramie.
8. *Ibid.*, 14 July 1869, page 2.
9. The South Pass *News*, 27 October 1869, page 6. American Heritage Center, *op. cit.*
10. *Ibid.*, 9 April 1870, page 1.
11. Article by Gerald O. Kelver, in *Single Shot Rifle News*, Volume 45, Issue Number 3 (May-June, 1991).
12. Huseas, Marion McMillian. *Sweetwater Gold, Wyoming's Gold Rush 1867-1871.* Cheyenne Corral of Westeners, International Publishers, 1991.

This photograph of the "Freund's Sporting Outfits" store is unidentified, but possibly is the earliest Freund gun store in Denver, Colorado at 24 Blake Street. Frank W. Freund (in white hat) still appears rather youthful here. (Freund Family Collection)

CHAPTER THREE

THE DENVER YEARS

pparently, Frank and George Freund spent some time in Denver during 1869, as George's catalog issued in Durango[1] refers to the Colorado Armory as having been established in Denver that year. It is logical that the brothers spent some time there in the winter of 1869 looking for a store site, and even may have leased one in late 1869. Denver was enjoying a boom as an oufitting town in the early 1870s, and the brothers established their store at 24 Blake Street between F and G Streets. The *Daily Rocky Mountain News* on October 2, 1870 included the following mention in a listing of premiums awarded at the late fair:[2]

CLASS D
Freund Brothers, display of sportsman goods, diploma.

The Freund brothers received from the Sharps Rifle Manufacturing Company three sporting rifles, on November 14, 1870. They were the first Sharps rifles shipped to the Freund brothers, but were returned to the factory for some unstated reason.[3]

In 1871, the Freunds were listed in the *Rocky Mountain Directory and Colorado Gazetteer.* They had listings under:

Fishing Tackle (Dealer in)
Freund & Brother, Blake between F and G,
and
Gunsmiths
Freund & Brother, 24 Blake.[4]

The second Freund gun store in Denver, at 31 Blake Street, with its famed seventeen-foot-long, 200-pound rifle-sign attached at the second-story corner. The steps at lower right lead to a shooting gallery; fully-mounted bighorn sheep and bear can be seen in the right and left windows, respectively. Frank W. Freund stands in white hat, with brother George at the left, at foot of steps. The white dog is believed to be "General", Frank's mascot. Sign in window at right reads, "Rifles & Shot Guns Made to Order." (Freund Family Collection)

Interior view of Freund's second gun shop in Denver, taken in 1873. "General", Frank Freund's dog, lies to the right, behind the first counter. (Private collection)

Interior window view facing Blake Street, with its mounted bighorn sheep. Taken in 1873. (Private collection)

During 1871 the Freunds received eight additional sporting rifles from the Sharps Manufacturing Company. Two rifles were shipped on September 7th, two on September 30th, two on October 17th, and the last two rifles were shipped on October 26th, 1871.[5]

In 1872 the brothers moved their sporting goods store to 31 Blake Street in Denver. Apparently, advertising hasn't changed much over the years, as the following appeared on December 11, 1872, in the *Daily Rocky Mountain News:*[6]

One day later, on December 12th, the brothers received the following write-up in the same paper:

FREUND BROTHERS

News for Sportsmen

One of the best arranged, thoroughly convenient and most fully stocked business houses in this city, is that of Messrs. Freund & Brother, manufacturers, importers and dealers in guns, pistols, ammunition, gun materials, cutlery and sportsmen's articles of all kinds, at No. 31 Blake Street— opposite the American House— in this city. These gentlemen are now fully established in their new quarters, and are showing a line of goods equal to anything ever before opened in the western country. The interior arrangements of this store are fully appointed and artistically finished, presenting a pleasing aspect, and displaying the mammoth stock in the most advantageous manner.

One side of the store is devoted to the larger firearms, fowling pieces, rifles, etc., and we find here the latest improvements and finest finish of the noted gunmakers. There are Winchesters, Remingtons, Sharps, Greeners, and other makes, of all descriptions, finished to suit every taste; guns with gold, silver, and nickel-plated mountings; double and single barrel breechloaders; snap action breechloaders— new; in short, a full line of every description and all qualities, warranted perfect in every particular. Messrs, Freund & Brother pay particular attention to this line of their business. On the opposite side of the store are cases filled with revolvers— army and navy— knives, flasks, belts, fishing tackle, game bags, field glasses, riding whips, ammunition, and, in short, everything that the sportsman can possibly desire. The showcases are filled with pistols— revolvers, single-barrel, Derringers and others— together with a full line of pocket and belt cutlery— dirks and Bowie knives— fishing lines, trout flies, and a thousand and one articles which go to complete a stock of goods of the character of which we are writing.

Sportsmen and parties generally desiring to procure an outfit can find everything at the establishment of Freund & Bro. that they can desire— guns, ammunition, and all the accessories of a thorough equipment. We take pleasure in referring all persons to this establishment, for the reason that only the best and latest improved materials are kept on hand, and, further, that the stock is sold at the lowest eastern prices.

Messrs. Freund & Bro., in addition to their stock above referred to, keep a full line of natural specimens, buffalo heads, elk, wolves' heads, and all animals and breeds natural to the country, which are prepared by Mr. R. Borcherdt, a taxidermist of great skill, and who was for a long time connected with Barnum's New York museum. Visitors to Denver will find it to their advantage to examine the specimens with which the store abounds.

In addition to the wholesale and retail trade, this house keeps experienced workmen constantly employed in repairing work. Their shop is well appointed and work is fully warranted. The location of the establishment is easily designated by the sign of the large gun which is suspended in front of the block, and which was manufactured by Messrs. Freund & Bro. It is a double-barreled gun, in wood, seventeen feet long, weighs two hundred pounds, and is admirably proportioned, being the largest sign of the kind ever made.[7]

Interior view of Freund's second Denver gun shop, 1873. Note the profusion of Colt revolvers and other pistols, and ammunition displayed in glass showcases on and behind the counter. (Author's collection)

Another interior view taken in 1873, showing racks primarily of shotguns, full-mounted bear standing at center, and mounted animals and heads above showcases. (Private collection)

During 1872, Freund & Bro. were shipped the following rifles and carbines from the Sharps Rifle Manufacturing Company:

On April 22, 1872	5 Military rifles
June 6, 1872	2 Military rifles
	2 Sporting rifles
	5 Carbine rifles
July 30, 1872	2 Sporting rifles
August 6, 1972	5 Military rifles
August 31, 1872	5 Sporting rifles
Sept. 3, 1872	4 Sporting rifles
Oct. 16, 1872	10 Military rifles
Nov. 12, 1872	6 Sporting rifles
Dec. 7, 1872	4 Sporting rifles

Of the above fifty rifles and carbines shipped in 1872 to Freund & Bro., the four sporting rifles sent on December 7th were the last Sharps rifles shipped from the company to the Freund brothers while in Denver. All told, they account for a total of sixty-one Sharps rifles shipped to Freund & Bro. while they worked in that city, as follows:[8]

34 sporting rifles
22 military rifles
 5 carbines
___
61 total

From then on, whatever Sharps arms the brothers purchased came from other dealers. Apparently it was either less expensive that way, or perhaps they received more liberal credit terms from the large wholesale gun houses in the East than they got from the Sharps Rifle Manufacturing Company.

One of those large eastern dealers was Schuyler, Hartley, and Graham of New York City, whom it appears from the records acted as a warehouse (and perhaps banker in one form or another) for the Sharps Rifle Manufacturing Company.

Among the other eastern dealers who supplied guns to the western gun dealers and gunsmiths were Benjamin Kittredge and Co. of Cincinnati, E.S. Harris of New York City (Sharps' only general agent), J.P. Lovell & Sons of Boston, and Spies, Kissam & Co. of New York City.

The five carbines shipped to the Freund Brothers in Denver on June 6, 1872 make for an interesting study. These five carbines were early New Model

Another rare interior view of Freund's gun shop, looking down the center aisle and out toward Blake Street, 1873.
(Author's collection)

1873 corner window view from interior of Freund's Denver gun shop, showing bear and other animal mounts. Note monkey on counter top holding business cards.
(Author's collection)

1869 carbines, with serial numbers in the percussion range. (The one owned by the author bears the serial number "112128".) These percussion serial numbers have been verified by Dr. R.L. Moore, Jr., (owner of the factory records) as having been shipped as 1869 carbines, and all appear in the records of the Sharps Rifle Manufacturing Company.[9]

The Freund Brothers were instrumental in forming a Schuetzen Club in Denver, as evidenced by the following notice which appeared in the *Daily Rocky Mountain News* on February 28, 1873:

Incorporations Filed.

Also articles of incorporation were filed for the organization of Denver Deutscher Schuetzenverein, by M. Sigi, F.R. Frotzascher, R. Kroeck, William Whist, F.W. Freund, G. Freund, Albert Borcherdt, Rudolph Borcherdt, Ernst Phisterer, August Gueck, Edward Schwalbe, Richard Borcherdt, H. Brook, and C.A. Jachmus.

The object of the society is to diffuse useful information amongst the members; to practice after the target; to acquire a proficiency in the use of firearms; and for social amusements. The trustees in the society for the first year are: M. Sigi, F. R. Frotzschar, and R. Kroeck. The place of business is Denver.[11]

In 1873 Frank and George Freund entered an exhibit in the 8th Territorial Fair of Colorado, in which Frank W. Freund was awarded a handsome medal for the "Best Colorado-made Gun in 1873." It was given by the Colorado Industrial Association, and inscribed to F.W. Freund. It is an indication of how comfortable

Obverse and reverse of the silver medal awarded to F.W. Freund at the 8th Territorial Fair of Colorado, in 1873. (Freund Family Collection)

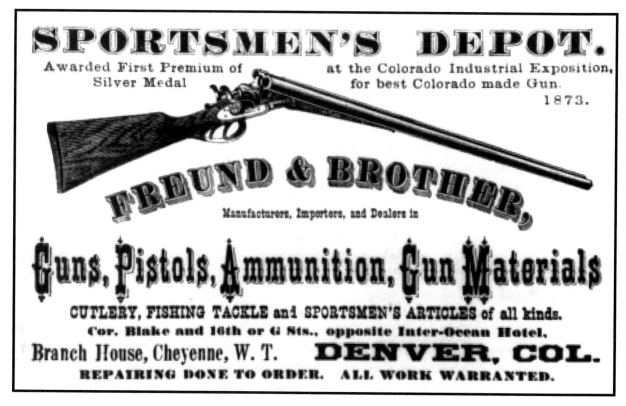

The Freund brothers' business card for their second Denver "Sportsmen's Depot," at 31 Blake Street at Sixteenth. Note mention of the silver medal won at the 1873 Colorado Industrial Exposition. (Author's collection)

the brothers' relationship was at the time, that Frank changed the inscription in their advertising to read "Freund Bro's."

After the brothers split up their partnership and George took up business in Durango, both continued to use the cut of the medal that read "Freund Bro's." Ironically, the gun for which Frank won this medal has never been identified.

With the store in Denver, and another one in Cheyenne, Frank probably was not able to spend much time on sales, and this aspect of the business would have been attended to by George.

On October 13, 1873, Frank applied for the first of his patents. This was an improved sear and sear-spring that enabled Frank to install double set-triggers in breechloading guns having the lock to center and a low hammer (*see* chapter seven).

The Freund Brothers were not only in the gun and sporting goods business, as evidenced by this ad which appeared in the *Daily Rocky Mountain News* on June 2, 1874:[12]

—◆—

THE ORIGINAL
HYGIENIC INSTITUTE
212 Sixteenth Street
Opposite American House
After being closed for repairs, and the
addition of several European baths,
is now open.
Remember that this is the Original
Hygienic Institute
Freund & Bro.

—◆—

Again, in the June 6th, 1874 issue of the *Daily Rocky Mountain News* the brothers' ad appeared:

—◆—

Turkish bath 75 cents, at the Hygienic Bath Establishment,
212 Sixteenth Street, opposite the American Hotel, next door
to Freund Bro's gun store.

—◆—

In 1874, Frank W. Freund applied for three more patents. The first was for an improvement in metallic cartridges, and was applied for on February 4th. The second was for an improvement in breechloading firearms, applied for on October 19th. The third was for sights for firearms, and was applied for on October 19, 1874 (refer to chapter seven for information on these patents).

In 1875, Frank W. Freund applied for three patents, all on March 19th. The first was for an improvement in breechloading firearms, and was for changes to the Remington Rolling-Block system. The second was for an improvement in pistol-grip attachments for the stocks of firearms, which at the time was mainly applied to Remington Rolling-Block arms. This patent was later to be involved in a court case between Frank and the United States government. The third patent applied for was an improvement on guard levers.

It is to be noted that of the seven patents granted to Frank W. Freund during the Denver years, none of them applied to Sharps rifles. All of the Freund patents involving Sharps rifles were applied for from their locations at Cheyenne, Wyoming Territory.

A fine example of the high-class work that F.W. Freund was capable of is a fancy Remington Rolling-Block rifle once owned by Frank P. Sargent, a well-known single-shot rifle authority and collector. His description of it (from a letter dated November 26, 1948) is interesting:

> One is a fine heavy Remington Rolling Block Model and [it] appears to me this action or at least the frame may have been furnished to him [F. W. Freund] by E. Remington Co., not case hardened but left soft for his beautiful fine line engraving.... It is engraved in scroll writing, "Improved by F.W. Freund, Denver, Col., Pat. July 28, 1874, & Pat. Apld for." on the upper tang. In [the] same fine engraved scroll is "E. Remington & Sons, Ilion, N.Y. U.S.A." This Remington name engraved on the tang in old slanting script letters is different to me, as I have eight or ten other Remington rolling-block rifles and [on them] the name is pressed into the metal with a marking die or stamp and block letters used.
>
> This rifle is chambered for the .44 [caliber] bottle-necked shell, 2¼ inches long, and so [it] used the very early .44-90 B.N. in the first 2¼ inch length....
>
> The action is fitted with very fine double Schuetzen set triggers, two triggers spaced widely apart, made by Freund and which I have never seen furnished on a rolling-block model by [the] Remington factory. I have two Remington rolling-block factory jobs in single set only, and Remington Hepburns with Schuetzen double set.
>
> The action is a straight grip but Freund made up a trigger guard and pistol grip in steel and put a fine checkering job on [the] grip forming the pistol grip, as sharp as a file. The trigger-guard reverts back to the muzzle loader shape. He reshaped [the] nose on the hammer so when the hammer was cocked and the breech block rolled open for cleaning, the hammer at this point presented a flat surface ½ to ⁵/₈ inches long, parallel with the bore and presumably so [a] cleaning rod could not cut and scrape on the corner of the hammer as on standard Remington rolling-block models. The breech block is changed also. The checkered extension to open the action with the thumb is a separate piece of thin tool steel fastened to the face of the block with three small flat head screws spaced equidistant around the firing pin. The outer end is gracefully shaped and finely checkered making a thinner and neater looking end than [on] the standard Remington. Also maybe he [Freund] hardened this piece to get a longer life or tight headspace, and of course, it could be replaced if it was ever needed to reduce the head space after long use. He used the standard cross

screw in the block to retain the firing pin... so evidently [he] did not have in mind a front assembled firing pin idea. His extractor slides out in a groove in the side of the threaded barrel shank, but rides in a groove in the block twice as wide as on standard Remington allowing a hook on the bottom of the extractor twice as wide and so twice as strong against breakage. My standard Remington extractors take hold of the cartridge head at a place like 8 or 9 o'clock on a clock dial. He [Freund] seems to be like Harry Pope and knew the trouble in standard articles and brought out his own versions with corrections in them.

This Remington rolling-block also has the extra long [Freund] patented open rear sight. It is on the order of our later Rocky Mountain rear sight only this one has a spring blade about eight inches long and gives a very easy spring action in moving the elevator back and forth. It is built up very high at [the] crotch and calibrated with finely engraved numerals from 100 to 500 yards. There is only "0" in [the] center of the slide and as you move it along the elevation it fills in for 100, 200, 300, 400, and 500 yards... the sight is so high that a special black sole leather gun case which came with it has a large bulge formed in it at this point to clear this rear sight.

The barrel on this rifle is a very unique double octagon, with a second set of octagon flats milled 22½ degrees from [the] other set. I have never seen this [feature] on any rifle before and no one I have shown it to has seen it before. I think this was done to do away with an octagon flat on [the] barrel top between sights and [thus] prevent glare. On my Sharps Borchardt Express Model [the] barrel flat is matted to prevent glare. Freund had this idea of his to get the same result."[14]

The popular rifles of the day which were suitable for the largest western game were the Sharps, Remington, Springfield, Whitney, Maynard, and other big-caliber single shots, which had greater range and power than the repeaters. One such rifle that came from the Freund shop in Denver is mentioned in the book *Hunting at High Altitudes*, edited by George B. Grinnell. In the section written by Roger D. Williams is described an adventure in the Black Hills. Williams went to the hills in 1875 with one of the "outlaw" or "sooner" parties, which included the hunter and scout Moses Milner, better known as "California Joe." (This was not the Professor Walter P. Jenny expedition in the summer of 1875, for which "California Joe" served as a guide.)

Williams was armed with a fine, single-shot Springfield sporting rifle which he called a "needle gun," a name often incorrectly applied to that action.

Jim Baker, famed mountain man, scout, and hunting guide. According to Freund family records, Frank was very proud of his friendship with Baker.
(Private collection)

The rifle was .45 caliber, had a heavy octagon barrel, double-set triggers, and a curly maple pistol-grip stock. It was chambered for a case holding 107 grains of powder, a rather stiff load for a Springfield action. In an encounter with a wounded deer, Williams, to his dismay, broke the stock of this fine Freund-Springfield. He considered it the best hunting rifle he ever owned.[15]

The panic of 1873 naturally affected the western United States, and the Denver area was no exception. Yet the Freund brothers were craftsmen, and as evidenced by their advertisements, were not leaving any stone unturned to pick up a little extra income.

However, in April of 1875 the Freund brothers addressed a letter to the Sharps Manufacturing Company which revealed the business slump in Denver, at least in sporting goods and guns. In it the brothers asked for an extension of time on a note against them held by the company. Their letter closes with "business is very dull here, and it is impossible to meet obligations."

Then, in the May 20, 1875, issue of the *Daily Rocky Mountain News* there appeared a brief news item:

———◆———

Freund & Bro., 381 Blake Street, have sold their Sportsmen's Depot to John P. Lower, of the firm of Gove & Co., whose card appears this morning in our advertising columns. Mr. Lower is a clever gentlemen and a capable workman, and the News *takes pleasure in commending him to the public.*

———◆———

The street number 381 is correct, as it was the third store for the Freund brothers in Denver. The first was located at 24 Blake Street between F and G Streets, and the second was the Sportsmen's Depot in the two-story building at 31 Blake Street on the corner of Sixteenth Street. The store sold to John P. Lower at 381 Blake Street was in a one-story building in the middle of the block. John P. Lower had worked for and become a partner to Carlos Gove of Denver, before taking over the Freund Brothers store.

The sale of their Blake Street store marked the Freund's farewell to Denver. The brothers moved back to Cheyenne, where they had kept a store open since first entering Cheyenne in their head-of-track days with the Union Pacific railroad.

1. "Colorado Armory" catalog number 389, issued by George Freund from Durango, Colorado, courtesy of John Hartman. (*See* chapter eleven)
2. Western History Department, Denver Public Library.
3. Author's correspondence with Dr. R.L. Moore, Jr., of Philadelphia, Mississippi, owner of the original Sharps Rifle Manufacturing Company records.
4. Western History Department, Denver Public Library.
5. Dr. R.L. Moore, Jr., correspondence.
6. Western History Department, Denver Public Library.
7. *Ibid.*; *Daily Rocky Mountain News*, December 12, 1872, page 4, column 1.
8. Western History Department, Denver Public Library.
9. *Ibid.*
10. *Ibid.*
11. *Ibid.*
12. *Ibid.*
13. *Ibid.*
14. Barsotti, Part I, 1957.
15. *Ibid.*
16. Western History Department, Denver Public Library.

JOHN P. LOWER,
Guns, Rifles, Pistols, Ammunition

SPORTING AND FISHING TACKLE,
FIELD AND SPY GLASSES, INDIAN GOODS, &C.,

381 Blake Street, Denver, Col.

The Sportsmen's Depot store at 381 Blake Street, Denver, that the Freund brothers sold to John P. Lower on their departure for Cheyenne in May of 1875. Note that it is located in a one-story building. (Western History Department, Denver Public Library)

CHAPTER FOUR

THE CHEYENNE YEARS

1875

The return to Cheyenne was a logical move for the Freund brothers, as they had helped establish the town back in 1867. From all available information, they had continued to keep their shop open in Cheyenne from 1867 until their return in 1875. To the Freund brothers, returning to Cheyenne had to be like going home again, for no doubt they still had many friends and business acquaintances in Wyoming.

An advertisement which appeared in the Cheyenne *Daily Leader* on June 28, 1875, announced:[1]

Wyoming Armory
and
Pioneer Gun Store
FREUND BROS.
Eddy Street......Cheyenne, Wyo.
Guns, pistols, and ammunition of
every description at lowest prices.
Repairing done in the best style
and warranted.

The discovery of gold in the Black Hills occurred at the time of the Freunds' return to Cheyenne, and those participating in the rush to the gold fields used Cheyenne as a base or jumping-off point to the mines.

Gold prospectors were slipping into the Black Hills in ever-increasing numbers. This was the land that had been confirmed by formal treaty to the proud Sioux Indians. However, the Custer Expedition into the Black Hills had

The Freund brothers' "Wyoming Armory" in Cheyenne. Frank W. Freund is standing in the doorway with his dog, General, at his feet. George Freund stands at Frank's left, next to the crossed rifles in the window. Note profuse window advertising, and the Freunds' famed rifle sign. (Wyoming State Museum)

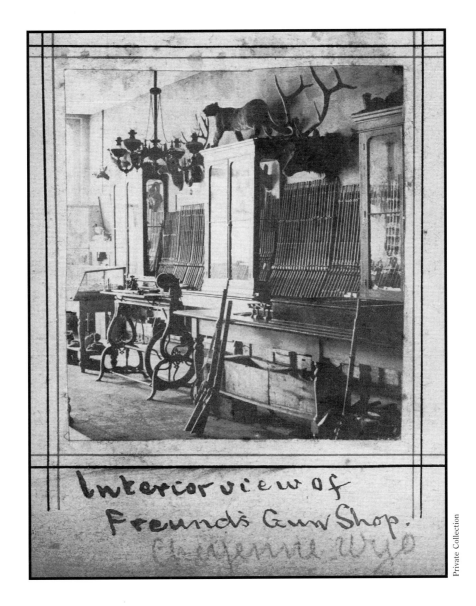

Interior view of Freund's Gun Shop. Cheyenne Wyo

put an end to that agreement by violating its terms.

Scout Charley Reynolds had carried Custer's official report of the expedition to Laramie City. The report had contained the news that the Black Hills were so rich that there was "gold among the roots of the grass."

Indian opposition to this rush onto their lands provided a real element of danger for white men entering the Black Hills area.[2] The papers of course played on the drama, and news items such as the following from the July 12th issue of the *Daily Leader* afforded the Freund brothers some valuable free advertising:[3]

———◆———

CHEYENNE AS AN OUTFITTING POINT!

As all who intend going to the Black Hills need fire-arms and ammunition for their own protection and to supply the commissary department, we are often asked whether these goods can be obtained here. We refer them to the Wyoming Armory and provision store of the Freund Bros. on Eddy St., who keep a large assortment of guns and pistols and ammunition of all kinds.

They are first class gunsmiths, and repair fire-arms in the most workmanlike manner.

———◆———

The rush to the Black Hills had to be one of the primary reasons for the Freund brothers' return from Denver to Cheyenne. Frank W. Freund had made a trip back to Germany after his father's death, and had received a good inheritance. He went to the Dakota Territory during the Black Hills gold rush and bought property. However, he lost both his property and inheritance when the at first easily-obtained gold played out.[4]

The first patent for Frank's work on Sharps rifles was applied for on December 3, 1875. The patent was not issued until January 2nd, 1877. It was U.S. patent number 185911, titled an IMPROVEMENT IN BREECH-LOAD-ING FIRE-ARMS. It concerned a gas check and straight-line firing pin (*see* chapter seven).

No doubt Frank probably had some ideas and possibly did some experimental work on Sharps rifles while still in Denver, but to date no work done by the Freunds to Sharps guns has turned up from that period.

1876

Frank's work on Sharps and other firearms took up a great portion of his time in 1876. He applied for seven patents during the year, including one for priming cartridges (number 184854), one for the disassembly of revolving fire-arms without tools (number 183389), another for improvements on front sights (number 189721), and four patents for improvements on breech-loading fire-arms (numbers 180567, 184202, 184203, and 216084). *See* chapter seven for a review of these 1876 patents.

A sworn statement from a professional hunter on April 11, 1876, gives us the first evidence that Freund's improvements to the Sharps rifle were being readily accepted by men in the field:[5]

Territory of Wyoming, County of Laramie

J.A. Meline of said county and Territory, being by me duly sworn according to law, deposes and says that he is a hunter by profession, and has been engaged in such business for twenty-two years— that he is well acquainted with the rifle known as Sharps' breech-loading rifle, and has used it exclusively for five years— that he has had them made to order for him at the factory, and always considered it the best rifle in use, finding but one objection to it— that there is a difficulty in forcing in the cartridge after the rifle has been fired a few times, and by reason of this trouble or difficulty he has known many accidents to occur, such difficulty being more observable where the rifle has a small bore and a long cartridge used.

This deponent further says that he has personal knowledge of the following accidents, as having occurred from the said defect in the rifle. While attacked by the Cheyenne Indians in the fall of 1874 on the north fork of the Smoky Hill River, one Charley Brown, a hunter, lost his life by his inability to force a cartridge into his rifle (which was one of Sharps' breech-loading rifles) soon enough. The Indians succeeded in taking his life before he could put his rifle in a condition to fire, the rifle being found after his death with a portion of the cartridge exhibited out of the gun, showing the fact of his difficulty in placing it in.

Also that one McLaughlin, a hunter of experience, was severely injured in the hand and lost one finger, by reason of the explosion of a cartridge, while he was in the act of forcing it in, it requiring the aid of a stick.

That one Henry Campbell and another James Campbell, both hunters, were injured in the hand and eye, while in a similar act as above mentioned. And that he has knowledge of many others injured thereby, and that he experienced considerable trouble with the rifle in that respect.

This deponent further says that he has lately been shown an improvement in the said Sharps breech-loading rifle, made by F. W. Freund of Cheyenne, Wyoming, it being an improvement of the breech block which avoids the danger of all such accidents, and that he admired the said improvement very much, for which said improvement this deponent is informed that the said inventor, F.W. Freund, has applied for a patent. And this deponent says that it is his opinion that no man using rifles will want any other kind after once seeing this improvement, that great credit in due that inventor, which all hunters will appreciate.

J.A. Meline

Subscribed and sworn to before me this 11th day of April A.D. 1876
E.P. Johnson, Notary Public

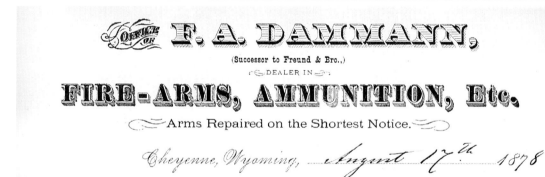

Letterheads used by the Freunds in Cheyenne. The earliest, top, was hand-embossed on plain paper; next shows the silver medal awarded in 1873; third is that of F.A. Dammann, "Successor to Freund & Bro."; the last illustrates Freund's patent rifle action and names "C. [Clotilda] Freund", Frank's wife, as the firm's manager.

In the Cheyenne *Daily Leader* on November 12, 1876, the following advertisement appeared:[6]

———◆———

WYOMING ARMORY
(illustration of a Winchester rifle)

FREUND BROTHERS
Have always for sale a large assortment of
SHARPS, WINCHESTERS
and all the latest
Sporting and Military Arms
AMMUNITION, ETC., ETC.
Colt's and Smith & Wesson pistols
and everything in that line
AT THE LOWEST PRICES

———◆———

Frank had been making periodic business trips back to New York City and other places in the East. On those trips, he also had paid visits to the Gasperrini family, where he played cards. The Gasperrinis were French, and Frank spoke that language as well as English and German. It was a large family, and before long Frank was playing suitor to one of the daughters, Clotilda. They were married by the mayor of New York City, in his office on December 18, 1876. Frank was thirty-nine years old, and Clotilda was twenty-one.

1877

Frank and his new bride returned to Cheyenne about the 20th of January, 1877. On that same trip back East he had mixed business with pleasure, and received a rifle action from the Sharps Rifle Company. He intended to alter it with his patents, and thus prove to the Sharps company that they needed his improvements to provide a more acceptable gun to the general public— especially in the West.

On January 29th, Frank wrote to the Sharps Rifle Company that he "had just returned home and found business very good which may delay my sending you the action for a month or more." Apparently the Sharps Rifle Company had asked Frank to pay for the action given him in December of 1876, for he subsequently sent them the following letter:[8]

Cheyenne, Feb. 15th, 1877

Sharps Rifle Company
Bridgeport, Conn.
Gents,

Enclosed find Post Office Order five dollars & eighty-two cents ($5.82), the amount of your invoice for the Sharps Breech Assembly which please place to our credit and acknowledge the receipt of same.

Yours Respectfully,
Freund & Bro.

On February 20, 1977, Frank sent a sample of his improvements along with this letter:[9]

Cheyenne, Wyoming
Feb. 20, 1877

Sharps Rifle Co.
Bridgeport, Conn.
Gentlemen.

I sent you by today's express the model of your rifle which you furnished me while East and to which I have added my improvements to wit: The breechblock and frame as constructed as to force the cartridge into its proper position on closing of the breech. The firing pin so formed as to carry the hammer to the half-cock by either opening or closing the breech and also with a gas check which will break the force of gas that may escape from a punctured cap when the arm is fired. You will observe that the breechblock has been slightly increased in length and so arranged that the cartridge has been forced over and is then prevented from falling back. This however is not absolutely essential and may be avoided if desired by giving the lever a little more swing. The cuts in the top of the breech for the shell extractor to fall into are done away with in my breech action. And another extractor may be put upon the right side of the rifle without any interference with the firing pin, as would be the case under your system of breech action. This will I think prove an important feature in the use of heavy charged rifles and will afford an equal outward pressure on both sides of the cartridge shell. In adding these improvements, I have made some minor changes in the form of the breech frame which are of course not necessary to make the workings of my improvements complete. But are made simply as suggestions and in case you should conclude to make the breech frame adapted to my improvements, think they will give the arm a more symmetrical appearance. In order to add my improvements to your present system the only changes required are as follows:

The lengthening of the nose of hammer, the breech block cut down substantially as shown by the model, the space between the rear of the breechblock and frame filler as shown by the detachable plate, a slight change in the form of the firing pin, and the trigger plate lengthened and which in connection with the pin or lug in the breech frame forms a guide of the breech block. All this may be done with little or no changes in your present milling process. And at a very small expense the improvements may be attached to your present system of breech action. To make the half cocking movement work easier, I have inserted a screw in the lockplate as shown in the model which breaks the force of the main spring after reaching the safety notch and which you will find makes the movement work sufficiently easy to meet all objections. This screw however may be made adjustable as to make the hammer rebound to the safety notch or not as desired. I send you also an extra hammer and firing pin, the firing pin is one of your highest with a face slightly changed as shown. This firing pin may be used on any of your rifles by only removing the sharp edges from the face of the hammer, as shown by the hammer sent you. No change is made in the locks with the use of this firing pin & hammer.

In correction I will add that I am satisfied that you will have no difficulty in milling out the rear portion of the frame should you conclude to make it solid. The model which I send you I desire to keep and after you have made another by it for yourself to keep, please return the same to our address at this place. State also what you will make me firing pins for in accordance with the extra sample which I send you.

Yours Respectfully,
F.W. Freund

P.S. I send you some copies showing the facility of forcing the cartridge home, what is wanted.

Inasmuch as Frank heard nothing from the Sharps Rifle Company regarding their receipt of the action he had send them, he sent the following letter to the company:[10]

Cheyenne, March 18th, 1877

Sharps Rifle Company
Bridgeport, Conn.
Gentlemen,

I sent to you Feb. 20 the action with the improved Breech and also a Hammer and Firing Pin, separate, and a letter concerning the matter.
If you have acknowledged the receipt of them, your letter must have been lost. And you will oblige me, if you will inform me in your next letter whether you received the above mentioned Breech &c.

Respectfully yours,
F. W. Freund

The Sharps Rifle Company did comply with this latter request, on March 23rd.[11] In the meantime a friend of the Freunds who was a large gun dealer in Dodge City, Kansas, F. C. Zimmerman, sent the following letter to the company recommending Freund's action:

I have heard and seen an improvement (or patent) of your Rifle by Mr. Freund in the breechblock. This is a very valuable improvement on your rifle, especially for hunters & Frontiersmen. This breechblock does away with the objections of your Rifle and if you invest in it you never invested in anything better. Suppose you wish to know why the present Breech Block is an objection, very well I will explain it to you. Since my experience with your rifle I have seen 20 to 25 men which had Cartrs. explode in their face some of them lost their eyesight in one Eye in the following way: if the Hunters reload their Cartrs, they often expand the shell so that the Cartrs. won't enter the chamber by about 1/2 inch or less. They have nothing to force the Cartrs. to their proper place and take the first thing they find; this is in general the wiping rod. With this they Force or drive the Cartr. to the place and there it happens that they explode which never can happen with Freund's Improvement. So are his other improvements of value. It just so happens that one of the victims of an exploded Shell steps into my store which happened only a few days hence (his name is S.J. Kensert) he bought a 44/90 gr. Rifle of me about 2 months ago. The other day he was practising with reloaded shells, one would not go to its proper place and would drive it in as before and said exploded, his whole face is full of powder and one Eye is sore. Another victim lives on a Ranch which lost the eyesight of one Eye, his name is J. Springer in the same way, and so did others suffer. Your guns are liked best of all for that objection it would be perfect. Mr. Freund is a splendid mechanic and a Gunsmith, I know him since 1857. We worked together in Germany, France, and England and together in America in the best gun shops, he always showed to be a Genius and I am glad to hear that it is him who made this improvement on your Rifle. I did not write him for several years but now I will congratulate him on his Improvement he so successfully made on your celebrated Rifle.

Whatever his shortcomings as a letter writer, Zimmerman was highly respected as a dealer and distributor of Sharps rifles. The company, therefore, had to answer the letter in such a manner as to not offend him. In answer to Zimmerman's glowing letter, the company replied on April 16, 1877:

> *We have a model of Mr. Freund's patent here, and are considering the propriety of adopting it. It undoubtably has some merit.*

From their letter, it would seem that the Sharps Rifle Company was considering the Freund improvements from a manufacturing standpoint. However, they did not inform Freund that they were thinking about its adoption.[12]

There is no record of Frank or George placing any orders for Sharps guns or parts since their arrival back in Cheyenne from Denver in June of 1875, until February 8, 1877. On that date they ordered parts again. Orders for parts were received by Sharps on 2/5, 3/16, 5/5, 5/31, 6/12, 6/16, 8/20, 9/6, 11/21, and 12/20/77, according to company records.

The Sharps records do not show any complete rifles being ordered from

Cheyenne during 1875, 1876, or 1877.[13] (Readers should keep in mind that the interpretation of the Sharps company records is quite a task. Dr. R.L. Moore, Jr. is to be congratulated on his efforts. The author also advises that not all of the Freund brothers' letters to Sharps reside in any one collection, and that any omissions that may occur in this text be overlooked.)

Frank's next effort was to take up the cause for the .40 caliber cartridge *vs.* the .44 and .45 caliber cartridges. An excerpt from his letter of May 8, 1877, ordering an eight-pound octagon barrel in .40 caliber states:[14]

> *We want this barrel for our new breech mechanism as the .40 caliber does much finer shooting than the .44 or .45 caliber. This is our experience and likewise that of our hunters in this section of the country.*

A dapper-looking Frank W. Freund, age thirty-nine, in 1876.
(Freund Family Collection)

Frank did not lose much time in stating his case for the .40 caliber, as he again wrote Sharps the following day:

Cheyenne, Wyoming
May 9, 1877

Sharps Rifle Co.
Bridgeport, Conn.
Gentlemen:

In our letter of yesterday we stated that we would have something further to say upon the use and superiority of the .40 calibre rifles. And we will now give you our unsolicited opinions on this subject:

We have sold of late quite a number of the .45 calibre of your make, but when put to test by those who have used the .40 calibre, they complain that the ball travels too slow, shoots too much lead and requires greater elevation with less accuracy in shooting. Those complaints are not the complaints of the few, but of many, and too from hunters. Who have had the practical experience of years in the use of firearms. All admit that the cartridge works freer and easier in the .45 calibre than in the .40. But, now that the breech system is so modified as to avoid this difficulty, no objections can be raised on that account. While the modified system by action of the breech is especially of great importance when applied in the .40 calibre gun. It is likewise valuable to all other calibres, and all unite in saying that it should be applied to all the calibres alike.

There is, however, no rifle now in use that will stand the .40 calibre 90 grain charge and at same time by operators without cleaning after each shot. Your rifle as it is now made will of course stand the charge as will some others. But after firing a few shots the cartridge cannot be got into the chamber unless it is by clubs in as with others. Likewise the Remington does not make the .40 calibre 90 grains for two reasons: 1st their breech will not stand the charge and secondly the cartridge cannot be got into the chamber as is the case with other guns.

It might be interesting, perhaps, to give you a little history of an accident that occurred to a man at Red Canyon, Wyo. last fall while using the .40 calibre gun of the Remington make. The gun had been re-chambered by someone in Denver, Colorado to take the 90 grains and while shooting it the breech block was blown out and the operator instantly kilt. Some of his deceased complained to Remington and in reply they stated that they would be in no way responsible for accidents occurring to their

.40 calibre rifles, which had been re-chambered for the 90 gr. charges.

Now in our opinion the .40 calibre with our breech system applied to your gun does away with those objections. We have changed some guns of your make to our improved breech system and thus far they give entire satisfaction.

Therefore, we feel justified in saying that with the .40 calibre as well as the other calibres of your make changed to our system of breech mechanism, it will give ample facilities for forcing the cartridge into the chamber and so combine in this .40 calibre more accuracy in shooting, and in time will force other guns to a back seat: For this simple reason that other guns are not operating with the .40 calibre & combined with the required resistance. And as in the past, the charges of powder will be increased in the future, rather than diminished with this calibre reduced instead of increased.

Your gun with our improvements thereon is pronounced by experts and those having experience to be the best gun in the world. Hoping to hear from you soon. We remain,

Respectfully,

Freund & Bro.

Then, just two days later Frank wrote again:

Cheyenne, Wyoming
May 10, 1877

Sharps Rifle Co.
Bridgeport, Conn.

Gentlemen:

We wish some long bullets for the gun barrel we ordered from you on the 8th inst. Now can you furnish us with a bullet for the .40 calibre in proportion to the long .44 cal. Creedmoor bullet. We want them for long range shooting and heavy charges, what do you think would be the best size. Please reply to this before shipping the gun barrel & bullet moulds that we may add to the order and ship by freight.

Respectfully,

Freund & Bro.

The next day still another long letter followed:

Cheyenne, Wyoming
May 11, 1877

Sharps Rifle Company
Bridgeport, Conn.
Gentlemen:

Since writing you the other day we have thought over the matter of chambering the .40 calibre rifle barrel which we ordered from you. And we want you to make the chamber fit as closely as possible to get the shell almost in the chamber by hand. It is our object in having the chamber fit close to the shell to avoid the expansion of same, which is so common with all guns now in use. It is our purpose especially to prevent the enlarging of the "neck" of the shell. As you know, the chamber that fits closely has little or no chance of expansion and in proportion to its expansion is due to loosely fitting chambers. Thus the power of the gas, too is reduced in a still greater degree the closer the chamber fits to the shell. With the chamber fitting as close as we indicate, no difficulty will (and the bullet will fit right) be experienced in extracting the shells. With the ample facility of forcing the cartridges in the chamber, should they go a little tight. The imperfections or non-roundness of the shells gives us no concern whatever. For when the breech is closed the shell will be swedged into its proper size. In the course of our business we have quite a demand for your rifles with our improvements thereon. Now what you charge to make alterations necessary to add our improvements to your gun. The alterations to be as follows, the breech block modified as shown in the model sent, guided and a steel plate to the rear of same attached with two screws to frame, and firing-pin with gas check on rear as shown in model. Two extractors; the firing pin and hammer constructed so as to carry the hammer to half cock. Either a new hammer or the nose of the old one lengthened as shown in the photograph sent you. What we want is the cheapest way in changes on locks or frames as necessary.

Send please an estimate of cost for doing the work indicated above as soon as possible.

Respectfully,
Frank Freund

P.S. We sent you today some views showing one section of the interior of our store. Reply if a new model is desired. Let us know and we will send you one at once.

F & Bro.

Then three days later, still another letter:

Cheyenne, Wyo.
May 14, 1877

Sharps Rifle Co.
Bridgeport, Conn.
Gentlemen:

In ordering the .40 calibre rifle barrel several days ago nothing was said about the cost of same. Therefore we wish now to inform you upon that point: We don't want a barrel that will cost any high figures on account of its having been tested and sighted for long range for we can do that ourselves. We simply want a good barrel and don't care about paying any "fancy" prices for same. We have been getting very good barrels from Remington at from $12.00 to $14.00. But for convenience of the thread and to save a little extra work we order from you, if it can be had at reasonable figures.

Respectfully,
Freund & Bro.

Then, nine days later another letter was sent to Sharps:

Cheyenne, Wyoming
May 23, 1877

Sharps Rifle Company
Bridgeport, Conn.
Gentlemen:

Your letters of the 17th inst. are at hand contents duly noted. In relation to the chambering of the .40 calibre barrel ordered from you I would say that I want you to make the tool for chambering the same and after using it for this barrel send it to me. In making the tool follow the directions given you in my last letter on this subject: that is make the tool for chambering of such a size as will make the chamber fit as closely as possible to a perfect seal and not drop into the gun as is the case with your system of chambering, and especially let the neck of the shell fit absolutely tight. In the matter of the estimate, I asked you to give, I would say that I did not expect estimate without a perfect model of the required changes, but thought you might give an approximate estimate. I will make, how-

ever, an exact model embracing only the changes necessary. As this model will be constructed, I would like your lowest estimate as to the cost of changing to my breech system 50, 100, 250, and 500 of your rifles. I want the estimate on each separate number of guns. The new model will embrace a slight change in the position and construction of firing pin so as to give a greater surface of resistance on the right side of the breech block than in your old style of breech. This of course is not absolutely necessary. But it can be done just as easily. It is now no longer a matter of my opinions as to whether my breech system is a success or failure, for military men high in rank and of great experience pronounce this arm with my system as perfect in all respects as it is possible to construct any arm. The parties to whom I allude have used your arms and know their defects and merits and I am now at work changing your rifles of which they are the owners to my breech system. Your arm as I modify it must take the lead in time and to make it do this is my purpose to introduce them to the attention of all. And if necessary cause a sufficient number to be altered and placed throughout the country where they will do the most good. In other words I propose to so introduce the arms with my improvements as will warrant any manufacturing company in making the necessary arrangements to manufacture the arms entire. The arm merits as you well know all I claim for it and with my system of breech action attached, your old style of guns will take a stand among the things that "were." For with the new movement I retain all the excellent features of your old system and attach with the same prices additional and valuable functions. The number of guns I have given you on which to make your estimate does not constitute by any means the entire number to be changed. For there are thousands of your guns now in daily use throughout the West and a large majority of these will want the improvements added to their guns. Which can't affect anything but a good profit to the manufactures. An arm with this improvement is just what the people in this section of the country are looking for and to manufacture the remodelling of the old style of arms is sure pay, for it is simple reason that the gun is in their possession until paid for.

  We sent you a model over three months ago and nothing of importance has been said on your part concerning it. Now while I am willing to give ample time to consideration to this matter, I don't want to yield to unnecessary delay. All I want in this matter is what is fair and right and if you fail to give us then we will have to look to other parties for it. If you don't want the improvements I would prefer that you would so state and if you do want it I wish also to know that.

I sent you firing pin so constructed as to carry the hammer to half cock and asked you what you would make 100 like it for. But have heard nothing from you. This half cock combined with the other improvements is pronounced by military men to be the safest arm for military use in the world.

Respectfully,

F.W. Freund

P.S. If you conclude to make the gun with my improvements it will of course be not necessary for me to make you a new model as you can make one to suit yourselves.

Respectfully,

F.W.F.

On May 21, 1877, Freund & Bro. wrote a five page letter to the Hull and Belden Company of Danbury, Connecticut (mechanists, tools, machine work and iron and steel forging), explaining the merits of their patent changes in the Sharps rifle.

On May 24, 1877, F. A. Hull, Jr., the president of the Hull and Belden Company, wrote to the Sharps Rifle Company as follows:[15]

Enclosed find a letter from Freund & Bro., Cheyenne, referring to modify some of your rifles. We send this to you thinking that it may be of interest to you. The gentlemen are partial in the line, know the needs of the shooters in the West. Possibly their ideas are some of them worthy of consideration. Please write them a reply to their letter. You can do this much better than us.

Pres. F.A. Hull, Jr.

By this time, the Sharps Company was getting fairly well fed-up with Frank W. Freund and his "improvement." Their disgust began to reveal itself in their answer to Hull & Belden:

We have already had enough correspondence on this subject with F. & Bro. to fill an octavo volume. We fail to see the extreme importance of the attachment as alluded to by F. & Bro. Still are perfectly willing to do the work if they will give us enough of it to make it an object to make tools

& c. and to satisfy us on the matter financially.

In answer to Freund's letter of May 23, the Sharps company wrote:

> *We have been looking into the matter of changing guns somewhat on the principle of the model which you sent us. It would be useless for us to take this unless we could have as the lowest calculation 500 at one time, & we think 1000 as few as we could start on. The cost of alteration wd be some where in the neighborhood of $5.00 each. We do not doubt but that the matter could be done mostly by hand work for all what wd be required.*

Freund probably thought that the price for alteration was too high, as for once he didn't answer this letter!

The company, however, was still answering letters from writers who wanted to know when they were going to adopt Freund's improvement. The position of the Sharps Rifle Company is fairly well summed-up a in a letter to one W. N. Bronson of Hot Springs, Colorado:[16]

> *In regard to the Freund improvement of Sharps rifles we have to say that we do not consider it so valuable as do Messrs. F. We have never adopted it and probably shall not. The improvement consists simply of an arrangement of the slide by which a shell or cartridge may be forced into the chamber of a gun. We claim that no cartridge should be forced in our rifle, that does not go into the chamber without any force other than of the fingers of the man using it. If the cartridge be forced in, the extractor will not be able to throw it out, & it must be forced out by the wiping rod.*

The remaining correspondence of the Freund brothers for 1877 is concerned with ordering parts, paying off their debts, and other miscellanous problems.

1878

Clotilda Freund was having difficulty adjusting to life in the West. She had been born on Walker Street in New York City, and as a child Clotilda would spend her Sunday afternoons rolling a hoop down to City Hall Park, where for two cents she could buy a hot ear of corn. Her father was a tailor, and as soon as she was old enough she had been taught to help him.

The houses that Frank owned in Cheyenne were one-story duplexes. Each had three rooms, with an extra shed joined to the main part of the house. Frank connected two of the duplexes together to make a more attractive and liveable home for his family. One of the kitchens was used as the maid's room after the first child arrived.

Winters were terrible in Cheyenne, quite a departure from those Clotilda had known in New York City. Frank advised her that she could no longer wear the flimsy underwear she had been used to back East, and he ordered what he believed would be a "suitable" kind for her— long red mens' flannels. After a lengthy dispute and some tears, Clotilda was allowed to compromise on the flannel, but in creamy white, and in a weight not quite as heavy.

The storms in Cheyenne were severe. Clotilda told her children of an incident during one of those blizzards, when it had snowed the whole night. In the morning, and against Clotilda's protests, Frank insisted on going to a store that was not too far away. In many places the snow was higher than one-story houses. Frank started out; then along about five o'clock in the afternoon he arrived back at the house weak and faint. He had never even found the store.

In better weather the family took their evening meals at the Inter-Ocean Hotel. Clotilda would walk down and meet Frank when they ate at the hotel. Francis E. Warren, later senator from Wyoming, and his wife also took their evening meals at the hotel. Clotilda and Mrs. Warren became good friends, and would visit each other often. The Warren's daughter later married General John Pershing of Spanish-American and First World War fame.

"General"

One of Frank's prized possessions in Cheyenne was his dog, "General Freund," who was an excellent hunting dog and very intelligent. At one time one of Frank's friends admired the dog so much that he finally prevailed upon Frank to sell him the dog. The new owner of course took General to his home, which was in Denver, but within two weeks' time the dog was back at Frank's house in Cheyenne, 106 miles away. No one ever knew how the dog did it, but Frank refused to consider parting with the dog again. (No mention was made of Frank refunding his friend's money!)[17]

The author believes that it also was during 1878 that Frank hunted buffalo. In the book *Powder River Country*, by Margaret Brock Hanson, there is a chapter from the works of Edward Burnett titled "Freund's Castle." The following quotation is from that article:

> *This big sandstone butte on the South Fork of Powder River, ten miles north of Salt Creek, is still called [by that name], but no one in the country knows why this name; so you boys have a little patience with me and I'll tell you why it is so named and all about Freund, the gunsmith, who improved on the Sharps rifle for the buffalo hunters.*
>
> *The Sharps rifle invented and patented by Christian Sharps was used extensively in the Civil War. It lived up to the name stamped on the barrel, "Old Reliable", heavy of barrel, strong breech block that inspired confidence; not a saddle gun at all, but just the gun when a stand was got on a bunch of buffalo, a very accurate shooter that was wanted as the buffalo hunter did not figure on using more than one cartridge a hide. Though I have heard that on the Southern range shot guns and buck shot were sometimes used. Buffalo hunters were not out for sport.*
>
> *Freund's camp was on the river bottom right there where the South Fork of Powder River bridge is; where it was easy to find, as the castle could be seen for miles around. He had his dugouts and drying ground. They made the melted lead and 5% tin almost red-hot in a dutch oven— buffalo-fat fire above and below.*
>
> *The next camp, down river, was Jim White's, at the spring 4 miles below old Fort Reno, and between them they worked this locality. It was Jim White's men, who in a joking way, named this butte "Freund's Castle."*
>
> *Freund's Castle does look something like a castle in the distance, though it's gigantic and must cover 100 acres. You can climb up it on the east end. It's sandstone formation. The cattle trail from Oregon passed here.*
>
> *Freund had worked in the buffalo range, and in gun matters he was quite a genius. The first thing he did was to put in another extractor. The old style rifle had only one extractor and when there was fast shooting and the rifle got dirty and the cartridge hard to get out, this extractor would sometimes break and that meant the ramrod had to be used, which was slow work. The hunters did carry extra ones but by the time they got*

it in, their stand was spoilt. Then Freund went to work on the breech block, which he cut down on one side and brazed a plate on the other so that it worked as a wedge and forced the cartridge into the chamber. These were great improvements, but I consider that his nicest work was on the sights. His patent front sight was a combination of three sights, two of them day sights— cloudy weather, red bead, sunny weather, black bead, and a silver one for night which enabled a hunter to do good work if the night was not real dark. The hind sights were wide spread and quick to catch the bead. These patents he put on a gun for $30 to $40. His workshop was in Cheyenne.[18]

From the above, Mr. Burnett claims that the sandstone butte was named "Freund's Castle" by Jim White's men. In early 1878, the partnership of Jim White and Frank Collinson split up, and White headed for Wyoming. In September White formed a partnership with Oliver Hanna, and by October Jim White had moved up to Montana, where he was murdered by unknown parties in 1880. It is believed that the most probable time for White's camp to have been near Freund's Castle would have been during mid-1878.

No other direct mention discovered by this author names Frank W. Freund as a buffalo hunter.

In early 1878, Frank again inquired about the status of his conversion of the Sharps action.[19]

Cheyenne, Wyo, Jan. 26, 1878

Sharps Rifle Co.
Bridgeport, Conn.
Gents:

We sent you nearly one year ago, a Model for breech improvement on your gun. At the time we sent it to you, you informed us that you were not prepared to answer the matter, just then. But wanted time to look into it. We would like to know if you have come to any conclusion on the subject. If you have not, please let us know about what length of time you require to further inspect it. If you have concluded and can make no use of it, please return us the Model.

Respt. Yours,
Freund & Bro.

On February 5th, E.G. Westcott wrote that he expected to be in Cheyenne in

March, and would discuss the improvement in person. Frank Freund replied to Westcott six days later:[20]

Cheyenne, Wyo. Feb. 11th, 1878

Mr. E.G. Westcott

Pres. Sharps Rifle Co.

Dear Sir:

Your favor of 5th is received. With references to the Model. You can keep it for the present. We shall be pleased to see you out here. As we think we can talk matters out better than we can write them. We understand, there is to be a board of Officers, convened in March to examine the merits of different breech loading guns. Would it not be a good idea to put in one of your 40/c with our improvement? As it has two (2) extractors and a superior breech movement. We think for accurate shooting and for the most flatness in trajectories the gun should get away with all others.

Respt. Yours,

Freund & Bro.

Evidently, Westcott never made the trip west. On March 12, Westcott wrote again: "We are too busy now to fit up to alter the 30 systems with your patent, for long range rifles."

This letter ended the correspondence between F.W. Freund and the Sharps Rifle Company on the matter of the Freund improvements. The last letters in the company letter books concerning the improvement appear in August 1879, after a burst of advertising activities by Freund in Western newspapers. Much favorable publicity was given to Freund, and his statements that "the Sharps will take a stand among things that 'were'." The Company position in the matter still remained the same:

Mr. Freund won't live long enough to hurt the reputation of our guns. There are thousands & tens of thousands of people who know better, who feel about it as you do, & whose experiences have been the same as yours. Mr. Freund kindly sent us a paper with the article marked. The so-called Freund improvement is simply a device for forcing the cartridge into the rifle & extracting the shells. The rifle is no better with the "improvement" than without it, and no worse.[21]

Later in this chapter is reproduced another letter from Clotilda Freund, urging

Sharps to put some of their idle equipment to work and to make a product they could sell.

Apparently a fine Freund rifle owned by General George Crook ended up with a broken stock. They wrote the Sharps Rifle Company the following:[22]

Cheyenne, Wyo, Feb, 27th, 1878

Sharps Rifle Company
Bridgeport, Conn.
Gents:

Enclosed, we send you the action of Gen. Crook's Rifle to put a new Pistol grip stock on. The Rifle stock was broken by Gen. Crook on the last campaign into the Big Horn and sent to us to be restocked. He will be here at Cheyenne shortly and will want his Rifle to use. Please make it without delay. And if you have any charges, send the bill to us. We will not make any charges to him.

Respt. Yours,

Freund & Bros

On March 19, 1878 the Freunds received a bill from Sharps for $16.00, for the repair to General Crook's rifle.[23]

On the 15th of the following month, the Freund brothers again wrote Sharps explaining to them why the Westerners preferred octagon barrels:

Cheyenne, Wyo., April 15th, 1878

Sharps Rifle Company
Bridgeport, Conn.
Gents:

Yours of the 8th is received and contents noted. In ordering the rifle as we did "Octagon" that is the only way we can use it for our purposes as we find more accurate shooting can be done with it owing to the fact that at times hunters and others desire to take a rest with the gun. And this can be done better and more sure with a flat than a round surface. We do not suppose that the cost of the gun will be more than the extra cost long barrel. We would like the sides of the breech where you cut the plate and insert wood, to be solid. Instead of cut for wood.

Respt. Yours,

Freund & Bro.

However, the big news in April 1878 was the following letter to Edw. G. Westcott, President of the Sharps Rifle Company:[25]

Cheyenne, Wyo. April 4th 1878

Mr. Edw. G. Westcott
Pres. Sharps Rifle Company
Dr. Sir:

> *Enclosed we send you Draft No. 26417. for $17 15/100. This closes our old a/c to date. We have sold our entire stock of goods to Mr. F. A. Dammann, of this place. Any goods he may want of you. We recommend him as a perfectly reliable man to do business with. We are still engaged in the manufacturing and repairing, and are kept pretty busy, altering Sharps Rifles to our new Breech movement.*
>
> *Please send us the following order—*
> *6 Lever Springs, 3 Set Triggers*
> *3 Hammers, 3 Breech Blocks*
> *3 Hind Stocks Creedmoor Butt.*
> *3 Main Springs & 3 Guides for Hind Sights*

also one (1) 45/90 Rifle Model 1878, 34 inch Octagon Barrel. Creedmoor weight. Not to be an expensive but plain gun. Please send the goods to Mr. Dammann, for the purposes of opening trade with him. You can do so. And we will take them from him. Sometime in February we received a letter from you stating that you would be here in March. Do you still entertain this idea of coming? If so please let us know about how soon. We think you will lose nothing by coming out.

> *Yours Truly,*
> *Freund Bros.*
>
> *In looking over your Bill again, we find it only for $14 85/100. Please credit us for Balance $2.40 & oblige.*
> *F & B*

F. A. Dammann had purchased all of the stock of the Freund Brothers store. Frank and George would continue to work in the shop, continuing with their manufacturing, converting, and repairing of firearms. They would receive all their parts through Dammann. They did not miss the chance to goad Sharps that they were too busy altering Sharps rifles to take care of the store. With

Dammann taking care of the store, this may also explain how Frank found time to go buffalo hunting in 1878.

Also note that they ordered a .45/90 Sharps rifle Model 1878 with a 34-inch octagon barrel in this same letter. To the best of the author's knowledge, this is the first complete rifle ordered by the Freund brothers from their Cheyenne location.

The Freund Brothers and Dammann received the following rifles from Sharps in 1878:

5/24/78	1 - 1878 Sporting Rifle .45/90
8/29/78	1 - 1878 Sporting Rifle .40 cal.
9/9/78	4 - 1874 Sporting Rifles
	1 - 1874 Business Rifle
	3 - 1874 Long Range Rifles
Total:	10 Sharps Rifles from factory in 1878.

In addition to the 10 rifles, they received parts orders on 3/19, 4/20, 4/22, 6/17, 7/1, 7/29, 8/22, 8/29, 9/21, 10/1, 10/14, 10/19, 10/30, 12/10, and 12/26.[26]

On October 5, 1878, Frank applied for a government patent for an "Improvement in Breech-Loading Fire-Arms." It was an improvement on the Ballard type of gun, and was granted U.S. patent number 211728.

1879

In the year 1879 the first child, a son, was born to Clotilda and Frank, and named William F. Freund. William joined the U.S. Navy when he reached maturity, served on the USS *Kansas*, and was captain of the rowing team, which was quite sucessful in winning many trophies. William died of appendicitis while at sea, at the age of 47 years.[27]

In the meantime, F. A. Dammann continued to operate the store, while Frank and George continued to operate the shop. It is interesting to note that there were not any patents applied for in 1879, so both Frank and George must have been busy converting Sharps rifles to Frank's patent.

F. A. Dammann seemed to be having a problem with the ordering of Sharps parts, as evidenced by the paragraph that followed an order for parts on January 14, 1879:[28]

By reading your letter over again, I cannot perceive the necessity of delaying my order, as my expression Breech Bl was certainly understandable to you, for you understood that I wanted 4 Breech Blocks without something. I called this something a slide & firing pin, you certainly understood what firing pin means, if you had reflected a moment you would have known what I meant for slide, for I combined the words slide & firing pin. This slide is a piece of steel covering the gas tube & firing pin w both in rear of slide or Breech Bl.

You may use your own expressions in your circular, but that don't signify that your customers should understand such without specific illustrations, but my terms were plain and audible enough to be understood.

Now please understand that the slides or Br Bl should be without firing bolt and the strip of steel in the rear of Br Bl slide.

Send goods by Express as soon as possible.

Respectfully,

F.A. Dammann

Please send me a set of illustrated circulars, containing all parts and pieces which I might hereafter order

He finally received these parts in an order shipped on February 14, 1879. On this same date of February 14, he had written Sharps again:

Have you forgotten the gun parts I ordered some time ago? Please send these articles as soon as possble as I am in need of same.

On February 8, the Sharps Rifle Company offered Dammann Model 1874 military rifles at $16.00 each. There is no record of Dammann or the Freund brothers taking advantage of this offer. They sold a large number of 1874 Sharps military rifles, but they probably purchased them from one of the large Eastern wholesale houses.

On February 19, 1879, Dammann ordered sixty sporting Model 1874 rifles made up. These 1874 rifles were sent to N. R. Davis & Co. of Cheyenne. N. R. Davis & Co. were traders in early Cheyenne, and apparently had the resources to pay for the sixty rifles. They probably then released these rifles on an individual basis as Dammann could pay for them.

On April 26, he ordered:

500 cartridges 40/90/370 lead patched ball
500 cartridges 40/70/330 B.N. lead patched ball

These were shipped from Sharps on May 1, 1879 which is probably faster action than you could receive today from Cheyenne to Bridgeport, Connecticut, if everything had to go by mail or freight.

In an order written on May 19, 1879, he writes:

> *Concerning the 2 Creedmoor Rifles which I have on hand yet, I must say that I cannot sell them in the state they are now, but I may dispose of them without these sights. Therefore inform me in your next what reduction you will make if I send you the scale Vernier & Peep sight also the spirit level windage front sight back.*

On May 20, 1879, Dammann ordered 5 rifles with English Stocks, but was told that the Sharps Rifle Co. cannot make "Freund Improvements." This implies that Dammann ordered the 5 rifles with the Freund conversions on them, and that the company was still set against making any of these changes.

The above is the last correspondence that the author or Dr. Moore have been able to locate between Dammann and the Sharps Rifle Company, until this final letter of September 6, 1879:[29]

> *Cheyenne, W.T., Sept. 6, 1879*
>
> *Sharps Rifle Company*
> *Bridgeport, Conn.*
>
> *Dear Sirs:*
>
> *I herewith notify you that I have sold all my interest in the Wyoming Armory, known under the firm name of F.A. Dammann to Mrs. C. Freund, wife of Mr. F.W. Freund. My successor to pay all my debts, contracted heretofore by me and also to collect all outstanding debts due the firm.*
>
> *I am authorized by Messrs. Freund Bros., managers of the firm, to say that they will be able to pay all debts as they go along. Thanking you for the many favors you have ever shown me during our short business relations, I hope you will carry them over to my successor.*
>
> *Very Respectfully,*
> *F.A. Dammann*

On the same date, Clotilda Freund sent a draft for $100.00 to Sharps, along with a note stating that more particulars would follow.

Her next letter was on September 17, 1879, and reads as follows:[30]

Cheyenne, Wyo. Sept. 17, 1879

Sharps Rifle Co.
Bridgeport, Conn.

Dr. Sirs:

Your statement of Mr. F.A. Dammann's a/c came to hand and comparing with my Bank, find that there is only a Balance due yours of $150.72. Please find enclosed Draft on N.Y. No. 17389 for the Amount, and send me a receipt for Balance in full.

I take great pleasure in informing you that all Bills contracted by Mr. Dammann have been promptly paid. Send by Mail as soon as possible 6. Sharps rear Sights for Sporting Rifle Complete. The sights are to be scaled graduated.

Yours Respectfully,
C. Freund
F. Br. Manager

Notice that Clotilda had the title of Freund Brothers' manager on this letter.

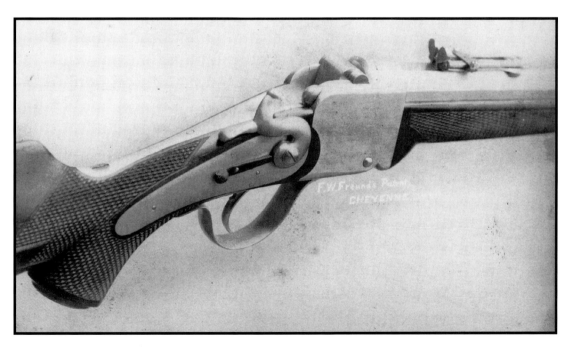

Frank W. Freund's "falling-block" rifle, probably an intermediate model before Freund's Wyoming Saddle Rifle. The back of the photograph states "low hammer and safety slide specially adapted for horseback use." The front of the mounting card is marked "C.D. Kirkland. Cheyenne, Wyo." (Freund Family Collection)

On September 19, 1879, Clotilda again wrote Sharps requesting that the following rifles be supplied:[31]

Cheyenne, W.T. Sept. 19th, 1879

Sharps Rifle Company
Bridgeport, Conn.

Gentlemen:

Please inform me by return mail, whether you can furnish us the following described Rifles, and how soon.
3 Sharps Rifles, oct., double trigger, 26 in long, 8½ to weight chambered 70 grain, cal 40.
3 Sharps Rifles 28 in long, 9½ to weight, chambered 90 grain.
All with checkered pistol grip & checkered front stock.

Very Respectfully,
C. Freund
per F. Bro.

P.S. Please inform us to what price you can furnish us the altered 45 cal. Carbine and also the 40 cal. Carbine & 45 Military. We refer to the guns made up of old materials & new barrels.

This letter was processed on October 6, 1879.

On September 22, 1879, an order was received by Sharps from Clotilda Freund, but E.G. Westcott refused to fill the order until the Dammann account was paid in full.

Apparently, their correspondence must have crossed in the mail, as the order for the six rifles of September 19, 1879, was processed by Sharps on October 6th.

Clotilda Freund answered this letter of October 6, stating that the 90 grain rifles must weigh 9½ pounds with the barrels being 28 inches in length, as stated. She also advised that one of each could have an English walnut stock, but all of them must have the steel butt plate:

The parts which you may leave in soft state are the following:
Receiver, slide, firing pin, extractor, lock plate, hammer, and butt plate.

Obviously, the Freund brothers had in mind applying some of their fine Germanic engraving on some of those gun parts.

On October 15th, Frank A. Dammann entered the picture again, writing the correspondence for Clotilda and signing her name with the initials F.A.D. below the signature. This continued until November 6, 1879, which is the last letter of which the author has knowledge of Dammann writing for Clotilda.

The Sharps records for 1879 show complete rifles being shipped to N.R. Davis Co. in Cheyenne. The records show parts being shipped to the Freund brothers and Dammann on 2/14, 3/19, 4/8, 5/1, 5/24, 11/12, 11/15, and 11/12. The records are very sketchy in the period after May 24th. A reason for this may be the inability of F.A. Dammann to order parts on credit, from approximately this date until the Freund brothers again took control of the store's business.

1880

The first and only patent that Frank and George applied for together was on March 9, 1880. It was for a sight for firearms, and is the patent for the More-Light Sight, having a diamond-shaped hole below the sight notch. It was granted United States patent no. 229245.

Apparently in March 1880, the Freund brothers and Clotilda were experiencing financial problems, as evidenced by the following:[32]

Freund Bros. Office Wyoming Armory

Cheyenne, Wyo. Ter., March 2, 1880

Sharps Rifle Company
Bridgeport, Conn.

Gents:

Enclosed please find draft for $150.00, all I can send you today. My sales at present are very slow, and most of my customers live a distance, and it takes quite a while to receive remittance from them. I will however send you the Balance as soon as possible. Please acknowledge receipt of Draft. What will you charge to make a Rifle of the following description, Oct. Barrel 40 cal, 26 inch length, Double trig., Pistol Grip, checkered Butt Plate (English Walnut is required), Pistol Grip, and good stock nicely checkered. The Breech frame is to be without the Bulges on side, but straight as in the drawing. Even if you have to make it by hand. The goods are going to a place where it will do the Sharps Company as well as myself considerable good.

Respt. Yours,

C. Freund

On March 29, 1880, the company wrote back stating the balance was $246.00, and terms were cash every 80 days since December 15, 1879.

Cheyenne, Wyo. Ter., April 3, 1880

Sharps Rifle Company

Gents:

Enclosed find Draft on New York for two hundred & fify (250) Dollars. Credit me the balance of old Account, and what is over on new a/c by order below. And acknowledge receipt of same.

By Freight as follows

All Bottle Neck

1 M Ctgs.	*40 Cal*	*70 Grain Powder &*	*330 Gr. Lead*	
1 M "	*40 "*	*90 "*	*"*	*370 " "*
1 M "	*44 "*	*90 "*	*"*	*500 " "*
1 M "	*45 "*	*100 "*	*"*	*Straight 500 " "*

By Express

1 Sporting Rifle 40 cal, 90 Gr., octagon 28 in Barrel, Double trigger, pistol grip, Front & hind stock and Butt plate checkered, Weight of gun 9 1/2 pounds, Model 1874. Would prefer the stock English Walnut. 1 do. do. 26 barrel 10# weight of gun. The bend of stock in these guns as the Sporting Rifle.

Would prefer the pieces in soft State if Convenient, but the barrels browned. And if you cannot send English Walnut, Please select a nice Stock of American Walnut.

By Mail

6 Sporting rear Sights, Complete.

1 doz. Extra slides for Military rear sights. As by enclosed drawing. You can send same time the 6 Sporting rear sights with the above, which was Short in Bill of Nov. 12, 1879. Of which we gave you notice same time as received. Please send the sights or credit us the Amount. And Oblige.

Respectfully yours,

C. Freund[33]

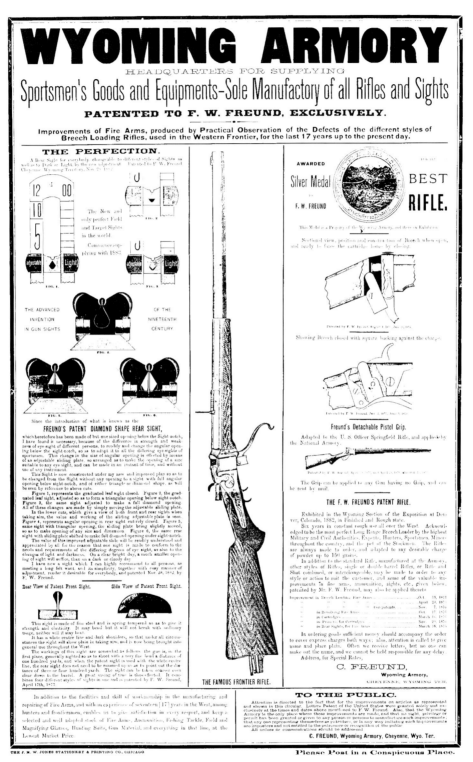

"Wyoming Armory" broadside, 1883; Clotilda's name at bottom. (Courtesy Dr. R.L. Moore, Jr.)

Sharps replied to Clotilda that they would advise her to order all ammunition from U.M.C. The company also advised her that they could not make up any Model 1874 rifles, ever again.

However, on April 14, 1880, they offered to Clotilda Freund 30 complete actions that were cut out to lighten the action so that heavier barrels could be used, at $15.00 each. Total weight of each was to be less than ten pounds. They sent a sample and remarked that it was the last chance to get Model 1874s.

Sharps again wrote on April 24, 1880, that they could supply the foregoing in the "soft state" at $23.00 each, and this was the last chance for the Model 1874.[34]

Clotilda wrote back on May 12, 1880, in reply to Sharps' offer:[35]

Cheyenne, Wyo. Ter., May 12, 1880

Sharps Rifle Co.
Bridgeport, Conn.

Gents:

Yours of 24 inst. duly received also the bridge frame & lock came to hand since writing to you. I don't know exactly what to do with the bridge frame or guns in the state you offered them to me. And therefore have to decline your offer. If you would get up the guns with stock & hammers & etc would be pleased to have another offer from you. Any material you have of 1874 such as springs, screws, parts, locks, & stock, receivers think of if you wish to sell, you might make out a statement & send to us. I have at my disposal from parties quite a number of Sharps Rifles, Model 1874. As to the Cartridges, using the UMC. This will only help Freund's Improvement.

Respectfully,

C. Freund

Clotilda again ordered paper patches on October 2, 1880, and sent a check for $22.85 to bring their account up to date.[36]

It was in 1880 that Frank W. and George dissolved their partnership, with Frank buying out George. No definite records for that transaction exist, to the author's knowledge. Most historians seem to think that the breakup was over financial difficulties. Whatever the problem was, the brothers seem to have parted as friends, for they had several business dealings after George moved to Durango.

Apparently Clotilda must have had a good opinion of George, as she introduced him to her sister, Ida Gasperrini, whom George later married (*see* chapter six).

Sometime after the first of October, 1880, and before September of 1881, Frank and Clotilda changed their business stationery. Until then, they had been using stationery with the Sharps Model 1874 depicted on it. After October, 1880 they switched to stationery with a cut of the Freund's patent camming action, and with the business name "Wyoming Armory, C. Freund" (*see* page 56).

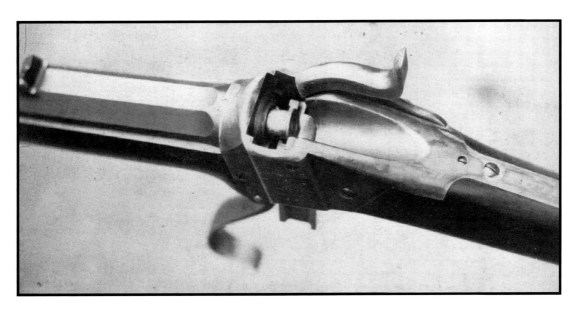

Freund patent altered rifles showing double extractors (top photo); rounded top on the camming block (lower photo). Note also the Freund patent rear sight and slim-waisted tang. (Photos by J.S. Mitchell, Eddy Street, Cheyenne, Wyoming Territory; Freund Family Collection)

1881

The following article was published in the Cheyenne *Weekly Leader* on December 22, 1881:

C. Freund

Wyoming Armory— Firearms, Ammunition, Sportsmen's Goods, Equipment, Etc.

We desire that special attention be given to the subject matter of this article.

We have a chance to expand, for here Cheyenne leads the world.

An industry is in our midst, that will 'ere long cause the name and fame of our city to be heralded unto the utmost parts of the earth.

We have amongst us a citizen who is a superior representative of the practical genius of the nineteenth century. The world delights to honor her great discoverers and place their name upon the very pinnacle of fame's proud temple. Permit us to step to the front. We herewith proudly present the name of Mr. Freund for a position in that luminous cluster of James Watts, George Stephenson, and Benjamin Franklin. Mr. Freund is an already eminent mechanic, to whose great skill and original genius, we owe the greatest improvements yet developed in firearms. By the constant study of every piece of mechanism that came in his way, through many years of enlightened activity, he acquired a large amount of sound practical knowledge, which upon being applied, has resulted in the discoveries that will immortalize his name and bring imperishable fame to the city of his adoption. We have not the space necessary to a complete description of all of Mr. Freund's improvements, but the principal are as follow:

In breech-loading fire-arms pat'd.	*Oct. 13, 1873*
" " " " " "	*Apr. 24, 1873*
" " " " " "	*Nov. 7, 1876*
Revolving fire-arms pat'd.	*Oct. 17, 1876*
In cartridges pat'd.	*Mar. 16, 1876*
In primers for cartridges pat'd.	*Nov. 29, 1876*
In rear sights for fire-arms	*Mar. 16, 1876*
In front sights for fire-arms	*Apr. 17,1877*
Detachable pistol-grip— three patents	
In rear sights, pat'd.	*June 29, 1880*
Two patents in breech blocks	*Jan. 28, 1879*
	and June 3, 1879

The Freund improved rifle is the most perfect long range gun in the world.

Its distinctive features are: No clubbing home of cartridges after the gun has been fired, which has led to so many serious accidents; no rearward escape of gas to injure the eyes: no premature discharge by loading or carrying the rifle; no working loose of breech parts: and no flaying out of breech block. The rifle is extremely simple and its parts very durably constructed. The breech mechanism can be taken apart and put together again in a moment's time, without the use of instruments or tools. The breech block has a rocking motion, thereby forcing the cartridge home by the closing of the breech. It has double extractors, the lock is thrown back at rest or half cocked by the positive motion of opening or closing of the breech. The breech block has its resistance in a right angle with the line of the bore, and direct in rear of the charge. The inside of the barrel can be seen and cleaned from the rear, the firing bolt and breech block are so constructed as to prevent breaking of caps and rear escape of gas, and to exclude all dirt, thereby preventing obstructions and misfire. Single or double triggers can be used. These guns are made, usually, forty caliber, ninety grains, and forty-five caliber up to one-hundred and fifty grains of powder. They have a pistol grip and a cleaning rod of hickory wood in the wiper and a brush in the butt of the stock, and the new patent sights. The rifle is also made of the very best material, and in the most careful manner, is severely and thoroughly tested, and the price including all of above is one hundred dollars, but are made up to three hundred dollars, according to extra work and engravings, fine cases, fixtures, etc.

In ordering rifles the payment of twenty-five dollars should accompany the order from parties not known. Satisfaction is guaranteed in every respect, and illustrated pamphlets will be sent on application.

The new Freund patent field and target sights are the greatest invention accomplished in gun sights. Both front and rear sights combine the most essential points of the finest and most perfect sights heretofore made, and experts declare the invention to be beyond improvement.

In proof of the foregoing, we present a few extracts from letters received by Mr. Freund:

Gen'l. Phil Sheridan says: "The more I see of your patent breech block, the more strongly am I impressed with the practical ideas you have developed in your invention."

Gen'l. Crook says: "The improvements need only to be seen by practical men, to commend themselves to general use."

Gen'l. W.E. Strong, of Chicago, president of the Peslittigo Lumber company, says: "I have tested thoroughly patent sights recently put on my Winchester rifle, and in my opinion they are by long odds the best ever invented for hunting purposes. On my recent trip I was astonished and delighted at the perfect sighting of the rifle, and the certainty with which I could strike the object aimed at up to 150 yards, by taking a fine or coarse bead, without raising the graduated leaf. I have shown the sights to hunters and experts who pronounce them the most perfect ever seen."

There are hundreds more of these letters, but want of space precludes their publication. Suffice it to say, that after five years of constant rough use all over the west, Freund's patent long range rifle is acknowledged to be the most perfect breech loader, by the highest military and civil authorities, experts, hunters, miners, and sportsmen throughout the country.

These guns are sent to all sections of the world, orders being received from South Africa, India, and nearly every large city in Europe, and more are made and sent to New York city than anywhere else. Orders are now lying on Mr. Freund's desk from Englishmen who will call for their rifles next spring.

All of these orders are from persons who are willing to pay the large prices attached to the manufacture of rifles by hand, rather than be without them and this of itself should be sufficient guarantee of their superior excellence.

The question is often asked, "Why do not the leading manufactures of the United States, or some of them, adopt the Freund improvement?" The question is easily answered: by adopting another, their own would be thrown in the shade, and they act upon the idea that their own inventions must be held up to the world as being the best. These erroneous ideas, coupled with jealousy have thus far shut out the Freund improvement. Mr. Freund, however, does not complain, knowing full well that when the merits of his invention are fully understood, they will be adopted and used in preference to all others, as they are undoubtedly the best the world has ever seen. We trust that a company may soon be organized for the cheap manufacture of these rifles so that the public generally may reap the benefit of these great improvements.

Mr. Freund has a most extensive stock of all the leading fire-arms made, from the finest finish to the cheapest grade. His armory also con-

tains an immense stock of every variety of revolver, fishing tackle, pocket and hunting cutlery, etc., together with full assortments of every description of sportsmen's goods and equipments.

All orders will be filled at as low rates as can be given elsewhere and complete satisfaction guaranteed.[37]

Clotilda kept up her correspondence with the Sharps Rifle Company throughout 1881, until the last letter that the author has been able to locate which was dated November 5, 1881. Altogether, the author has been able to locate only four letters for that year. The first three refer to problems with sights that either were shipped or not shipped. The final letter has a suggestion for Sharps and Mr. Ed. Westcott:[38]

Cheyenne, Wyo., November 5, 1881

Sharps Rifle Company
Bridgeport, Conn.

Your last circular received. Please send me catalog of what you have for sale as you may have some materials which I may find useful: in Materials & Tools. You say the Sporting Rifle Model 74 will be made again & you have some of the old ones at hand. I have bought some 28 inch w/Triggers 40cal. from Hartley & Graham for $22.50 regular Model 74. The cheapest I can hear of & in the hands of Dealers. I cannot see how it could be possible for any one not to take up the opportunity of making a gun which everybody wants and which will give a good Profit unless the old ones did not pay or prove profitable—. If you have nothing to do & don't know what to do with your tools, why not make some of the Freund's improved in small quantities. You will soon find a Market & realize a good reward, you will find something to build up a nice little business & will if started in small scale & if run economically will meet with success in the End.

In making my gun you have no competition you also have to by yourself specially finish guns & could get your Price for them & if any opportunity shall come to get contracts for larger quantities you will have as good a Show as anybody. See the Ballard Co. how many guns the total of Sporting Rifles & there is a good Deal of work on that Gun. I respectfully submit this to your consideration, but am not at all anticipated in receiving favorable reply.

Respectfully yours,
C. Freund

In May of 1881, Angie Freund was born. When grown, she married John Keegan. Clotilda lived with Angie and her husband after her daughter, Jeanette, married. (Jeanette was the youngest child, born in New Jersey). Angie passed away on October 24, 1954. While Clotilda was living with Angie, the attic and second floor of their home caught fire. Some of the Freund records and perhaps some of the early Freund-made guns were lost in the fire. Frank's tool chest survived the fire because it was stored in the basement, although it suffered water damage.

Two other daughters were born to Frank and Clotilda while they lived in Cheyenne, although we have been unable to determine the dates of their births. One of them was named Elsie, who died when she was but eleven years old. The other daughter was named Clotilda, and she died in infancy. Frank offered the doctor a gun if he could save the little one's life, but to no avail.[39]

Frank applied for another patent on August 9, 1881. The patent was a refinement on George and Frank's sight patent of 1880. It featured a sliding plate on the opening below the sight. The notch could then be opened, partially opened, or closed.

1882

McKenny's Business Directory of the principal towns of Utah, Wyoming, Colorado, and Nebraska for 1882 lists on page 797:

Freund, C., Wyoming Armory, guns, pistols, and ammunition, Fergerson Street, Cheyenne, Wyo.[40]

Obverse and reverse of bronze medal awarded for rifles to F.W. Freund in 1883 by the American Institute in New York. (Freund Family Collection)

The following poem was written at the request of the friends of Frank W. Freund, and was dedicated to him and his family, then on a visit to the old house of Clotilda in New York City.

MY LITTLE FAMILY

by W.P. Carroll

I am lonesome, and gloomy, and sad here today,
And I'm thinking of those that are gone—
My wife and my babies are far far away,
While I linger and plod here alone.

It was long ago since, in the years that have fled—
and my heart travels back to that land
There I whispered to "Clo,"—never mind what she said:
But in answer she gave me her hand.

We have journeyed together since, out on the road
Leading down through this life to the tomb—
Which is naught but the door to a future abode,
In a land of perennial bloom.

There are little ones now, with blue eyes and black,
Our "Angie", with soft, dimpled cheek,
And Tillie, my Will, with his bright eyes so dark,
And another of whom I will speak.

"Tis Elsie, our darling, our baby so sweet,
The sunlight and cheer of my life,
The picture of one whom I'm longing to greet—
My "Clo", my first darling, my wife.

They may talk about riches, and land, and of gold,
But my pets are far dearer to me
Than fame, wealth and honor, and fortunes untold,
And the gems of the ocean and sea.

Since they left me and went to a far distant land
The flowers they droop and fade,
And are missing the care which that true little hand
Has accorded each blossom and blade.

My home is now dreary, and dark, and in gloom,
Since my dear ones have all gone away.
But I'm counting the hours when again they'll come home,
And I wish that the time was today.

And may the Great Ruler protect each and all
And bring them safe back here again,
My "Clo", and my "Will", and my cherubs so small,
To our home on this border-land plain.

Then united again— our own band—
We will go all the way to the shore,
And at last then we cross to that far-away land,
Together we'll walk evermore.

(Quoted from the Cheyenne *Sun*, Sunday, February 19, 1882)[41]

1883

The winters in Wyoming were still very fierce, and especially so for Clotilda. She and Frank went to visit friends at a ranch one weekend, traveling across the prairie with a horse and buggy. When it came time to head home, everyone advised Frank against starting across the prairie, but of course he insisted. So they started out with their small baby. The wind came up, and they drove through the blinding storm for miles. Finally, as they neared Cheyenne and the storm grew worse, the horse became frightened and started to run away. Frank called to Clotilda to jump. They both jumped, each thinking the other had the baby. The horse went dashing off with the baby alone in the buggy. The carriage broke loose and turned over, but underneath everything, completely warm and snug in the carriage robes, was the baby still fast asleep. The storm kept up for the next week, and the family out at the lonely ranch suffered great hardship. Food supplies were down, and they were forced to break-up furniture to burn in order to keep warm. Of course, all of this did not serve to pacify Clotilda about living in the West.[43]

In the *Residence and Business Directory of Cheyenne*, compiled by Johnson & Tuthill in 1883, the Freund brothers' advertisement appears on page 105 as follows:[42]

WYOMING ARMORY

Fire Arms and Sporting Goods
of all descrptions pertaining to the business in large variety at the lowest market prices. Sole manufacturing place for the
FAMOUS FREUND's PATENT LONG-RANGE FRONTIER RIFLE AND CELEBRATED NEW PATENT SIGHTS, the advanced invention in Gun Sights of the nineteenth century, and none are genuine unless stamped and numbered, which is only done by the patentee. The only establishment in Wyoming where Guns can be made and repaired in a superior and professional style, and equal in workmanship and merits to any place in the world.
For special rates, address
C. Freund, Cheyenne, Wyo.
Address, 369 16th Street

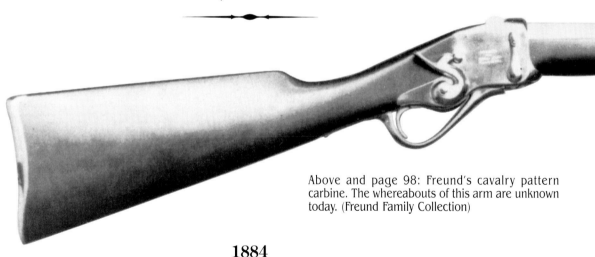

Above and page 98: Freund's cavalry pattern carbine. The whereabouts of this arm are unknown today. (Freund Family Collection)

1884

Clotilda never did get used to living in the West. She cried often, and especially when the song popular at the time, "Write Me A Letter From Home," was played. Whenever they were visiting and this song was played, Clotilda would cry and have to leave. All of this, no doubt, was having an effect on Frank.[44]

Frank applied for one more patent, number 496051, for a gun sight on October 18, 1884. This applied to a rear sight which was to be attached rigidly to the barrel, with a series of interchangeable sight plates that could be changed at will. It was the last patent that Frank W. Freund applied for, and it was not granted until April 25, 1893.

Among a group of old photographs of different F.W. Freund guns are two showing a working model of a military carbine. The action is essentially the same as that of a fine sporting rifle, and it is known to collectors as the "False-Hammer Freund," for lack of a better name and because, strangely enough, it does have a "false-hammer."

Nothing is known about the history of the false-hammer military model except that it was made by Frank Freund before he moved east in 1885. That it was made in Cheyenne would seem to be confirmed by the presence of a Cheyenne photographer's address on the photograph. It also is apparent that with this false-hammer action F.W. Freund had reached his goal of a perfect single-shot, falling-block action. As the letter from Congressman M.E. Post quoted below suggests, this military model was made by F.W. Freund, apparently in 1884 or 1885, to submit to the Ordnance Board for consideration.

At the bottom of this page are captions found on the photographs of the false-hammer military model rifle:

Rifle of Cavalry pattern, 28 inch barrel, Half Stock, Joint Cleaning Rods in butt of Stock.

The pattern and general style of this arm were suggested to me by the late Capt. Clark (of the 2nd Cavalry Reg't and Lieut-General Sheridan's staff U.S.A).

The Front Sight cannot be seen in the photograph on account of protection on each side, especially adapted for use on horseback.

A corresponding drawing of this rifle was sent to U.S.A. Headquarters before the construction of this arm, as photographed.

On the second picture the caption is shorter (both descriptions are pasted onto the pictures, and appear to have been clipped from a printed folder or brochure. Part of the printing has been erased for some unknown reason):

Cavalry Pattern. Bands superseded. Lock and Breech mechanism detachable instantly without ... instruments, and can be replaced as quickly.

The officer Freund credits with suggesting the general pattern of this rifle must have been Captain William Philo Clark, a young officer who served on the frontier for about ten years prior to his death in 1884. The Indians called him "White Hat", and as a lieutenant he played an important part at the surrender of the famous Crazy Horse in 1877. Captain Clark was well liked by the Indians, and brought in the remnant of the Cheyennes under Little Wolf after their epic march of 1878-79. He was an expert in sign language, and after his death his book, *Indian Sign Language*, was published in Philadelphia in 1885.

Exactly what did Captain Clark suggest to Frank Freund regarding the design for this rifle? Was the false-hammer his idea, to get around the usual Ordnance Board objection to a hammerless action? The three letters which follow are from the Freund family scrapbook, and give some idea of the origin and plans for this military rifle:

House of Representatives U.S. Washington, D. C.

May 23, 1884

Dear Mr. Freund,

I saw Captain Clark yesterday and had a long talk with him in relation to your matters. He tells me that he has written you fully on the subject. I am convinced that the only thing to be done under the circumstances, is that you make a carbine in accordance with his suggestions, and if it stands the test, as I have no doubt it will, the War Department will recommend an appropriation for funds to make say five hundred for a field test. With such recommendation there will be no trouble in securing the necessary appropriations.

Captain Clark tells me that some other arm will have to be substituted for the carbine now in use. I think it is a prize worth struggling for and would suggest that you make a big effort to secure same. I shall be glad to do all I can to assist you.

Very truly yours,

M.E. Post

Temple Court, New York

April 14th, 1885

Messrs. Richards & Co.,
24 & 26 4th Ave. N.Y.

Dear Sirs:

I am in receipt of yours of the 9th inst. asking my opinion of the

"Freund Rifle"

In reply, I beg you leave to say, I have , myself, personally and most thoroughly examined the "Freund Rifle" and sights, and am, I believe, very well acquainted with all the other leading makes of fire-arms, and from my experience and knowledge of such matters (gained by four years service in the field during the late rebellion) I feel no hesitation stating that the "Freund Rifle" military and sporting, and the Freund sights, are the best, most perfect and complete as any in existence.

As an illustration and comparison, the Freund Rifle and sights are as far superior to any other gun of the present day, as the first percussion capped gun of the past was superior to the smooth bore and flint lock.

The great and leading points of merit and advantage of the Freund Rifle are:

1st. Its great simplicity, as all its parts can be instantaneously detached and reassembled by anyone without the use of tools: because every part of the gun lock, stock, sights, and barrel, seems to dove-tail and fit in its proper place without screws, clamps, or any device so annoying to gunners. That fact of itself makes it a gun of especial value for military and frontier service, for no other gun in existence possesses that merit.

2nd. Its great safety and strength, on account of the solid breech block coming up squarely behind the cartridge renders it absolutely safe, while at the same time its peculiar action in forcing the cartridge home, combined with the double automatic extractor, admits of greater rapidity of firing.

3rd. The great benefit derived through the double extractor in equalizing contraction and expansion.

Mechanically, it has no rival, and from every standpoint, it appears to be beyond competition.

Not only is it the best, but in my opinion, if properly managed it can be produced ten (10) percent cheaper than any other first class gun on the market, and I will tell you why. Because it has but 31 pieces, all told—lock, stock, barrel, and sight; which is ten pieces less than any other gun of the same standard in the world. Read the following.

The total number of pieces in the different systems are as follows, not counting the sights as any sight may be used on either of the rifles:

Boshardt's Sharp's (sic) *61*
Springfield, U.S. System *60*
Sharp's (sic) *Michigan and other States* *60*
Remington, State of New York and others.......... *51*
Peabody-Martini, Turko-Russian *53*
Martini-Henry. British System *53*
Brown Standard Rifle *40*
Freund's Rifle (including sights)........................ *31*

Interior view of Freund's "Wyoming Armory" (probably the store located at 369 16th Street) in 1885, five years after the brothers dissolved their partnership. Note the sign advertising "Sharps' Rifles." (Freund Family Collection)

Besides being less in number, the difficult pieces in Freund's system are far simpler and easier to make and clean than those in either of the others, being in the aggregate lighter whilst at the same time stronger.

In my opinion, if it is handled with only ordinary business ability, it will undoubtedly out-sell any other rifle now made, as greatly as it now excells all other rifles in point of merit.

In conclusion, permit me to thank you for the value you seem to place on my opinion, and to say that I hope it may prove to be as of service to you.

Very Respectfully yours,

H.M. Munsell, Late Captain Company C.
99th Regiment Pa. Veteran Volunteers

* * * *

Hartley & Graham Arms and Ammunition
Agents Union Metallic Cartridge Co.
Bridgeport, Conn. P.O. Box 1760

New York, May 23rd, 1885

Mr. F.W. Freund
Denver, Colorado
Dear Sir:

We have examined your Patent Rifle (Single Shot) & consider it one of the best systems we have ever seen.

A good feature is the lock which can be taken off & apart without the aid of a screw driver and the Breech action is one of the strongest & perfectly safe for resistance of larger charges of powder. For simplicity & durability we do not think it can be excelled.

Yours truly,

Hartley & Graham

Was the Freund False-Hammer rifle ever submitted to the Ordnance Board for trial? If it was, what happened to it, and why did it fail? What seems most logical is that this fine action may have been perfected just too late for the 1880s. The Ordnance Department was earnestly seeking a suitable repeating rifle to replace the .45-70 Springfield. Single-shot rifles, and large calibers, were soon to be obsolete for military purposes.[45]

1885

There were times when Frank neglected the business of the gun store. Friends would tell Clotilda that they would go by the Freund armory, and it would be closed, with Frank in the back working on an invention.

It should be remembered that Frank was more interested in working on guns, inventing, or traveling in search of a new location, than being tied down to the life of a storekeeper.

The life of an inventor is either prince or pauper. Cheyenne had treated Frank very well. However, Clotilda was never very happy in Cheyenne and, due to her unhappiness, they finally did leave Cheyenne for the East, having resided in Cheyenne for almost ten years. Four of their eight children had been born there.

Frank sold his business, and they moved in the summer of 1885. But it was a poor move, and things never went as well in the East as they had in Cheyenne. Frank to his death always regretted having left Cheyenne.[46]

During the ten-plus years that Frank W. Freund worked in Cheyenne, he had his store in at least four locations:

1. The original store in 1867 was on the east side of Eddy Street.
2. They moved to the west side of Eddy Street, into the Whitehead Block, some time after that.
3. The next location was on Fergerson Street, between 16th and 17th Streets. (Fergerson was renamed Carey Street in 1910).
4. In 1883, they were located at 369 Sixteenth Street.[47]

Freund's cavalry pattern carbine, top view, showing camming action breech-block and "false-hammer" feature probably used as a cocking indicator. Single-shot, breech-loading rifled carbine capable of being field-stripped without tools, and having far fewer parts and simpler mechanism than other comparable arms of the 1880s. (Freund Family Collection)

1. Cheyenne *Daily Leader*, June 28, 1875.
2. Barsotti, Part I, 1957, pages 15-22.
3. Cheyenne *Daily Leader*, July 12, 1875.
4. Freund Family Collection.
5. *Ibid.*
6. Cheyenne *Daily Leader*, Nov. 12, 1876.
7. Freund Family Collection.
8. Author's collection.
9. *Ibid.*
10. *Ibid.*
11. Sellers, Frank: *Sharps Firearms*, page 188.
12. *Ibid.*
13. Correspondence with Dr. R.L. Moore, Jr., Phildelphia, Mississippi, owner of the original Sharps Rifle Manufacturing Company records.
14. Author's collection.
15. Courtesy Richard Labowski, M.D.
16. Sellers, *op. cit.*, page 189.
17. Freund Family Collection.
18. Hanson, Margaret Brock: *Powder River Country*, pages 116-119, courtesy of Mr. A.W. Burnett.
19. Author's collection.
20. *Ibid.*
21. Sellers, *op. cit.*, pages 189 and 190.
22. Author's collection.
23. Dr. R.L. Moore, Jr., correspondence.
24. Author's collection.
25. *Ibid.*
26. Dr. R.L. Moore, Jr., correspondence.
27. Freund family collection.
28. Author's collection.
29. *Ibid.*
30. *Ibid.*
31. *Ibid.*
32. *Ibid.*
33. Barsotti, Part II, 1958, pages 55-64.
34. Dr. R.L. Moore, Jr., correspondence.
35. Author's collection.
36. *Ibid.*
37. Cheyenne *Weekly Leader*, December 22, 1881.
38. Author's collection.
39. Freund Family Collection.
40. Courtesy Historical Research Collection, Wyoming State Museum.
41. Freund Family Collection.
42. Courtesy Historical Research Collection, Wyoming State Museum.
43. Freund Family Collection.
44. *Ibid.*
45. Barsotti, Part II, 1958, pages 59-61.
46. Freund Family Collection.
47. Courtesy Lee Jones, Cheyenne, Wyoming.

Frank W. Freund's large, oak wood gunsmithing box, shown with some of the gun parts and tools that were made by the master himself. Some are stamped "F.W. Freund Patent."

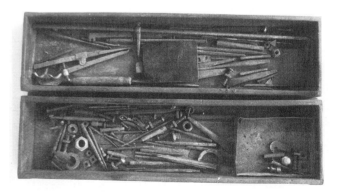

CHAPTER FIVE

THE
NEW JERSEY
YEARS

*D*uring the time the family was living in Wyoming, Clotilda Freund made a trip back East to visit her family, and to show off her children to their Eastern relatives. But on returning to the West Clotilda's recurrent homesickness was even stronger than before. Ultimately, Frank gave in to his wife's pleas, sold his gunsmithing business in Cheyenne, and they moved back to New Jersey. Despite reports that Frank had a very good job offer in California at the time, Clotilda and New Jersey finally had prevailed.[1]

The move east had to have been in the summer of 1885, as Frank started to advertise using his new address in *The Rifle* in August of 1885, and was continuous in his advertising throughout July of 1887.

One reason that Frank stated to his friends for moving east was to better manufacture his rifles. The East had better facilities for gun manufacturing, he said, and a large pool of skilled craftsmen to draw from. Frank picked Easton, Pennsylvania, as the site to set up his manufacturing business, and made repeated trips there to look at property available for purchase. Soon after locating the right property, he turned over the required funds to two men who claimed to be lawyers. They were to negotiate the land sale for him, and for the construction of a factory in Easton for the manufacture of Freund Patent rifles.

Instead, they absconded with Frank's money shortly after it was entrusted to them, and with them vanished Freund's hopes of large-scale manufacture of his invention. Following this terrible blow, Frank returned to New Jersey a broken man. He was past fifty years of age, and had lost everything, including the funds from the sale of his Cheyenne property and business. At home, Frank sat on a platform rocker with his head in his hands, and confessed to Clotilda that he would have taken his life were it not for the children.[2]

He did have some fine regular customers, but Frank never was able to re-establish himself as he had done in the West. Nor was he ever able to recover financially from the catastrophe. With a large family to support, and very much despondent, he turned back to the only trade he knew, that of the solitary gunsmith.

He advertised that he had spend a lifetime in gunmaking in the best firearm factories of the world, in such important cities as Vienna, Paris, and London, as well as in America. He called his patrons' attention to the fact that he had been located on the Western frontier for many years, during which time he had been successfully engaged in the repairing and manufacturing of fine rifles and sights.

Frank W. Freund was listed as a gunsmith in the Jersey City business directories from 1886 to 1908. During that time he continued the manufacture and sale of Freund's patent sights, "the best open sight for sporting rifles in the world", he claimed, and guaranteed his gunsmithing on rifles to be "substantial, most elegant and of unsurpassed wearing quality." His reputation for such work brought him many new and loyal customers. Among them was famed big-game hunter and civil servant Theodore Roosevelt, who enthusiastically endorsed Freund's rifles and sights.

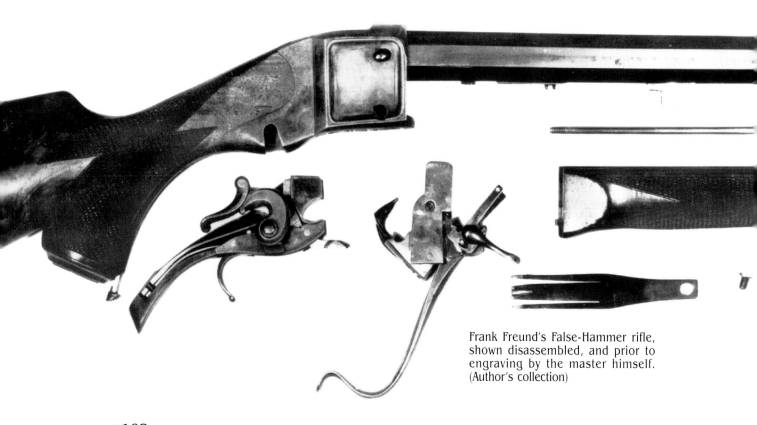

Frank Freund's False-Hammer rifle, shown disassembled, and prior to engraving by the master himself. (Author's collection)

Nevertheless, sometime around 1893 he closed his gun shop and accepted the position of foreman at the Brooklyn Navy Yard, where his talents could be more profitably employed. One day, while traveling to or from work on the streetcar, he fell and broke his ankle. He was in the hospital for quite a length of time, and then spent a long period at home recuperating. Frank did not return to the Navy Yard, as he would have had to do so in a lesser position. That the proud German gunsmith refused, saying that he "would not work for a bunch of tinkerers." Following his accident and recovery Frank worked out of a shop located in one of the rooms in his home, where he did custom work for sportsmen in the New York City vicinity.

Freund family records indicate that Clotilda was in the hospital at the same time as Frank, giving birth to daughter Stella, who was born on January 1, 1894. Stella was the seventh of their eight children; Jeanette would be the last child, born on December 4, 1897, when Frank was sixty and Clotilda was forty-three. It must have been around then, too, that Frank built the sporting model of his false-hammer rifle. The top barrel flat of this rifle was engraved with the inscription, "F.W. Freund Patentee & Maker. Jersey City, N.J."

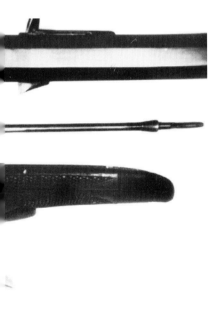

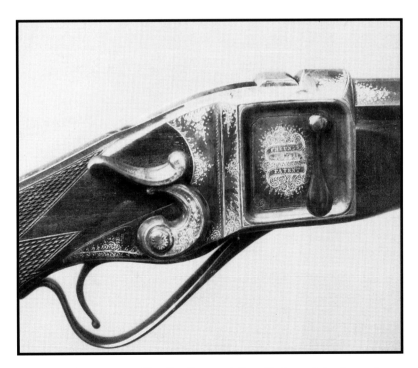

Freund's False-Hammer rifle after its completion. Photograph by Ayers, Jersey City. (Author's collection)

At that time Frank also built two rifles for Theodore Roosevelt, one having an extra barrel and forearm. Illustrated below is one of Roosevelt's four letters of endorsement written to Frank Freund. Three were written while he was a member of the U.S. Civil Service Commission, and one from the White House when he was President of the United States.[4] In a letter, Roosevelt wrote: "I have always used a slight modification of your sights." In addition, there are three 1876 Winchester rifles in the Cody Firearms Museum that formerly belonged to Theodore Roosevelt. All three have Freund sights, and were pictured in an article by Richard Rattenbury in the November-December 1982 issue of *Man At Arms* magazine. Other fine engraved Winchesters also exist with the preferred Freund sights.

COMMISSIONERS:
CHARLES LYMAN, *President,*
THEODORE ROOSEVELT,
HUGH S. THOMPSON.
WM. H. WEBSTER, *Chief Examiner.*
JOHN T. DOYLE, *Secretary.*

Address: "CIVIL SERVICE COMMISSION,
WASHINGTON, D. C."

United States
Civil Service Commission,
Washington, D. C.

June 2, 1892.

F. W. Freund, Esq.,

    70 Montgomery St.,

      Jersey City, N. J.

My Dear Sir:

      Many thanks for sending me the plate.   It seems to me to be just what I want, but I should like to ask if one thing could be done. Of course, understand that all this extra work I will pay extra for.   What I want to know is if this line down the rear sight could not be made of vermilion paint.   I have an idea that a line of vermilion would serve all the purposes of the white line and yet would not be so apt to glare or dazzle the eye and would not seem to fade into the front sight in certain lights.   If possible I want my rifles at 689 Madison Avenue before June 12th, as I will then want to take them out to the country to try them.   Hereafter I shall always deal with you directly and not through an intermediary.   Do you ever load cartridges, etc.?

      Sincerely yours,

          Theodore Roosevelt

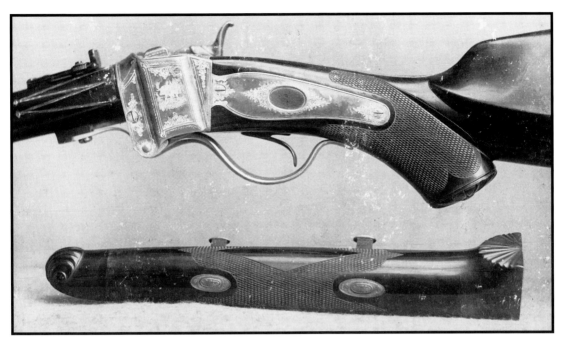

Theodore Roosevelt's custom Freund-built rifle. Notice the extra-fine "bank note" style of engraving, wood carving and checkering, and high degree of finish. (Freund Family Collection)

Another view of Theodore Roosevelt's custom Freund-built rifle, showing the special tools doubtless made by Frank W. Freund for removing the barrel from the receiver. (Freund Family Collection)

During the last quarter of the nineteenth century guns were inexpensive in the eastern United States. There were numerous manufacturers and gunshops competing for the gun dollar. The custom gunmaker's trade suffered as a result, and it was not easy for Frank to support his family on the income from his shop, alone.

Glowing reports of high prices and great prosperity in California appealed to Frank; and although he was sixty-seven years old at the time he packed his bags and headed for the Golden State in 1905. His oldest son, William, had previously joined the U.S. Navy, and it was William who bought Frank the train ticket. After working for two years in Los Angeles, Frank went up to Sacramento, then Seattle, and finally to San Francisco. In the latter city Frank obtained steady employment, and was able to send money home to his family in Jersey City. Then, in 1909, he wrote Clotilda that he was about to be made a member of the firm and looked forward to sending for his family.

But the firm failed, and soon Frank was taken sick with a cold, which at the time was thought perhaps to be influenza. He returned to Jersey City, the Masonic Order paying for his ticket. There he improved somewhat in health, and for about a year worked out of the shop in his home at 117 McAdoo Avenue, again building up his gunsmithing business. Frank advertised that he could do stocking, blueing, browning, and related gunsmith work. His youngest daughter, Jeanette, remembered too well the acid smell of the blueing and browning solutions that her father concocted in their kitchen. She also remembered that right up to the end of his life, he would go to and from the city, always with a gun in a case over his shoulder.[5]

In the closing years of his life, F.W. Freund used the business card shown below, front and back:

F. W. Freund

GUNSMITH
117 McADOO AVENUE
JERSEY CITY, N.J.

Stocking, Bluing, Boring, Damascus Browning.
Refinishing high grade guns, difficult repairing.
Through my long experience in this line both here and in Europe with the leading gun houses, using the very best of tools and material, the result is that I can guarantee all my work to be substantial, most elegant, and of unsurpassed wearing quality.
Terms are very reasonable.
Your patronage respectfully solicited.
Write or call.

Having spent a lifetime in fine art throughout gun making in the best fire arms factories in the world, such as Vienna, Paris, London, and in America, and having been located in the Western Frontier some years, engaged in repairing fire arms from almost all parts of the world, and manufacturing rifles and sights, have now permanently located East, for the reason of better facilities of labor and material, enable me to fill all orders at short notice, at the lowest figures. All work is done under my instruction, and goes through my hands before sending it to the customers.

Any person can put these sights on themselves, provided they are for standard Rifles. Can be sent by mail. It is impossible for any one to illustrate or to understand their full value unless seen on a rifle and tried under unfavorable circumstances.

Letters and testimonials from many eminent hunters and leading sportsmen from all parts of the country have been received, who having used these Sights, highly recommend them for their superior merits and fine workmanship, as well as the promptness and care with which each one has been suited to the eyesight and test; as well as other complimentary encomiums given in the past year in the leading sporting papers, all of which will appear in a new circular.

Thanking for the past liberal patronage, and hoping for a continuance of the same, I remain very respectfully yours,

{ SATISFACTION }
{ GUARANTEED. }

F. W. FREUND.

For Sale by L. P. Hansen 7 & Montgomery St Jersey City N.J.

Postcard-size advertisement for Frank W. Freund, undated but circa 1900. It apparently was distributed by L.P. Hansen of Jersey City, a celebrated Schuetzen rifleman of the time. (Courtesy John Barsotti, *Gun Digest*, 1958)

My Terms are very reasonable

For the accomodation of my customers in New York and vicinity who desire it, I will call at their residence on receipt of a postal card.

Soliciting your kind patronage, with thanks for past favors I remain,

Yours respectfully,

F. W. FREUND

254 McAdoo Ave.,   Jersey City, N. J.

Awarded
Silver Medal to
F. W. FREUND.
for Rifles & Sights
Denver, Col.
OMNIA VINCIT LABOR
COLORADO INDUS-
TRIAL ASS'N.

Trade card of Frank Freund showing his shop address in his residence at 254 McAdoo Avenue, Jersey City, New Jersey. (Author's collection)

Although he was getting older, Frank Freund still stood very straight and proud, and was a fine looking man, his eyes steel blue in color. He was about five feet eight inches tall, and weighed between 160 and 165 pounds. He sported a full mustache, which he darkened a bit, along with the hair at his temples.[6] Frank still spoke with a distinctly German accent, and when something pleased him especially, he would exclaim "Grahnde, Grahnde." He usually wore a wide-brimmed, soft felt hat—a lingering touch of his years in the West, for at heart he was always a Westerner and never really happy living in the East.[7]

The last photograph taken of Frank W. Freund, still straight and proud at age 71, before his death in 1910 at 73. (Author's collection)

Frank W. Freund never completely recovered from the sickness he had suffered in California, which the family later believed probably had been a stroke. Yet he went on working, even coming up with several new firearm inventions, and continued paying Masonic dues right up to the end. Receipts show that Frank paid his dues to the Cheyenne lodge up through the 31st day of December, 1910.[8]

By the middle of that year, though, Frank had suffered a second stroke which proved fatal. Adversity and age finally took their toll. Frank W. Freund died quietly at home on July 27, 1910, at the age of seventy-three years.

Fruend— On Wednesday, July 27, 1910,
Frank W., beloved husband of Clotilda
Fruend, age 73 years.
Relatives and friends, also Bay View
Lodge, No. 146, F. and A.M. and Hugh
De Pavens Commander, No. I.K.T.,
are respectfully invited to attend the
services at his late residence, 117
McAdoo Avenue, on Saturday, July 30,
at 8 p.m.
Interment Sunday at 2 p.m.

(Note mispelling of the Freund name)

On the afternoon of Sunday, July 31st, 1910, Frank W. Freund was laid to rest in Bay View Cemetery in Jersey City.[9]

When the Freund family first moved from Wyoming to New Jersey, Frank listed his address as 912 Bergen Avenue, Greenville. His next address was at 54 Stevens Avenue, which probably was his first shop in Jersey City. He then listed his shop at 78 Montgomery Street. His residence moved to 412 Jackson Avenue, then to 254 McAdoo Avenue (later renumbered 144), and finally to 117 McAdoo Avenue, all in Jersey City.[10]

Clotilda lasted until 1941, reaching the age of 87, living on a pension from their son, Will-

This photograph shows Clotilda Freund with her youngest daughter, Jeanette, about 1915, in Jersey City. (Author's collection)

New Rifle Sight.

THE BEST OPEN SIGHT FOR SPORTING RIFLES

In the World.

They are used on the **Plains** and in **The Rocky Mountains** in preference to all others.   Can be adjusted to any rifle if description of rifle is forwarded.   Send 2-cent stamp for illustrated pamphlet, and read explanatory testimonials and directions.   Discount to the trade only.   Price of sights, extra fine finished, $10 per set; plain finished, $5 per set.

F. W. FREUND,

912 Bergen Avenue,   Greenville, N. J.

After Frank W. Freund moved his family east to New Jersey in 1885, this ad appeared in *The Rifle* magazine until the summer of 1887. (Author's collection)

iam. He, the Navy sailor, had died of appendicitis and was buried at sea at the age of 47. Because William never married, his pension was paid to his mother.

Clotilda lived first with Jeanette, the youngest daughter, until Jeanette married, then she went to live with Angie, the oldest daughter, who was married to John Keegan. While Clotilda was living with the Keegans there was a fire in the attic and the second story of their house. It is believed by family members that most of the papers of Frank and Clotilda, along with some of Frank's model guns, were consumed in the fire. Frank's tool box, now in the author's collection, was in the basement during that fire and survived, suffering only minor water damage.

Without doubt, Frank W. Freund was a gifted inventor, a master gunsmith, and an expert engraver, but he was not a good businessman. It seems that, like many true artists, he did not have the inclination or temperament for the day-to-day, humdrum repetition of business. Perhaps if he had not "saddled the Sharps horse" but instead had gone after the attention of Winchester, as did John Browning, or had chosen the Rampant Colt, the story of Frank Freund's life, especially during the years following their move to New Jersey, would have had a very different ending.

Certainly, to his death Frank always regretted having to leave the West, especially Cheyenne. Many years now have passed since the Freund brothers headed west, following the tracks of the Union Pacific. Some of the towns they helped to found have returned to sagebrush, while others today are modern cities. But the frontier life that they knew has vanished forever, like the free-roaming buffalo. Up in Powder River country, near the site of Frank Freund's hunting camp, the great sandstone butte known to only a few as "Freund's Castle" stands alone as their monument.

But the fine hunting and sporting rifles they made are prized just as highly today by collectors, as they were by their original owners over a century ago. These products of the Freunds' Wyoming Armory are a just and fitting tribute to Frank and George Freund—the "pioneer gunmakers to the West."

1. Freund Family Collection.
2. *Ibid.*
3. *Ibid.*
4. *Ibid.*
5. *Ibid.*
6. Barsotti, Part II, 1958, pages 55-64.
7. Freund Family Collection.
8. *Ibid.*
9. *Ibid.*
10. *Ibid.*

George Freund as he appeared around 1880, about the time of his move to Durango, Colorado. (Freund Family Collection)

CHAPTER SIX

THE
DURANGO
YEARS

*T*here exists very little information about the early life of George Freund. Although his brother Frank always listed his birthplace as Heidelberg, Germany, in the 1910 census George listed his birthplace as Sovitz, Germany, which probably is an abbreviation for Schwetzingen, a small town about four miles distant from Heidelberg. In the records of George Freund's death notice, his date of birth is stated as May 6, 1843.

The 1910 census also lists the year of George Freund's immigration to the United States as 1865; however, the first time the author located mention of George was on page three of the Nebraska City *News* for January 19, 1867, as follows:

> *F.W. Freund and brother, Nebraska City gunsmiths and dealers in arms and ammunition.*

The above would lead one to suspect that George had met Frank back East, after Frank was discharged from the Union Army, and that the two traveled west together. Eventually they opened up their famed gun business in Cheyenne, Wyoming Territory.

Thereafter, George was to appear regularly as the latter half of the firm of "F.W. Freund & Bro." (or "Freund Brothers" or "Freund & Bro."), until 1880. Throughout their relationship Frank always seemed to take the lead, being the older brother and apparently the more skilled or inventive gunsmith, at least from the number of patents held in his name.

George seemed content to stay in the background, running the brothers' stores and not becoming involved in the deals and trips that Frank seemed always ready to undertake—establishing new shops, hunting buffalo, and traveling. Yet the relationship between the two brothers appeared to remain stable throughout their time in business together, and even after 1880 when George moved on to Denver, and then to Durango, Colorado.

The reason for the breakup of the partnership in 1880 seems to have been financial. Frank had married, and was starting a family. The boom in the goldfields of the Black Hills was over. The buffalo were gone, and the economy of the Cheyenne area was in a slump. The wide and wild frontier was fast disappearing.

Perhaps, also, George had reached a point where he was ready to strike out on his own into new territory. Apparently the dissolution of the partnership was very amiable, and they parted as brothers and friends. Records of contracts show George selling Cheyenne real estate to Frank and Clotilda, even after his move to Durango.

George either took Freund guns with him to Durango, or bought them from Frank, as one extant example is marked "Wyoming Armory, Cheyenne, W.T." on the sideplate and also has George's marking on the left side of the action: "Geo. Freund, Manufacturer Dealer of Fire Arms, Durango, Colo."

In the early 1880s, Durango was a new boomtown in the scenic mining and ranching region of southwestern Colorado. George Freund's Colorado Armory on First Street occupied a single-story red brick building, surmounted by a tall pole atop which was a large wooden rifle. Among the items to be found for sale there was the "More Light" sight, which had been jointly patented on June 29, 1880 by Frank and George Freund, the only patent to be issued in both brothers' names. George was an expert jeweler as well as a popular gunsmith during Durango's early years, and his Colorado Armory became a popular meeting place for local hunters and ranchers. Because guns played such a vital part in everyday life on the frontier, George did a large and profitable business.

In 1949 pioneer buffalo hunter Colonel Frank H. Mayer, who lived to be 103 years of age, recalled George Freund: "George came to Durango in the early Eighties, '81 or '82, and opened a gun store there. I was at that time ranching in Montezuma County, and having had trouble with extraction of swelled shells in both the Sharps' and Ballard's rifles, we got together and finally devised the double ejector principle on both the rifles." Mayer went on to call George Freund "a first class gunsmith." (Apparently Colonel Mayer was unaware of the Freund brothers' application of double extractors to Sharps rifles, as noted in their letter of February 11, 1878, to the Sharps Rifle Company.)

The first mention the author located of George's residence in Durango is from a newspaper announcing that he had arrived in that city on April 16, 1881.[1] Other writers have tried to tie George Freund in with Fort Lewis, the military post that was built seventeen miles west of Durango at about the time he moved there, but this author has been unable to establish such a connection. Nor has Duane A. Smith, professor of Southwest studies at Fort Lewis College in Durango, who has done extensive research on Fort Lewis for his book, *Sacred Trust.*[2]

Records of Durango real estate transactions and notices in the newspapers show that George leased a building and frontage on First Street, and opened a business during the latter part of 1881. He also received a permit to open a shooting gallery on October 1, 1881.[3] Through advertisements in the newspapers, as well as articles and notices, George's life can be traced through the 1880s, 1890s, and into the early 1900s. On June 11, 1890 George married Ida

George Freund's pioneer Colorado Armory, located on First Street in Durango, Colorado. George Freund stands in doorway, at left, wearing the hat. In addition to firearms, sights and gun accoutrements, George did a large business in hunting, fishing and mining supplies, as well as tobacco, books, jewelry and general notions. For years, his store was a popular Durango meeting place. (Courtesy Center for Southwestern Studies, Ft. Lewis College, Durango)

Gasperrini, the younger sister of Clotilda Freund, his brother Frank's wife. There was one child born to their marriage, Clarence Jethro Freund. Apparently, George and Ida Freund did not share a happy relationship. Ida was a great deal younger than George, and probably wanted something more out of life than being married to a frontier storekeeper. In any event, the couple were divorced within a year or so of their marriage. Thereafter, Clarence resided with his mother, who subsequently married a man named Whell and was living in Golden, Colorado.[4] It is believed that Clarence was adopted and afterward used the Whell surname, as nothing more is known of him by the Freund family.

Clarence Jethro Freund, George's only child, shown about 1896. (Freund Family Collection)

Artist's sketch of the interior view of the Colorado Armory, Durango, about 1889. (Courtesy John Hartman)

It appears that Ida had already left, when on June 21, 1891 the Colorado Armory exploded!

DESTROYED

FIRE IN A GUN STORE IN DURANGO YESTERDAY CREATED CONSIDERABLE EXCITEMENT THE EXPLODING CARTRIDGES MADE THE FIREMEN'S WORK DANGEROUS

Several people were more or less injured... A number of runaways caused.

Special to the News.

Durango, Colo., June 21.—A terrific explosion in the jewelry and gun store of George Freund on First Street, this city, at 6:15 o'clock this evening. The building was a one-story brick, with metal roof, and one of the most substantial in the city, having long been known as the Colorado Armory. The front and rear of the building were blown out and the roof completely demolished and scattered.

Immediately following the explosion, the entire building was enveloped in flames which for a time threatened to wipe out a whole block and prove disastrous, from the fact that the continual promiscuous shooting, small explosions and shooting of loose cartridges intimidated the firemen, who, by this time, were playing three large streams of water on the fire from a distance, and were in great danger of their lives.

UNDER CONTROL

This did not last long, however, for the brave chief, James Veitch, rallied the forces and urged them on the flames, which were soon under their control. A team of horses hitched in front of the building ran away down First Street, one horse being shot by a stray cartridge, carrying with them the inside window curtains.

The cartridges in stock were concealed under the shelving, and most of the discharges were up and down, thus saving, doubtless, many lives.

Hon. Adair Wilson was badly bruised by being knocked down by an unruly horse, but his injuries are not serious.

Mrs. Charles Tucker, wife of a jeweler living next door, sustained serious injuries from a falling wall, but she will recover.

The water had been turned off from the city mains all day for repairs, but fortunately the repairs were finished only one hour before and the supply was quite sufficient.[5]

EXPLOSION AT DURANGO

Flames Complete the Destruction
of the Colorado Armory Building

Durango, Colo., June 21 —(Special)—At exactly twenty-four minutes after six o'clock this evening an explosion took place in the Colorado Armory building, knocking out both the front, rear, and a portion of one side, and the roof, and in less than two seconds the building was one mass of flames which leaped high into the air, and for a time threatened many adjoining buildings. The firemen responded quickly and soon had three streams of water playing on the flames. Owing to the constant discharge of cartridges, it was extremely dangerous to get very near, but led by their chief, J.R. Veitch, the brave firemen risked their lives by getting close to the building and in fifteen or twenty minutes had it under control, but not until the entire contents were a total loss.

The building was a substantial one-story structure, being known as the Colorado Armory, owned by Phil Gerow, and valued at $14,000. It was insured for $2,500. It was occupied by George Freund with a large stock of guns and jewelry, whose loss is about $8,000.

The debris caused by the explosion was scattered for some distance.

A team of horses in front of the building ran away. Hon. Adair Wilson happened to be near by, and before he could get out of the way, was knocked down by the runaway team, but was not seriously injured. Mrs. H.A. Tucker, wife of a jeweler occupying an adjoining store, was hit by falling brick and hurt quite badly but not dangerously. Another man, name unknown, was quite badly cut about the face and neck.

Considering the location of the building, in the center of town, it is very lucky no one was killed. The large dial on the regulator in the front part of the store was the only thing left and the minute hand stands at twenty-four minutes past 6, showing the time the explosion occurred. The cause of the explosion is unknown.[6]

A.J. Andrews, a young boy at the time who did odd jobs for the gunsmith, such as sweeping out the store and running errands for him, recalled that George had just closed his business for the day and retired alone to his living quarters in the back of the store. There he had lit a fire in the stove. Heavy, dusty curtains separated the front of the store from the living quarters. According to Andrews, all at once the whole store seemed to explode, probably accounted for by the dust in the air and on the curtains.

Soon, however, George was back in business, as evidenced by local newspaper advertisements. They gave the new address for his Colorado Armory as 92 Union Block, whereas the old address had been on First Street.

Indications point to the fact that George did not do any converting of Sharps rifle actions at his store in Durango. Surviving specimens of his work done there do show some gunsmithing, such as altering muskets to carbines,

In this photograph taken about 1900, looking north up Main Street in Durango, George Freund's second Colorado Armory and City Jewelry Store are in the middle of the picture, at the corner of 9th Street. (Courtesy La Plata County Historical Society, Animas Museum)

George Freund's "skull and crossbones" registered trademark, stamped atop the barrel of a rifle either altered or sold by him at his Colorado Armory in Durango. (Author's collection)

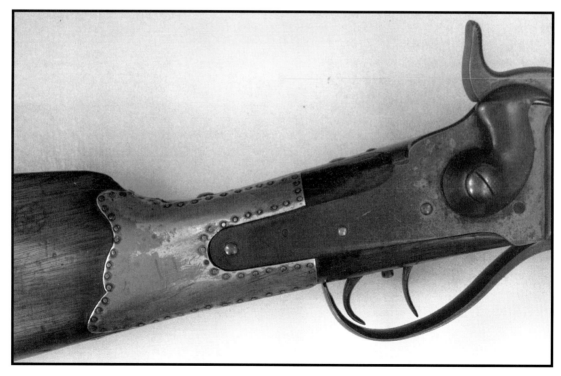

A fancy brass and tack repair to the wrist of a Sharps rifle, done by George Freund at his Colorado Armory in Durango. (Author's collection)

repairs to broken wrists, and installing sights, especially the Freund More Light sights. All conversions of Sharps rifles that the author has seen have shown the Wyoming Armory address in Cheyenne.

Several interesting differences show up in George's advertising. Although the silver medal won at the Colorado Industrial Association Fair in 1873 was awarded to Frank W. Freund alone, the advertising cut was changed while Frank and George worked in Cheyenne, so that it appeared to have been awarded to both Freund brothers. Later, in George's advertising in Durango, the medal is depicted as having been "'Awarded to Geo. Freund, Durango. Colo., For Breech-loading Rifles, Sights and Ammunition." The cut showing the reverse side of the same medal was changed to read, "Awarded to Freund's Best Colorado made Gun 1873", the word "Brothers" having been eliminated.

Shortly after settling in Durango, George Freund patented designs for two safety shells and a cap for blasting charges for mining. Then, in 1893, he was awarded design patent number 22406 for his "Smelter City" novelty silver spoon. But, other than U.S. patent number 229245 awarded in 1880 to both Frank W. and George Freund for their "More Light" sight, George produced no additional inventions or improvements related to firearms. George did, however, have the More Light sight patented in Canada and six European countries.

In the later years of George's life he continued to live and work in Durango. His shop goods drifted more into the jewelry and notion areas, with less emphasis on guns and ammunition, although he did some gunsmithing and

Illustration of George Freund's "'Smelter City" novelty silver spoon design, for which he was granted U.S. design patent number 22406 on May 9, 1893.

COLORADO ARMORY.

George Freund, Prop.

ESTABLISHED 1881.   **DURANGO, COLORADO.**

Headquarters for all Kinds of

Guns, Pistols, Knives, Sportsmens' Optical

GOODS, AMMUNITION, ETC.

Have constantly in stock, Winchester, Bullard, Remington, Ballard, Marlin and Sharp's Rifles of all sizes and styles. Shot Guns single and double barrel, of most approved American and Foreign manufacture. REVOLVERS—Colt's, Smith & Wesson, Marvin, Hurlburt & Co's and other leading manufacturers. Largest stock of

Pocket Coslery, Hunting Knives, Bowie Knives, Razors, Optical Goods,

Fishing Tackle, Powder, Shot, Cartridges, Shells, eto., in the Southwest

A Complete line of Gun Parts Promptly on hand, and repairing done promptly.

MY WORK DEFIES COMPETITION.   Send for price list and Circular.

GEORGE FREUND, Durango, Colo.

Freund's patent More Light Gun Sight, $5 to $10 per set. Miners' Knives a specialty — can be sent by mail. Supplies in this line at Byrne & Adam's, Cortez, Montezuma Valley. Key Fitting and Locksmithing to order.

THE FAMOUS HUNTER'S SIGHTS.

THE **ADVANCED INVENTION IN GUN SIGHTS OF THE NINETEENTH CENTURY.**

REAR VIEW

SIDE VIEW

The only original "MORE LIGHT" Patent Gun Sights.

Front and Rear Sights as they appear while taking aim. **PRICES:**

No. 1, Spring, on Step Notch, rear sight ------ $3.00
No. 2, Hinge Leaf, fine finish, rear sight ------ 5.00
No. 1, Steel Front Sight, white centre --------- 1.50
No. 2, Steel Front Sight, gold centre --------- 2.50

CAN BE SENT BY MAIL.

All orders should be accompanied with remittance to cover amount. Care should be taken to give name and address clear. Illustrated Circulars and Price List of Fine Arms, Ammunition, etc., and Patent Miners' Pocket Knife on application.
Address all communications to

GEO. FREUND, Patentee,
Colorado Armory Building, DURANGO, COLO., U. S. A.

MINERS, ATTENTION !

The most Complete and Successful Miner's Article ever Invented.

MINER'S POCKET KNIFE

REDUCED TO ¼ SIZE.

Can be carried in Vest Pocket. Knife with one large blade, cap-crimper, fuse-cutter and can-opener, $2.25. Without can-opener, $2.75, by registered mail on receipt of price. The entire Knife is made of Metal with Nickel Finish.

Address GEORGE FREUND,
Colorado Armory, Durango, Colo.

the making and installing of Freund's More Light sights. He also did lock and key work, jewelry repair, and sold sporting goods such as rods and other fishing equipment, and hunting-related supplies. As the years overtook George Freund he became badly crippled by arthritis, and had to use a crutch or cane to get around. The sight of the little frontier gunsmith hobbling about town was a common one among pioneer residents, although he remained active in the Masonic Order in Durango for the rest of his life.

On March 25, 1911 George Freund died at Durango's Oshner Hospital. He was 67 years, 10 months, and 19 days of age. Interment was in lot 2, block 6 of the Masonic plot at Greenmount Cemetery, and records of the funeral indicate that the Masonic Lodge paid most of his funeral expenses. Sadly, George's grave lay overgrown with weeds and forgotten for many years, until found in 1951 by a local historian named Paul Crawford.[7]

From the time George Freund first appeared on the Western scene at Nebraska City, soon after the close of the Civil War, he was destined to live out his life on the plains and in the mountains. He was a true pioneer in the settlement of the West.

George Freund's grave marker in Greenmount Cemetery, Durango, Colorado, as it appears today. (Courtesy Ed Sullivan)

1. Durango *Herald*, December 16, 1881.
2. Smith, Duane A., *Sacred Trust*. Boulder: University of Colorado Press, 1991.
3. Durango *Herald* (various dates).
4. Freund Family Collection.
5. *Rocky Mountain News*, June 22, 1891, page 1, column 1.
6. Denver *Republican*, June 22, 1891, page 1.
7. Barsotti, Part II, 1958, page 56.

George Freund's Colorado Armory letterhead. Note the presentation medal's inscription now reads, "Awarded to Freund's Best Colorado made Gun 1873." (Courtesy Richard Labowskie)

The actual 1873 medal awarded to Frank W. Freund at the Colorado Industrial Association Fair in 1873. Note the original inscription. (Author's collection)

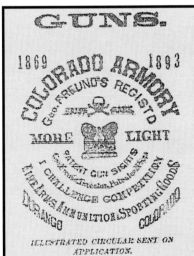

The following is a summary of the U.S. patents held by Frank and George Freund:

Patent Number	Date patent applied for/ granted	Description	Patent by	Location
153432	10/13/73 7/28/74	Breech-loading Fire-arm	F.W.F.	Denver
160762	10/1/74 3/16/75	Improvement in Breech-loading Fire-arm	F.W.F.	Denver
160763	2/4/74 3/16/75	Improvement in Metallic Cartridges	F.W.F.	Denver
160819	10/19/74 3/16/75	Sights for Fire-arm	F.W.F.	Denver
162224	3/19/75 4/20/75	Improvement in Breech-loading Fire-arm	F.W.F.	Denver
162373	3/19/75 4/20/75	Pistol-grips for Stocks of Fire-arms	F.W.F.	Denver
162374	3/19/75 4/20/75	Guard Lever and means for operating Breech Block of Breech-loading Fire-arms	F.W.F.	Denver
168834	3/19/75 10/19/75	Detachable Pistol-grip for Fire-arms	F.W.F.	Cheyenne
180567	6/1/76 8/1/76	Breech-loading Fire-arm	F.W.F.	Cheyenne
183389	9/27/76 10/17/76	Revolving Fire-arm	F.W.F.	Cheyenne
184202	5/11/76 11/7/76	Breech-loading Fire-arm	F.W.F.	Cheyenne
184203	2/23/76 11/7/76	Improvement in Breech-loading Fire-arm	F.W.F.	Cheyenne
184854	10/12/76 11/28/76	Improvement in Primers for Cartridges	F.W.F.	Cheyenne
185911	12/3/75 1/2/77	Breech-loading Fire-arm	F.W.F.	Cheyenne
189721	4/7/76 4/17/77	Front Sights for Fire-arms	F.W.F.	Cheyenne
211728	10/5/78 1/28/79	Breech-loading Fire-arm	F.W.F.	Cheyenne
216084	10/30/76 6/3/79	Breech-loading Fire-arm	F.W.F.	Cheyenne
229245	3/9/80 6/29/80	Sights for Fire-arm	F.W.F. & G.F.	Cheyenne
268090	8/19/81 11/28/82	Sight for Fire-arm	F.W.F.	Cheyenne
273156	10/18/82 2/27/83	Safety Shell for Blasting	G.F. & R.B.F. Reed	Durango
289768	10/9/83 12/4/83	Cap for firing Explosive Charges	G.F.	Durango
292642	6/23/81 1/29/84	Safety Shell for Blasting	G.F.	Durango
297375	12/27/83 4/22/84	Pocket Knife	G.F.	Durango
313414	8/19/84 3/3/85	Knife for Miners	G.F.	Durango
496051	10/18/84 4/25/83	Gun Sight	F.W.F.	Cheyenne
D22406	4/3/93 5/9/93	Design for Spoon	G.F.	Durango

CHAPTER SEVEN

FREUND & BROTHER

THE FREUND PATENTS

As a result of their combined inventive genius the two Freund brothers held a total of twenty-six United States patents. Frank W. Freund held nineteen patents in his name, and one in partnership with George; George held five in his name plus the one in common with Frank, and another in partnership with a man named Robert B.F. Reed, in Durango.

Frank applied for seven patents from Denver, Colorado, and twelve from Cheyenne, Wyoming, including the one held in partnership with George. All of the patents Frank Freund applied for (and was granted) were firearm or cartridge related.

All five of George's patents were applied for from Durango, Colorado, including the one held in partnership with R.B.F. Reed. All of the patents George Freund applied for from Durango (and was granted) were connected either with explosives for the mines, or for pocket knives. George also held a design patent for an ornamental spoon.

The following are synopses for each of the Freund Brothers patents:

No. 153432

Frank W. Freund

Applied for Oct. 13, 1873   *Issued July 28, 1874*

BREECH-LOADING FIRE-ARMS

This improvement relates to an improved slotted sear and forked sear-spring, which enabled Frank Freund to put double set-triggers in breech-loading guns which have the center lock and a low hammer.

Figure 1 is a side view, showing a portion of the arm in section. The firearm is not cocked.

Figure 2 is a side perspective view of the improved sear.

Figure 3 is a top view of the trigger box, or bottom piece of the stock with the improved sear and sear-spring and other trigger mechanism. The main-spring of the hammer is broken-away in order to have it not hide the mechanism.

Figure 4 is a perspective view of the improved sear and forked spring in proper relation to one another, but detached from the arm.

F. W. FREUND.

Breech-Loading Fire-Arms.

No. 153,432.   Patented July 28, 1874.

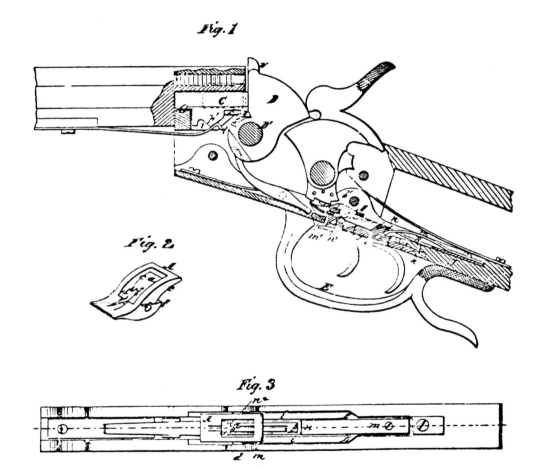

Fig. 1

Fig. 2

Fig. 3

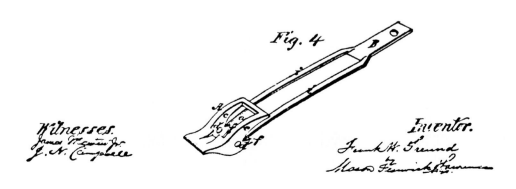

Fig. 4

Witnesses.

Inventor.

No. 160762

Frank W. Freund

Applied for Oct. 1, 1874   Issued Mar. 16, 1875

IMPROVEMENT IN BREECH-LOADING FIRE-ARM

The nature of the invention consists of a tail on the lower part of the hammer, along with a notch on the guard lever, enabling the hammer to be cocked before moving the breechblock. It also consists of a nose or cam formed on the tail of the hammer enabling the firing pin to be moved back with a positive action and held until released by a pin when the hammer is set free.

It additionally consists of a forked guard lever which receives the breech between its prongs and fastens rigidly upon a moveable pin of the breechblock.

The breechblock can be opened and closed by the guard lever or by a thumb piece, a matter of importance in localities where repair shops and gunsmiths are not readily available.

It also consists of the particular construction of a cartridge shell extractor, which is applied and held in position without a pin or other fastening.

This is quite an involved patent, which probably is the reason it never caught on with the sporting trade.

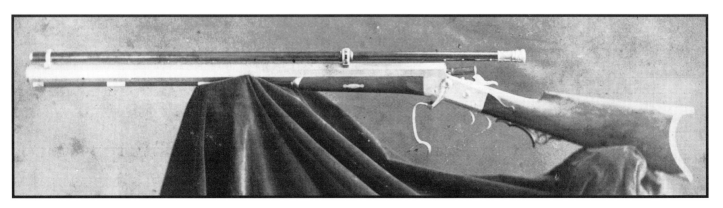

A Remington Rolling-Block rifle that was altered on the Freund patent no. 160762. (Freund Family Collection)

3 Sheets--Sheet 1

F. W. FREUND.

Breech-Loading Fire-Arm.

No. 160,762

Patented March 16, 1875.

Fig. 1

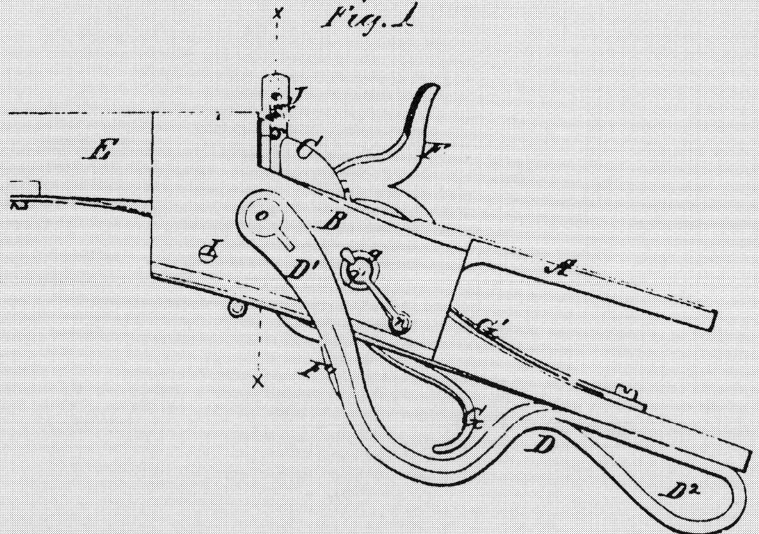

Fig. 2

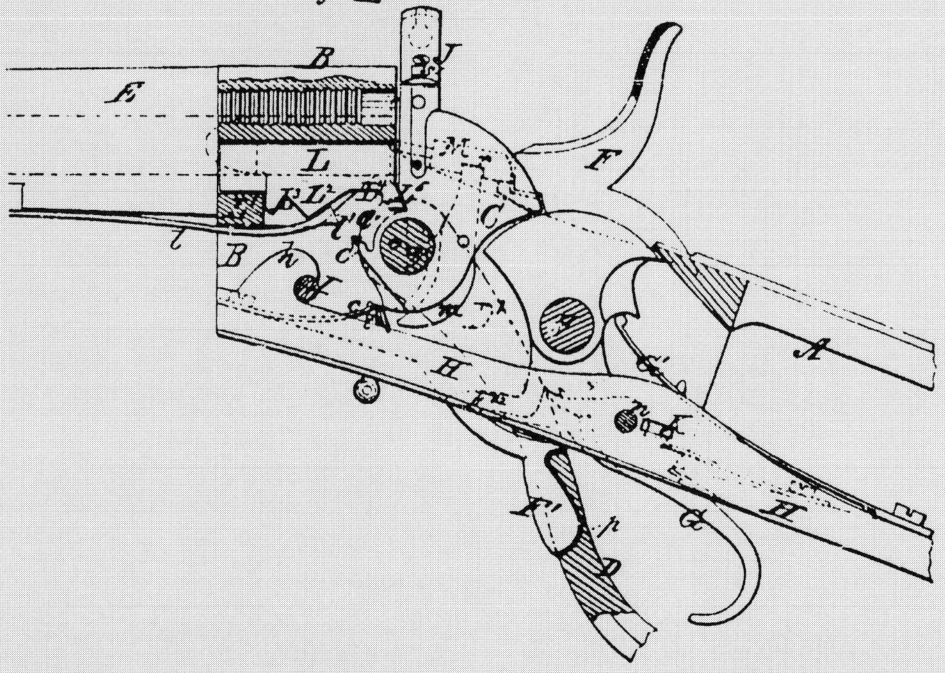

Witnesses.

Inventor:

F. W. FREUND.
Breech-Loading Fire-Arm.

3 Sheets--Sheet 2.

No. 160,762

Patented March 16, 1875.

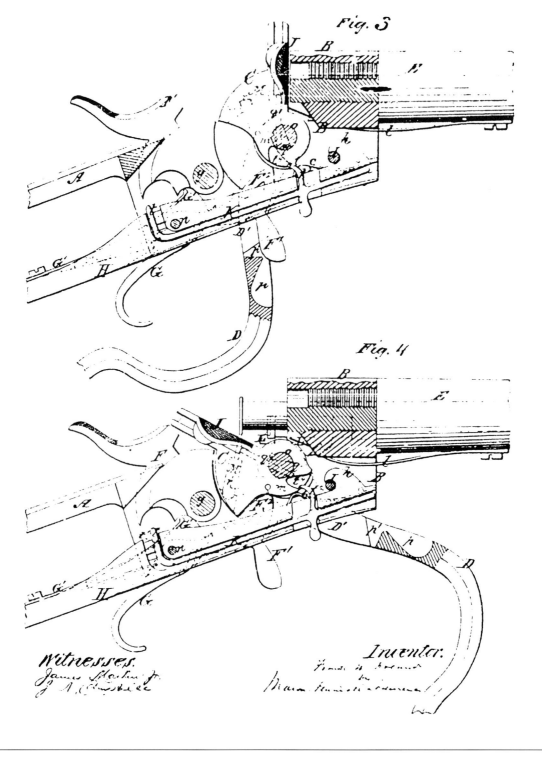

Fig. 3

Fig. 4

Witnesses.

Inventor.

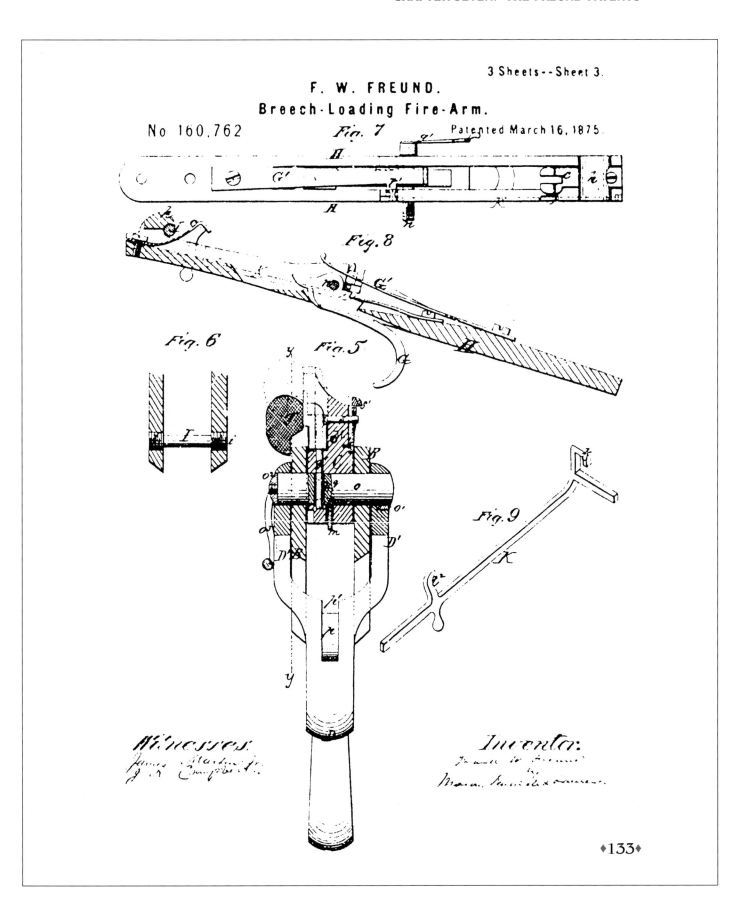

F. W. FREUND.
Breech-Loading Fire-Arm.

No 160,762   Fig. 7   Patented March 16, 1875.

3 Sheets--Sheet 3.

Fig. 8

Fig. 6   Fig. 5

Fig. 9

Witnesses

Inventor:

No. 160763

Frank W. Freund

Applied for February 4, 1874 *Issued March 16, 1875*

IMPROVEMENT IN METALLIC CARTRIDGES

This was an attempt by using a longer cartridge to move the charge of powder forward of the breech; as well as an attempt to construct a cartridge with a thicker or solid head, so that the explosion of the cartridge occured up within the barrel instead of back at the breech.

Figure 1 indicates a portion of the cartridge which is made thinner from the point *xx* to the mouth; while in the rear of point *xx*, the cartridge is made solid or thickened.

F. W. FREUND.
Metallic-Cartridge

No. 160,763

Patented March 16, 1875

Fig. 1

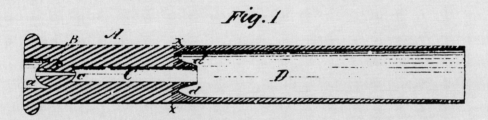

Fig. 2

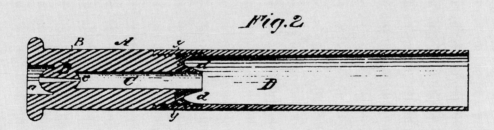

Witnesses.

Inventor:
Frank W. Freund
by his attorney

No. 160819

Frank W. Freund

Applied for October 19, 1874   *Issued March 16, 1875*

IMPROVEMENT IN SIGHTS FOR FIRE-ARMS

The invention relates to certain improvments in adjustable sights; and the object of the improvements is to prevent casual change in the sight after it has been adjusted, to hold the parts firmly together, and insure and facilitate accurate and smooth adjustments.

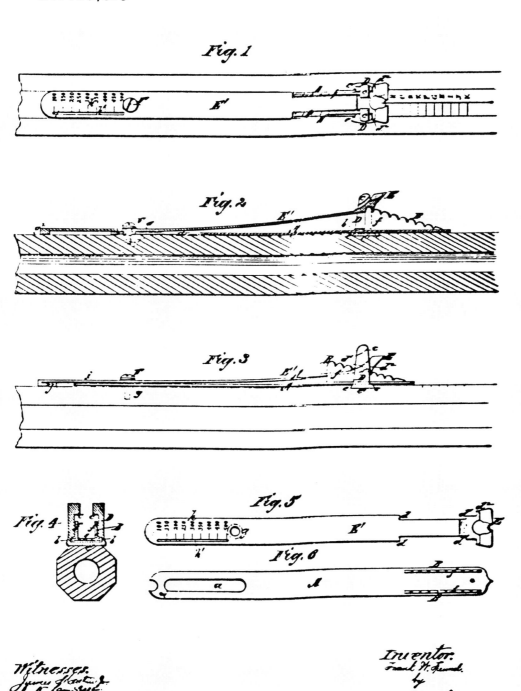

F. W. FREUND,
Sights for Fire-Arms.

No. 160,819   Patented March 16, 1875.

Fig. 1

Fig. 2

Fig. 3

Fig. 4   Fig. 5

Fig. 6

Witnesses.   Inventor.

No. 162224

Frank W. Freund

Applied for March 19, 1875   *Issued April 20, 1875*

IMPROVEMENT IN BREECH-LOADING FIRE-ARMS

The object of this patent was to make changes in the Remington Rolling Block action. It changed the cock, moving the slide forward or backward. It also prevents the full cocking of the hammer while the breechblock is open, or only partially closed, as well as prevents the breechblock from being opened when the hammer is cocked.

The pattern model of this patent is in the firearms collection of the Smithsonian Institution in Washington, D.C.

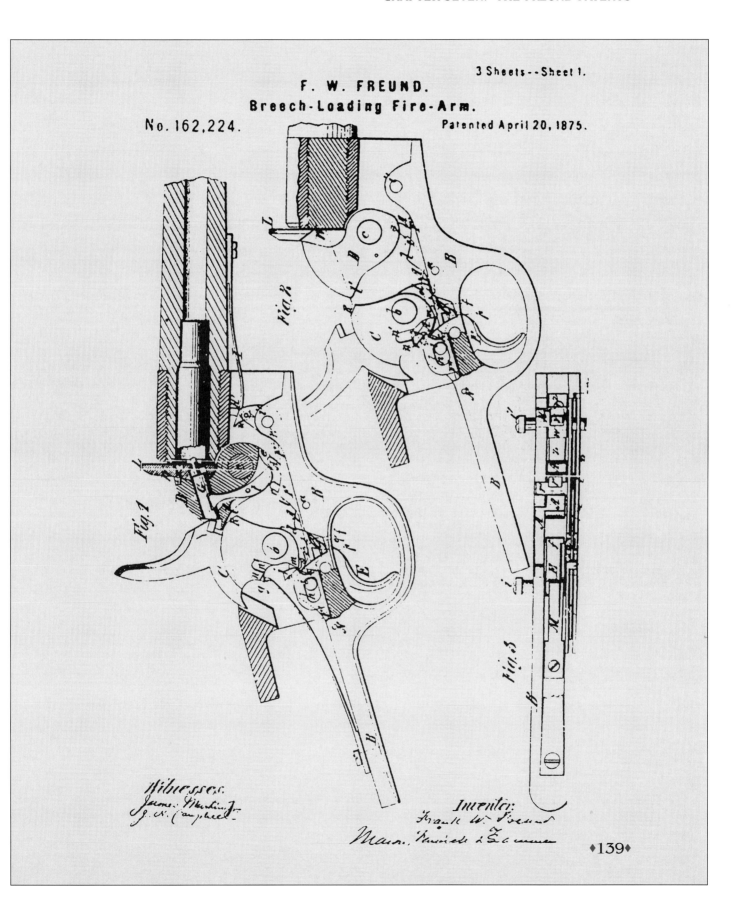

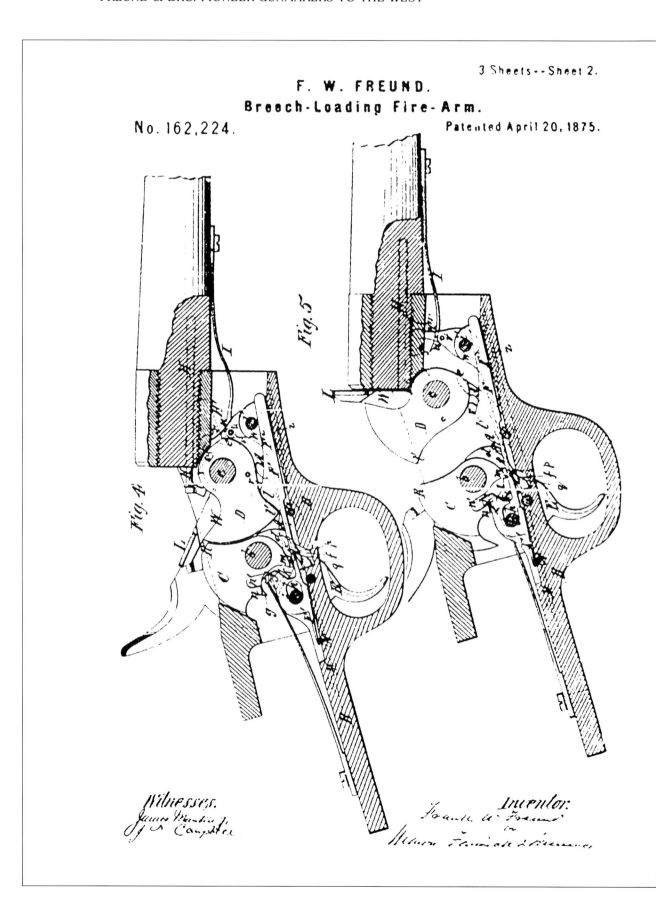

3 Sheets--Sheet 2.

F. W. FREUND.
Breech-Loading Fire-Arm.

No. 162,224.

Patented April 20, 1875.

Fig. 5.

Fig. 4.

Witnesses.

Inventor:

F. W. FREUND.
Breech-Loading Fire-Arm.

3 Sheets--Sheet 3.

No. 162,224.

Patented April 20, 1875.

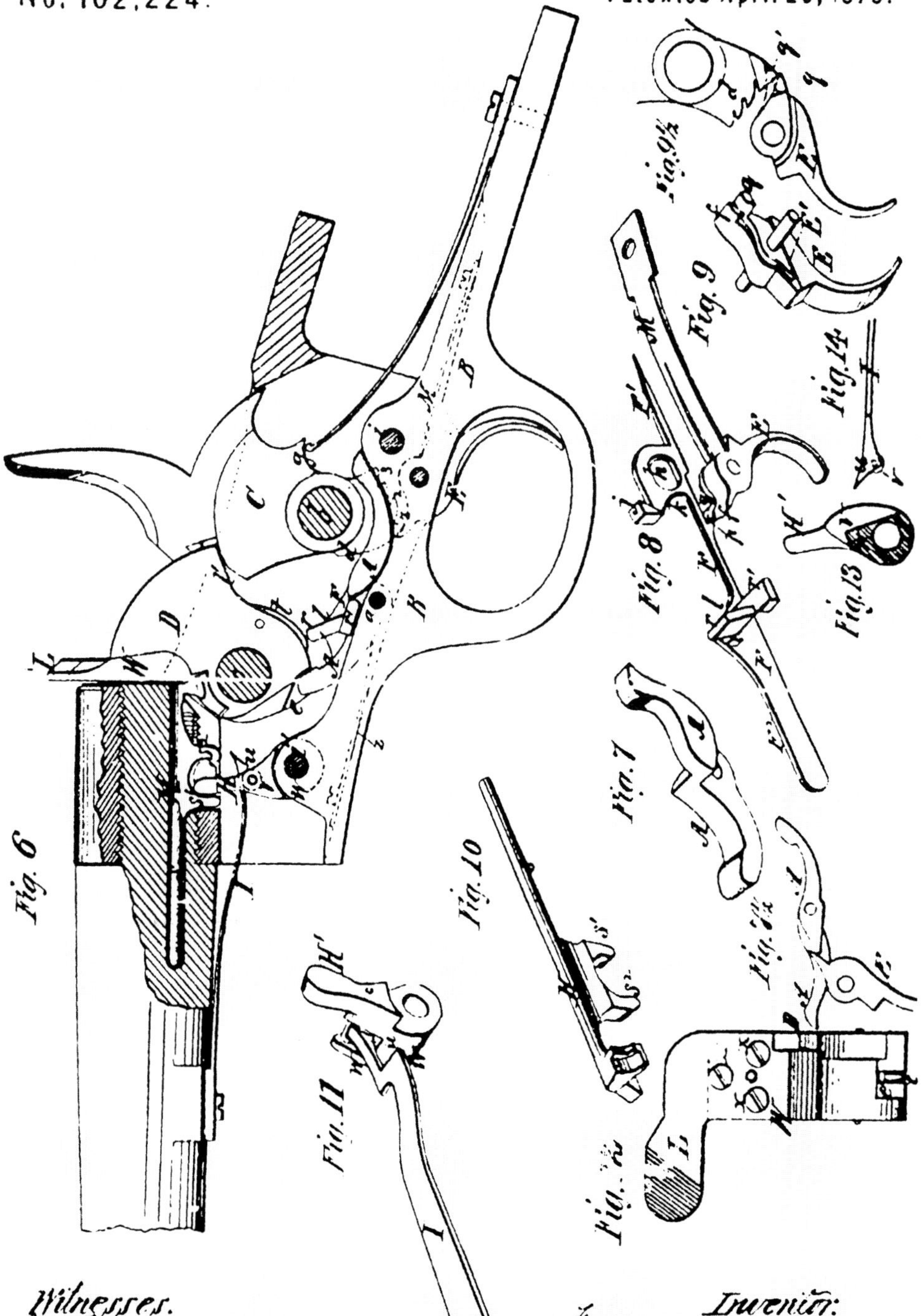

Fig. 6

Fig. 9

Fig. 14

Fig. 8

Fig. 13

Fig. 7

Fig. 10

Fig. 11

Fig. 12

Witnesses.

Inventor:

♦141♦

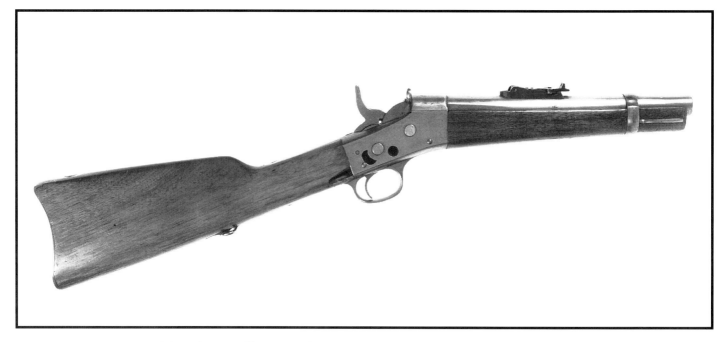

A pattern model Remington Rolling-Block rifle with the Freund patent no. 162224 alteration to its action, right side. (Smithsonian Institution collection)

Close-up view of the left side of the action of the rifle shown above, with "PAT. APL'D. FOR/1875/F.W.F." markings stamped at breech. (Smithsonian Institution collection)

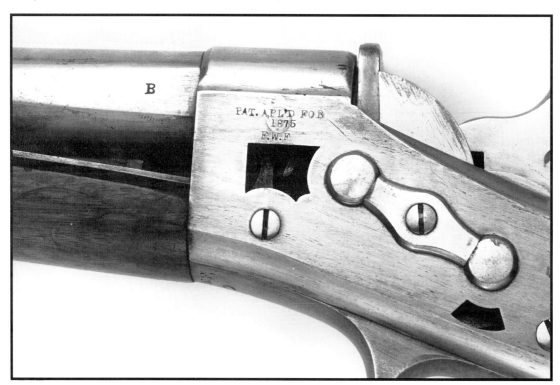

Right side and underside views of the U.S. Patent Office model (with original tag attached) for Frank W. Freund's detachable rifle pistol-grip, as shown in patent no. 162373. Inscription on presentation shield reads, "F.W. Freund. Denver, Col." (Courtesy Dr. R.L. Moore, Jr.)

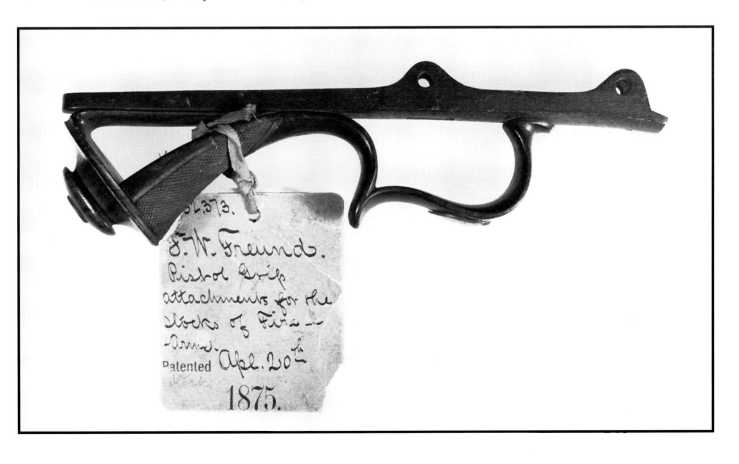

No. 162373

Frank W. Freund

Applied for March 19, 1875   *Issued April 20, 1875*

IMPROVEMENT IN PISTOL-GRIP ATTACHMENTS FOR THE STOCKS OF FIRE-ARMS

This patent had for its object the construction of a combined pistol grip and trigger guard, which in connection with the trigger plate of a firearm, could be readily applied to the ordinary gun then in use.

This device generally was applied to the Remington Rolling Block rifle. In the patent are shown a guard lever, pistol grip, and triggerguard made of one piece of metal.

F. W. FREUND.
Pistol-Grip Attachment for the Stocks of Fire-Arms.
No. 162,373

Patented April 20, 1875.

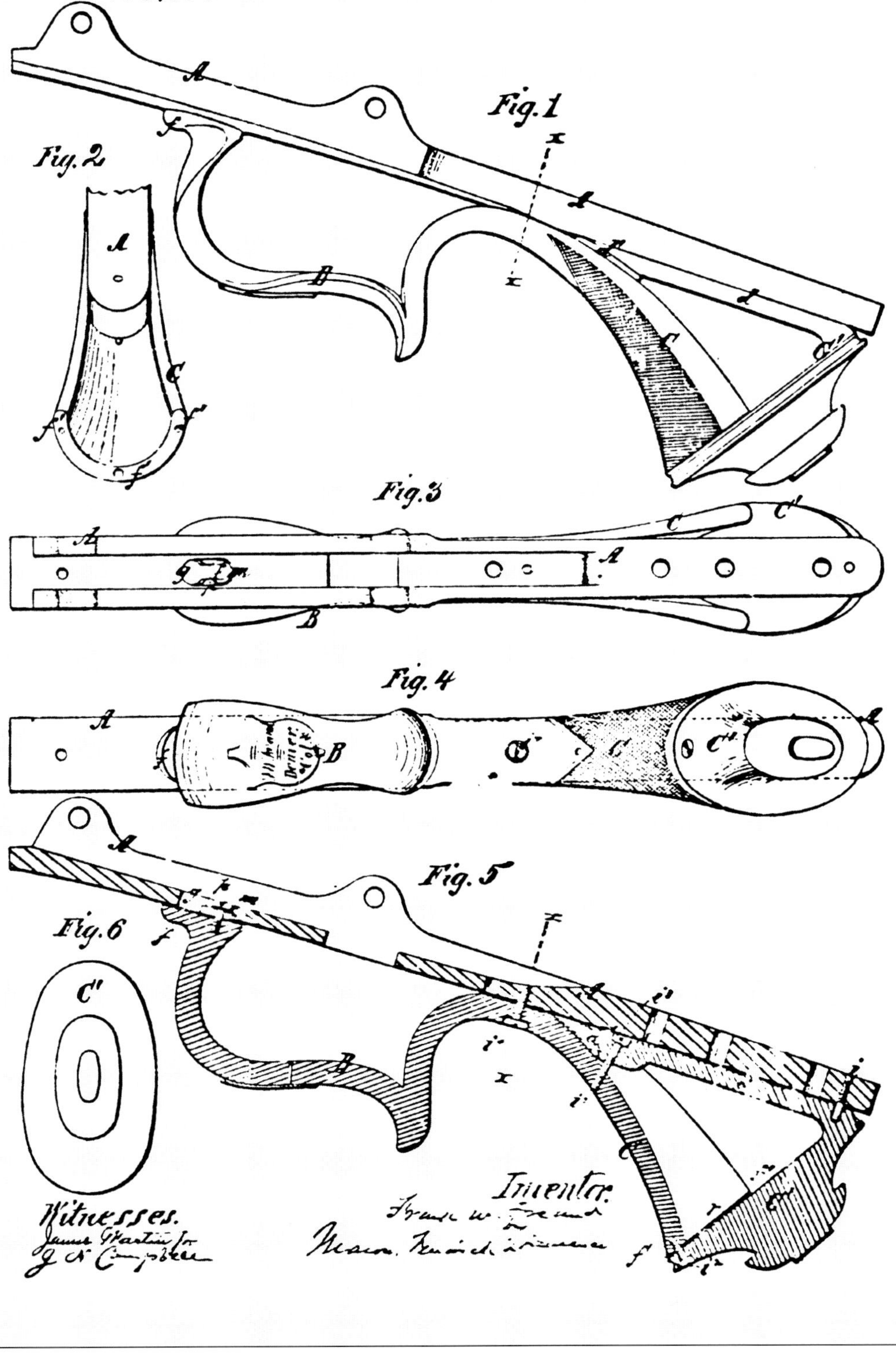

Fig.1

Fig.2

Fig.3

Fig.4

Fig.5

Fig.6

Witnesses.

Inventor.

No. 162374

Frank W. Freund

Applied for March 19, 1875   *Issued April 20, 1875*

IMPROVEMENT IN GUARD LEVERS AND MEANS FOR OPERATING
THE BREECH-BLOCK OF BREECH-LOADING FIRE-ARMS

This patent was applied for and granted on the same dates as patent no.
162373. It incorporates a pistol grip into the guard lever. The breechblock may
be operated either independently, or by the guard lever, as desired.

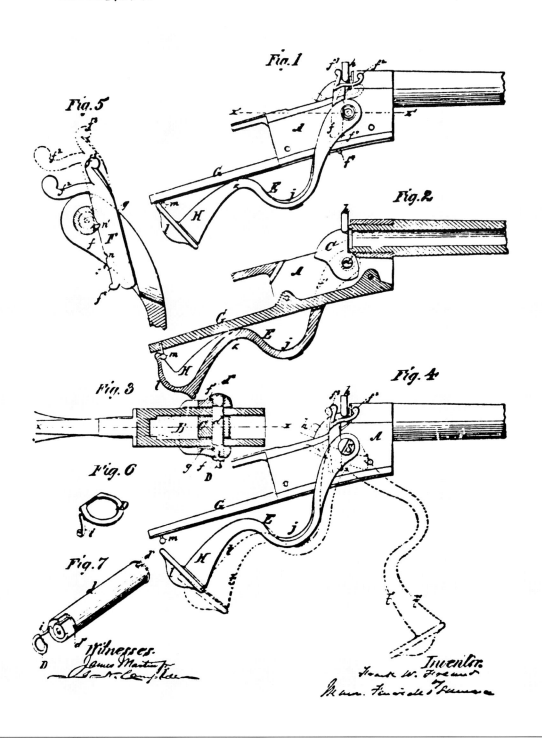

F. W. FREUND.

Guard-Lever and Means for Operating the Breech-Block of Breech-Loading Fire-Arms.

No. 162,374.

Patented April 20, 1875.

No. 168834

Frank W. Freund

Applied for March 19, 1875   *Issued October 19, 1875*

DETACHABLE PISTOL-GRIP FOR FIRE-ARMS

This patent consists of a grip for guns made separate from the stock, and adapted to be attached thereto.

This patent was assigned by Frank W. Freund to his wife, Clotilda, who in turn hired a firm of lawyers to pursue a claim against the United States for using a form of this pistol grip on the Officers Model of the Springfield rifle.

A later chapter of this book takes up that claim.

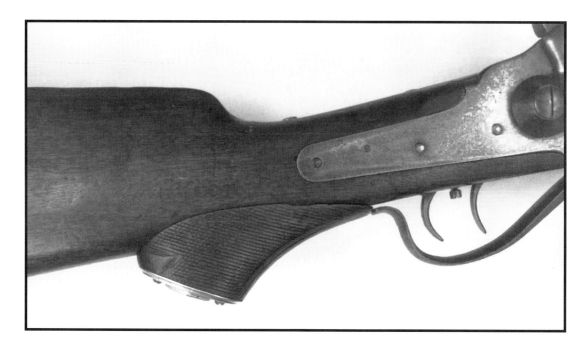

Close-up view of the right side at stock tang, of a Sharps rifle which has a Freund's patent no. 168834 detachable pistol grip attached. Butt cap has a relief-cast diamond which reads, "FREUND'S PATENT, Oct. 19, 1875." (Author's collection)

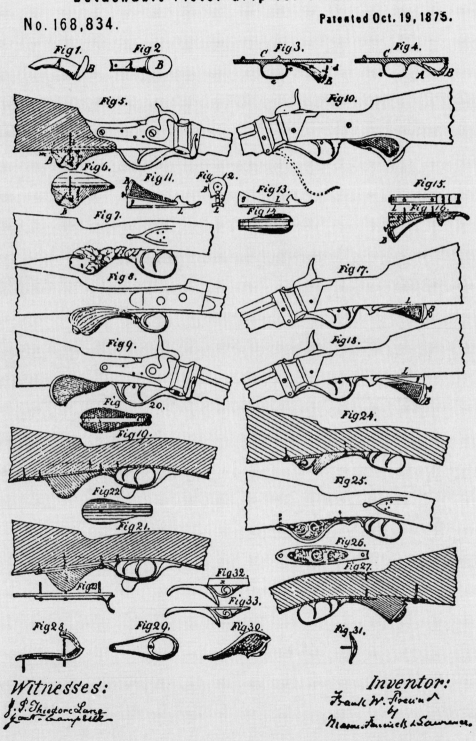

F. W. FREUND.

Detachable Pistol-Grip for Fire-Arms.

No. 168,834.

Patented Oct. 19, 1875.

No.180567

Frank W. Freund

Applied for June 1, 1876   *Issued August 1, 1876*

IMPROVEMENT IN BREECH-LOADING FIRE-ARMS

This patent is the one we normally associate with Frank W. Freund, for his change in the breechblock of the Sharps rifle.

The nature of the invention consists of a breechblock normally rounded on the top, and adapted to wedge the breechblock firmly into its closed position. The combination of the breechblock having a guideway, and the breech frame having a lug, forces the cartridge into the barrel.

This prevents one of the chief objections to the Sharps rifle: that after firing the rifle several times in rapid sucession, cartridges could not be forced into the barrel.

Nothing is noted in this patent concerning two extractors.

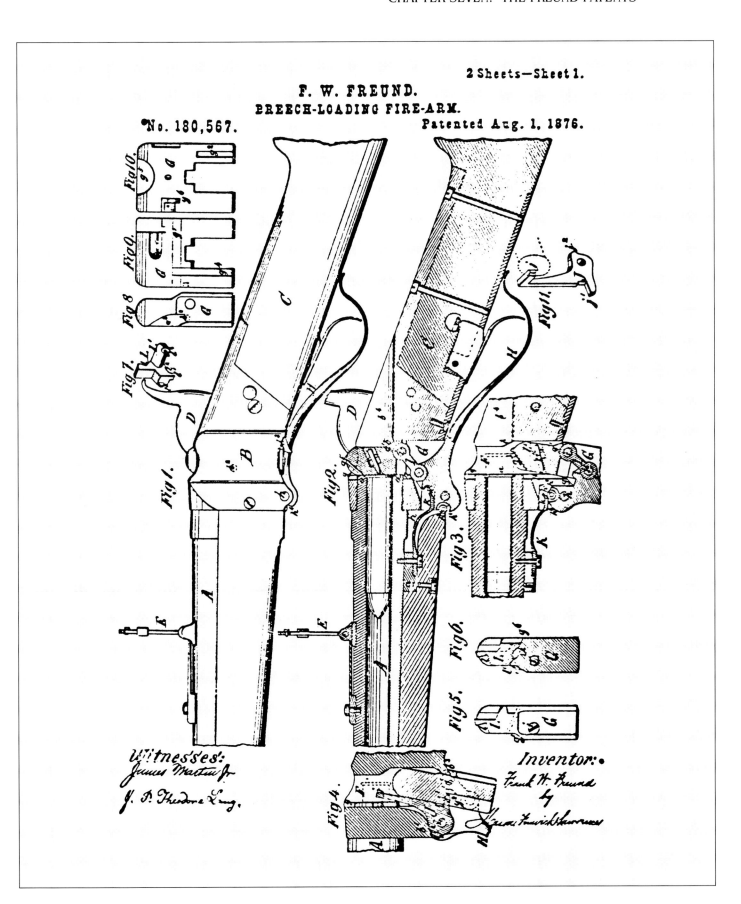

2 Sheets—Sheet 1.

F. W. FREUND.
BREECH-LOADING FIRE-ARM.

No. 180,567.

Patented Aug. 1, 1876.

Witnesses:
James Martin Jr.
J. P. Theodore Lang.

Inventor:
Frank W. Freund
by

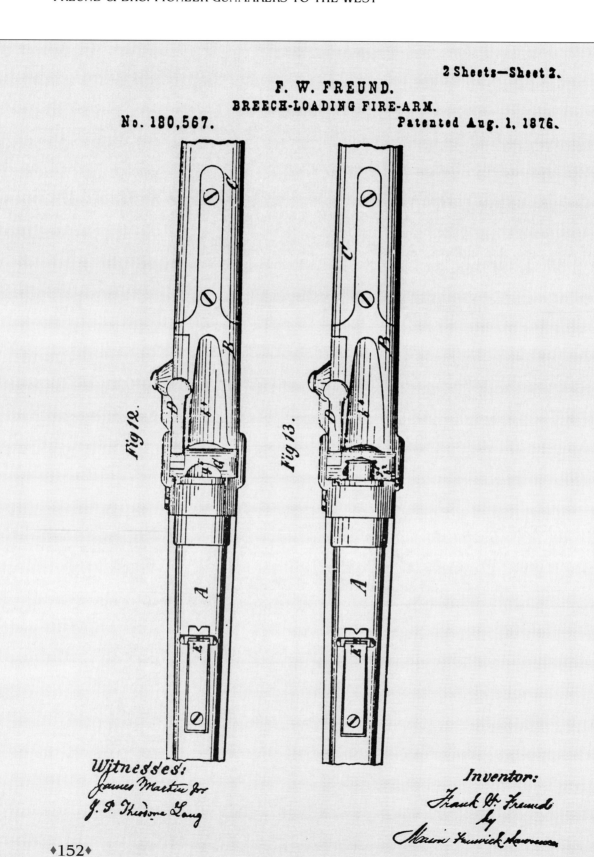

F. W. FREUND.
BREECH-LOADING FIRE-ARM.

No. 180,567.

2 Sheets—Sheet 2.

Patented Aug. 1, 1876.

Fig 12.

Fig 13.

Witnesses:
James Martin Jr
J. P. Theodore Lang

Inventor:
Frank W. Freund
by
Marion Fenwick Harmon

No. 183389

Frank W. Freund

Applied for Sept. 27, 1876   *Issued October 17, 1876*

IMPROVEMENT IN REVOLVING FIRE-ARMS

The object of this invention is to enable the operator to take the firearm to pieces, and to put it together again without the aid of tools. Another object is to avoid weakening the cylinder frame by drilling holes into it for the reception of pins and screws, and still another object is to give the firearm a neat appearance by means of smooth and well-polished surfaces, and by avoiding unsightly appearances by dispensing with the use of exposed pins and screws in holding the parts together.

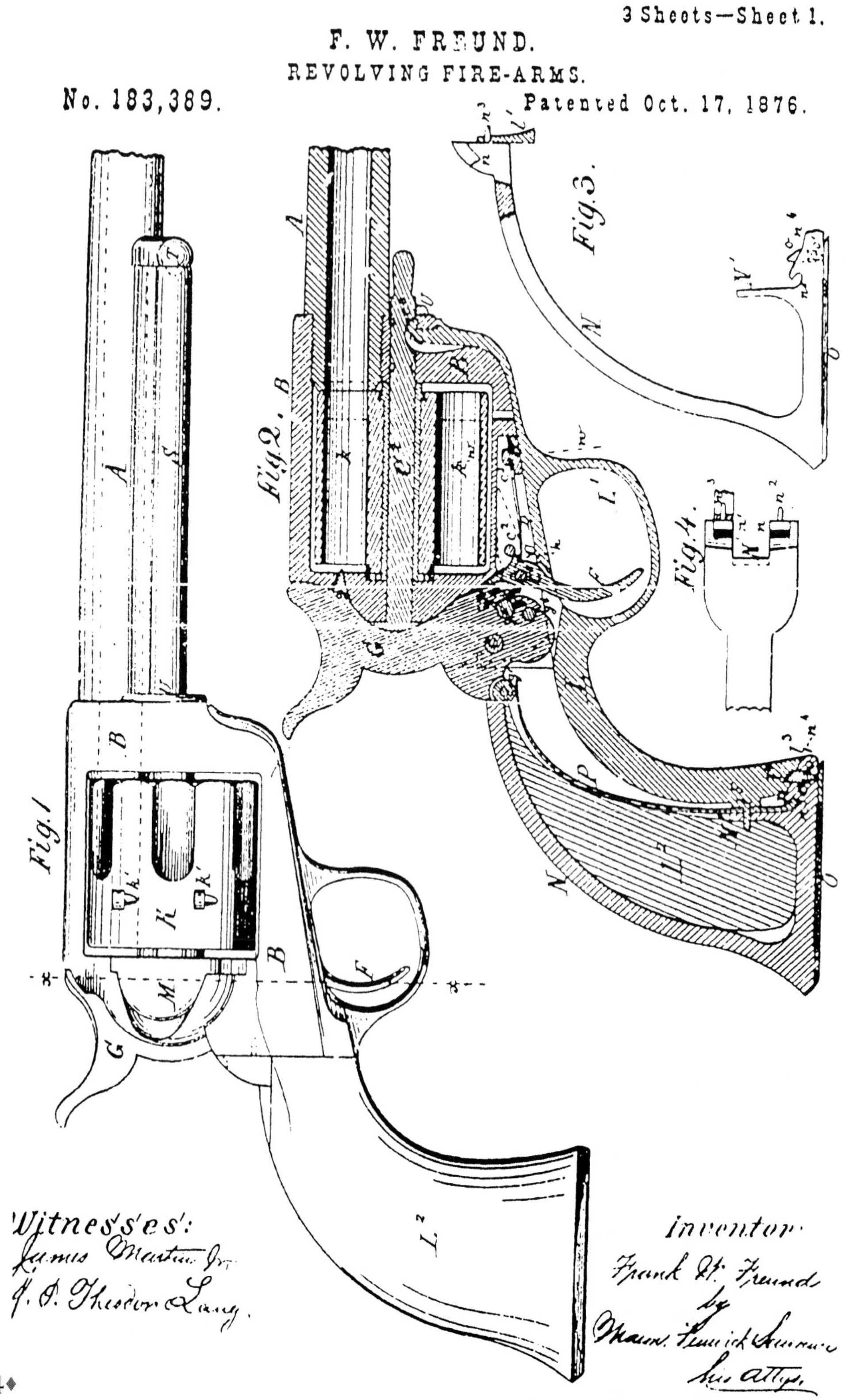

F. W. FREUND.

REVOLVING FIRE-ARMS.

3 Sheets—Sheet 1.

No. 183,389.

Patented Oct. 17, 1876.

Witnesses:
James Martin Jr.
J. P. Theodor Lang.

Inventor:
Frank W. Freund
by
Mason, Fenwick Lawrence
his attys.

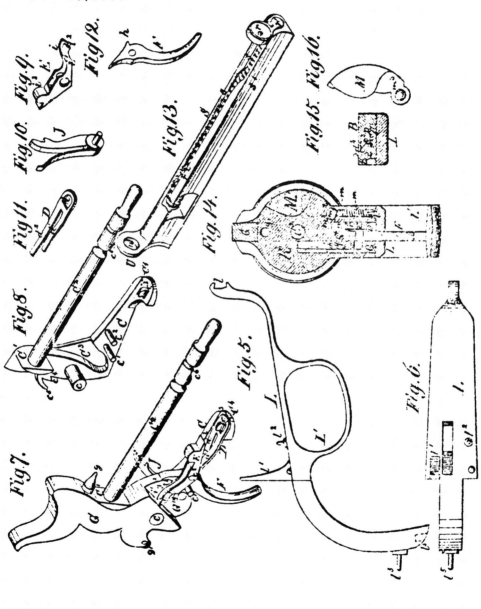

3 Sheets—Sheet 2.

F. W. FREUND.
REVOLVING FIRE-ARMS.

No. 183,389.   Patented Oct. 17, 1876.

Fig.12. Fig.4. Fig.10. Fig.11. Fig.8. Fig.7. Fig.13. Fig.14. Fig.5. Fig.15. Fig.16. Fig.6.

Witnesses:

Inventor:

F. W. FREUND.
REVOLVING FIRE-ARMS.

3 Sheets—Sheet 3.

No. 183,389.

Patented Oct. 17. 1376.

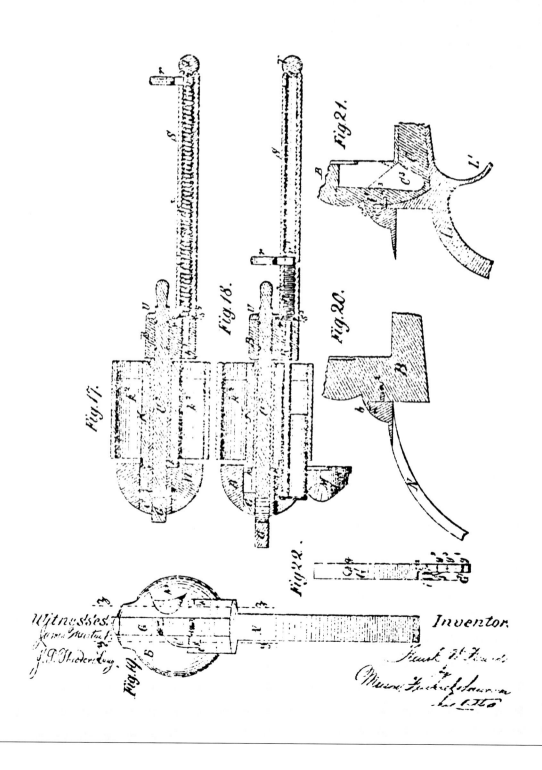

Fig.21.

Fig.18.

Fig.20.

Fig.17.

Fig.22.

Witnesses.

Inventor.

Fig.19.

No. 184202

Frank W. Freund

Applied for May 11, 1876   *Issued November 7, 1876*

IMPROVEMENT IN BREECH-LOADING FIRE-ARMS

The object of this invention is to furnish a firearm in which all the parts of the lock mechanism and other parts composing the arm are interlocked without the use of detachable pins or screws, and secured in permanent position by a final fastening device which can be operated by hand, the operation of which permits all the parts severally or together to be detached from one another without the use of tools or instruments. The firearm also may be taken apart without tools, such as screwdrivers; and owing to its simplicity of construction, may be taken apart and put together by anyone in day or nighttime without danger of losing the small parts.

Obviously, this is the patent on which the false-hammer gun that Frank W. Freund built in New Jersey was patented.

The next patent, no. 184203, issued on the same date, also is incorporated into the false-hammer New Jersey-made firearm.

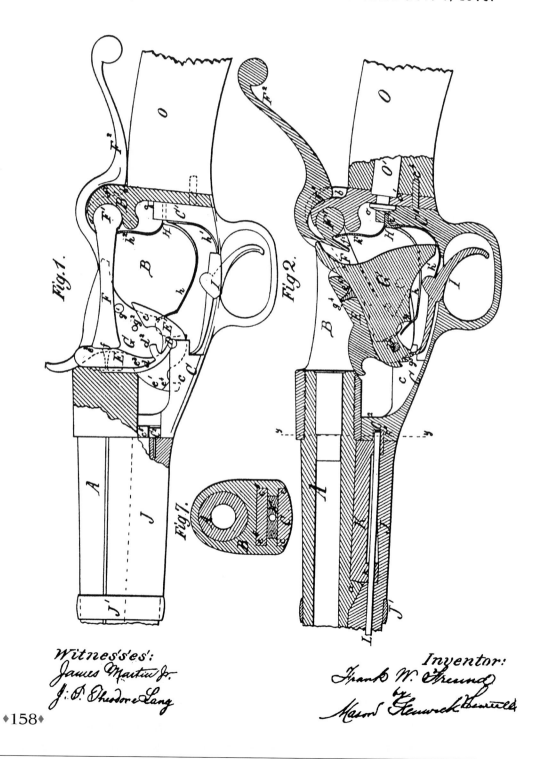

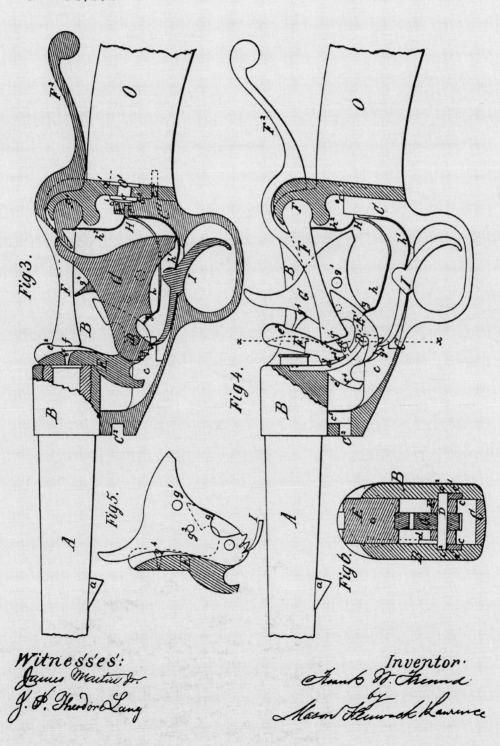

6 Sheets—Sheet 2.

F. W. FREUND.
BREECH-LOADING FIRE-ARM.

No. 184,202.   Patented Nov. 7, 1876.

Fig 3.

Fig 4.

Fig 5.

Fig 6.

Witnesses:
James Martin Jr.
J. P. Theodor Lang

Inventor:
Frank W. Freund
by
Mason Fenwick Lawrence

F. W. FREUND.
BREECH-LOADING FIRE-ARM.

6 Sheets—Sheet 3.

No. 184,202.

Patented Nov. 7, 1876.

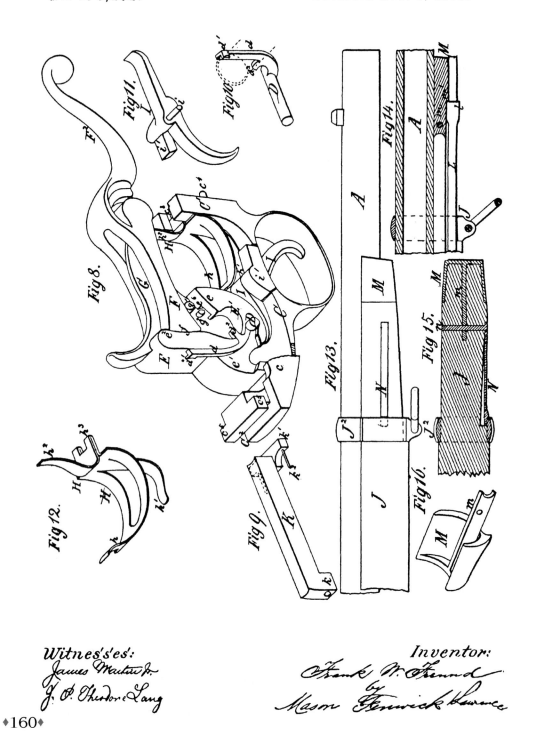

Witnesses:
James Mactire Jr.
J. P. Theodore Lang

Inventor:
Frank W. Freund
by
Mason Fenwick Lawrence

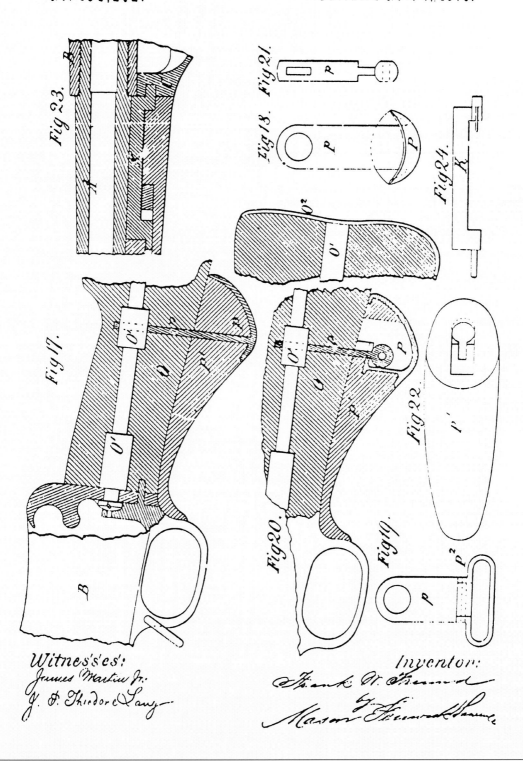

6 Sheets—Sheet 4.

F. W. FREUND.
BREECH-LOADING FIRE-ARM.

No. 184,202.   Patented Nov. 7, 1876.

Witnesses:
James Martin Jr.
J. F. Theodore Lang

Inventor:
Frank W. Freund
by
Mason Freund

F. W. FREUND.
BREECH-LOADING FIRE-ARM.

6 Sheets—Sheet 5.

No. 184,202.

Patented Nov. 7, 1876.

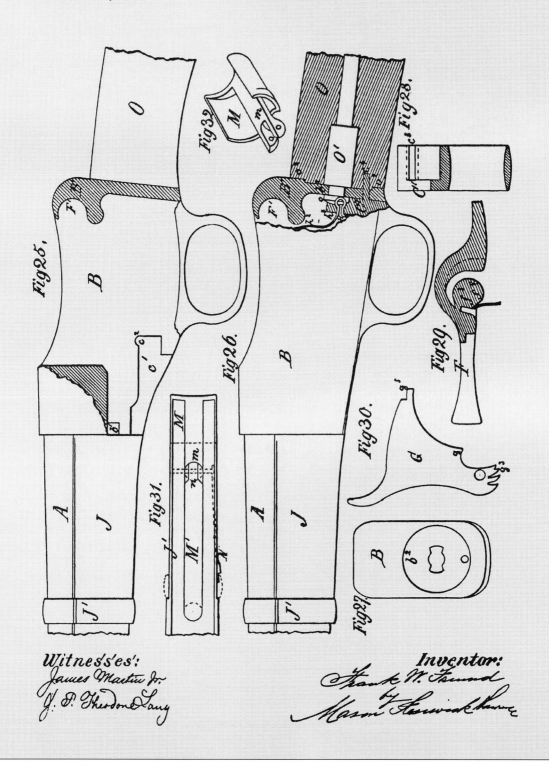

Witnesses':
James Martin Jr.
G. P. Theodore Lang

Inventor:
Frank W. Freund
by Mason Fenwick Lawrence

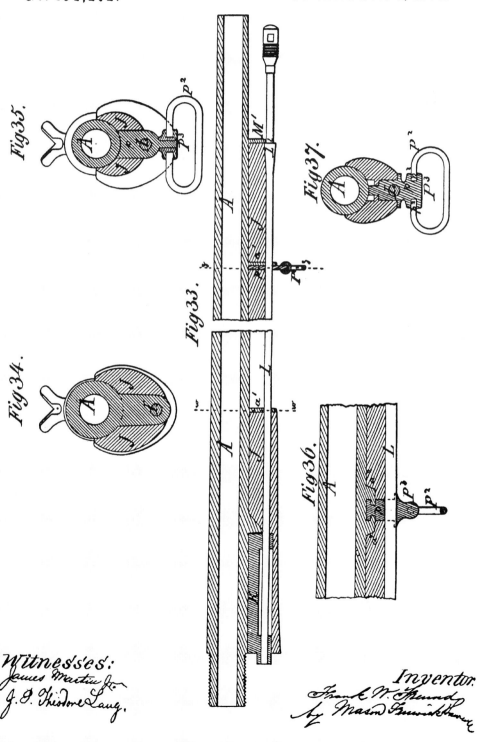

6 Sheets—Sheet 6.

F. W. FREUND.
BREECH-LOADING FIRE-ARM.

No. 184,202.

Patented Nov. 7, 1876.

Fig 35.

Fig 37.

Fig 33.

Fig 34.

Fig 36.

Witnesses:
James Martin Jr.
J. D. Theodore Lang.

Inventor.
Frank W. Freund
by Mason Fenwick Lawrence

Freund's patent cavalry pattern carbine, having features of patent no. 184202, right side. This arm was constructed to be field-stripped without the use of tools. (Freund Family Collection)

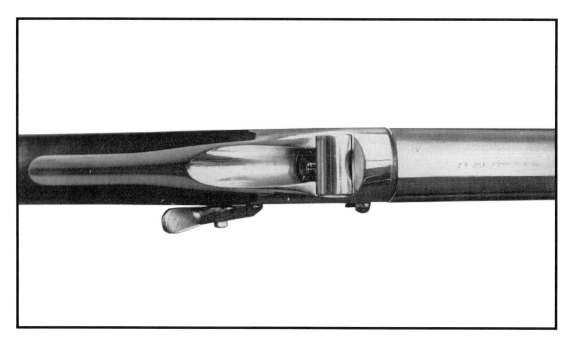

Top view at breech, of the cavalry carbine pictured above. Top of barrel is engraved, "Working model, Cal .45." (Freund Family Collection)

No. 184203

Frank W. Freund

Applied for February 23, 1876   *Issued November 7, 1876*

IMPROVEMENT IN BREECH-LOADING FIRE-ARMS

This patent consists of a provision for removing the parts of the breech and the lock mechanism, by simply removing and replacing them by hand without the aid of tools. They also may be inserted and fastened without the aid of tools. The triggerguard is made the means for opening, folding, and closing the breech-piece and its locking piece, and half-cocking the arm. Again, this patent is incorporated in the false-hammer gun made by Frank Freund in New Jersey.

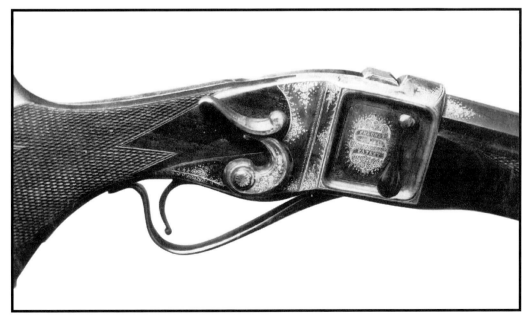

Close-up view at right side of action, of Freund's false-hammer rifle, built having features of both patents nos. 184202 and 184203. The triggerguard folds out and releases the breechblock. Although this rifle was completed in New Jersey, undoubtedly it was begun while the Freunds still resided in Cheyenne. (Freund Family Collection)

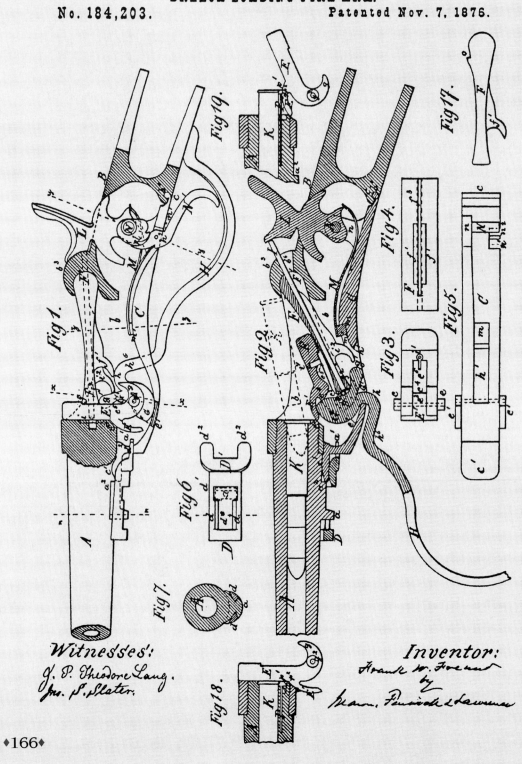

F. W. FREUND.

BREECH-LOADING FIRE-ARM.

No. 184,203.

2 Sheets—Sheet 1.

Patented Nov. 7, 1876.

Fig 19.

Fig 17.

Fig 1.

Fig 2.

Fig 4.

Fig 5.

Fig 3.

Fig 6.

Fig 7.

Fig 18.

Witnesses':

J. P. Theodore Lang.

Jus. S. Slater.

Inventor:

Frank W. Freund

by

Mann, Penwick & Lawrence

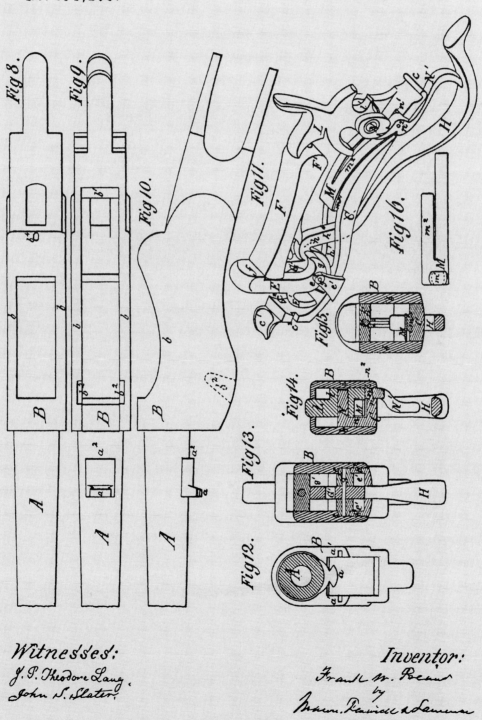

2 Sheets—Sheet 2.

F. W. FREUND.
BREECH-LOADING FIRE-ARM.

No. 184,203.

Patented Nov. 7, 1876.

Fig 8.
Fig 9.
Fig 10.
Fig 11.
Fig 16.
Fig 15.
Fig 14.
Fig 13.
Fig 12.

Witnesses:
J. P. Theodore Lang.
John S. Slater.

Inventor:
Frank W. Freund
by
Mason, Fenwick & Lawrence

No. 184854

Frank W. Freund

Applied for October 12, 1876   *Issued November 28, 1876*

IMPROVEMENT IN PRIMERS FOR CARTRIDGES

The nature of this invention consists of a removable cap or primer, which is adapted for use in conjunction with a cartridge for preventing the escape of gas rearward when the cartridge is exploded.

This avoids the consequences resulting from the escape of gas when the cartridge is fired, such as burning the operator's face and hands or wounding him by fragments of the cap or of the cartridge case.

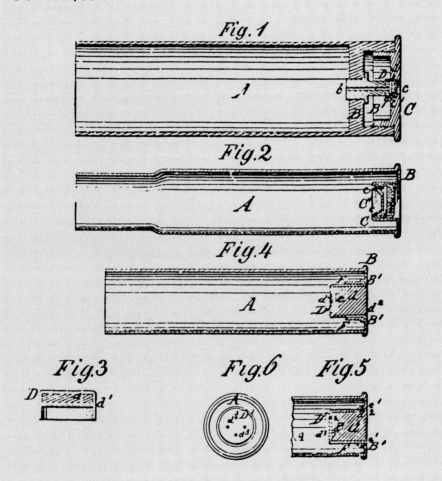

F. W. FREUND.
PRIMERS FOR CARTRIDGES.

No. 184,854. Patented Nov. 28, 1876.

Fig. 1

Fig. 2

Fig. 4

Fig. 3   *Fig. 6*   *Fig. 5*

Witnesses:
James Martin Jr.
J. P. Theodore Lang.

Inventor:
Frank W. Freund
by
Mason, Fenwick Lawrence
Attys

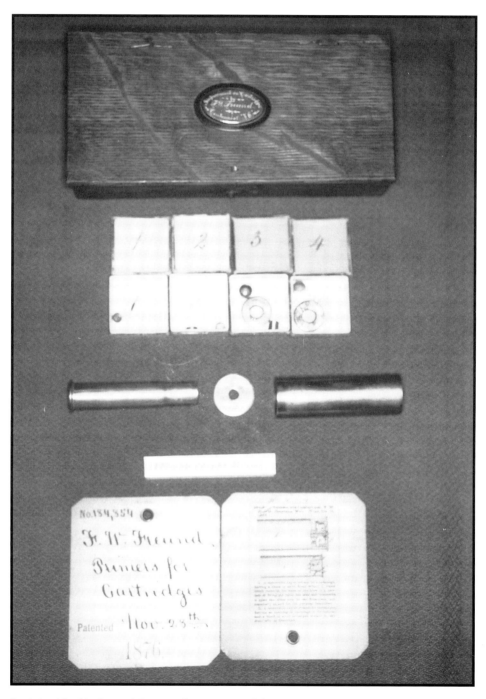

Contained in this fine oak box are the patent models and related tags covered under Freund's patent no. 184854, for improved cartridge primers. The plaque in the box lid (opposite page, bottom) reads, "Improvement on Cartridges/by/F.W.Freund./Centennial '76." (Courtesy Robert T. Buttweiler)

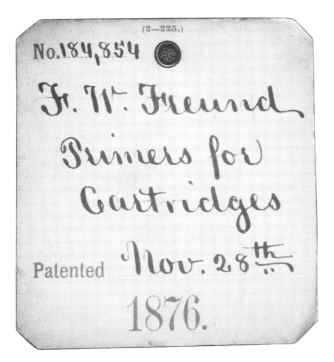

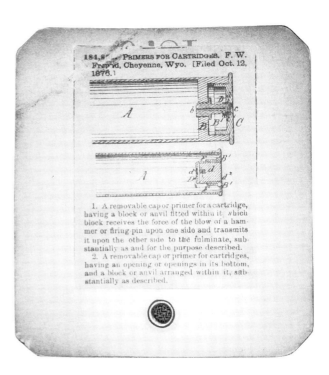

No. 185911

Frank W. Freund

Applied for December 3, 1875   *Issued January 2, 1877*

IMPROVEMENT IN BREECH-LOADING FIRE-ARMS

The object of this invention is to provide breech-loading rimfire or centerfire firearms with a gas check, which, when the arm is fired, will make gas-tight all rear joints or passages leading from the firing-pin to the rear of the breechblock.

This also will provide a gas check or firing pin actuator which will be caused to half-cock the hammer, when it is bearing upon the gas check or firing pin carrier, by the operation of lowering or raising the guard lever.

Another object is to have the firing pin thus arranged move in a straight line, and thereby overcome the diffculty and danger of clogging when the breech-block is lowered for reloading.

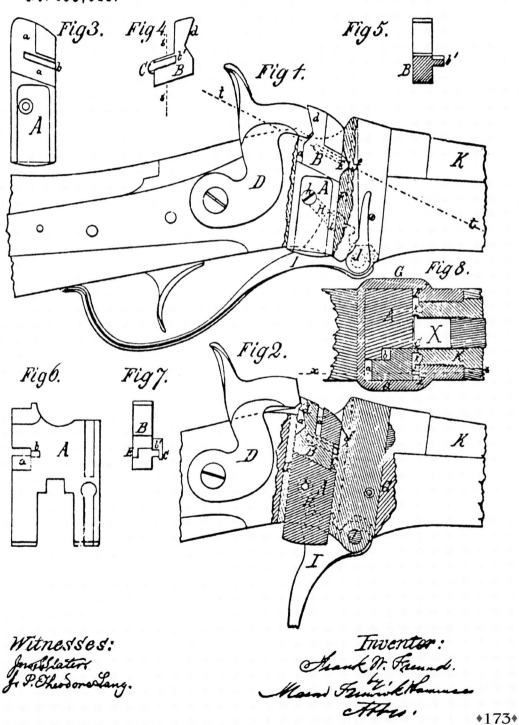

F. W. FREUND.
BREECH-LOADING FIRE-ARM.

No. 185,911.

2 Sheets—Sheet 1.

Patented Jan. 2, 1877.

Fig 3. Fig 4. Fig 5.

Fig 1.

Fig 8.

Fig 6. Fig 7.

Fig 2.

Witnesses:

Inventor:

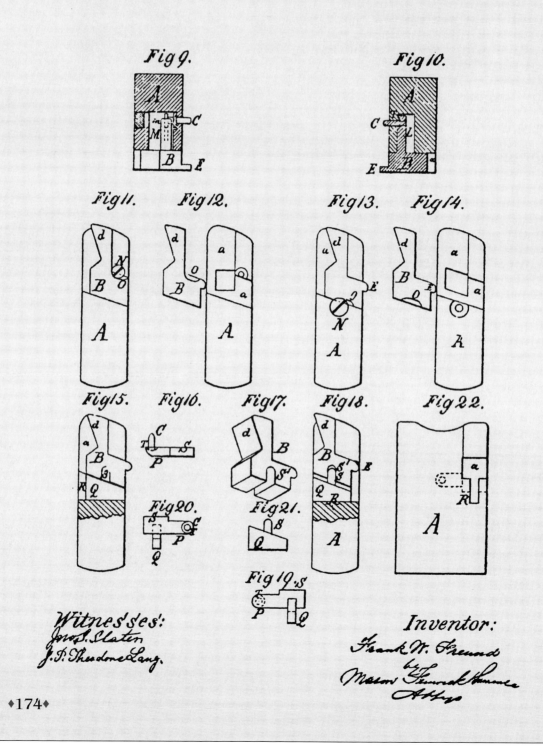

2 Sheets—Sheet 2.

F. W. FREUND.
BREECH-LOADING FIRE-ARM.

No. 185,911.   Patented Jan. 2, 1877.

Fig 9.

Fig 10.

Fig 11.   Fig 12.   Fig 13.   Fig 14.

Fig 15.   Fig 16.   Fig 17.   Fig 18.   Fig 22.

Fig 20.   Fig 21.

Fig 19.

Witnesses:
Jno.S.Slater
J.S.Theodore Lang.

Inventor:
Frank W. Freund

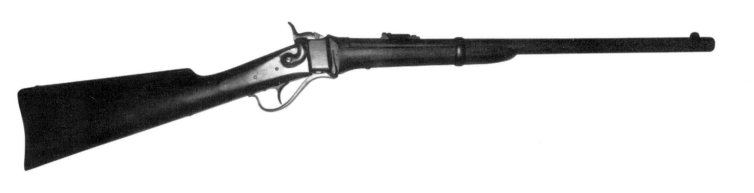

The patent model gun for Freund patent no. 185911, right side. This rifle is not chambered for a cartridge, nor does it have an extractor. The lockplate is engraved, "F.W. Freund/Cheyenne/Wyo. Terr./1875." (Courtesy Gary Gallup and Chris Schneider)

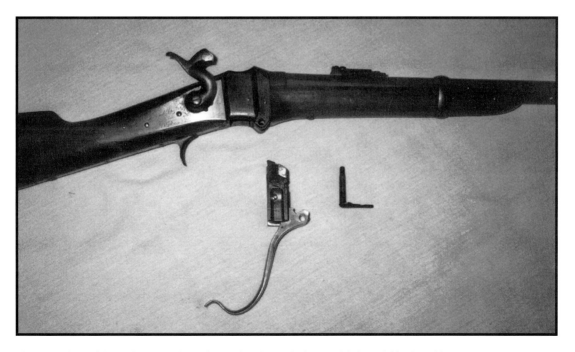

Close-up view of right side at action, of the rifle pictured above, with breechblock and lever removed. Hammer shows evidence of hand piecing. (Courtesy Gary Gallup and Chris Schneider)

No. 189721

Frank W. Freund

Applied for April 7, 1876   *Issued April 17, 1877*

IMPROVEMENT IN FRONT SIGHT FOR FIRE-ARMS

The object of this patent is to produce a front sight for firearms which shall be equally as efficient in dark and cloudy as in fine weather, and for aiming at dark as well as light objects; and to this end it consists of making the rear face of the sight two contrasting colors, preferably black and white, combined in such a manner that the body of the sight shall present one color, bounded on the top, side, or edge with a line of opposite color.

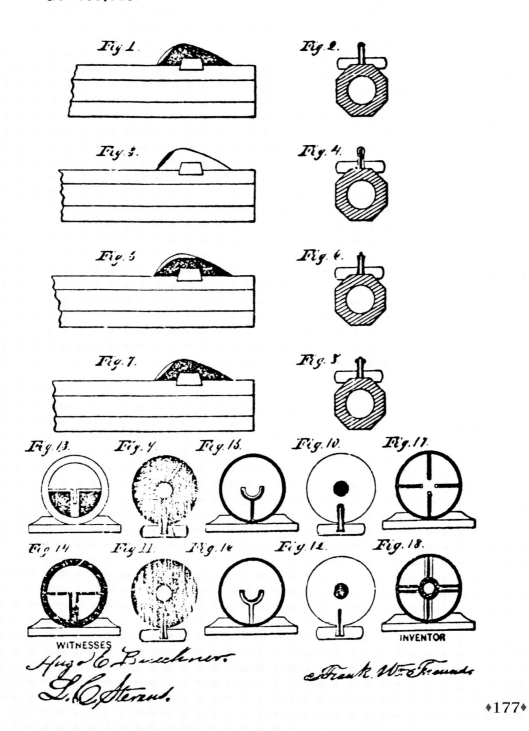

F. W. FREUND.

FRONT SIGHTS FOR FIRE-ARMS.

No. 189,721.

Patented April 17, 1877.

No. 211728

Frank W. Freund

Applied for October 5, 1878   *Issued January 28, 1879*

IMPROVEMENT IN BREECH-LOADING FIRE-ARMS

This patent applies to an improvement in the Ballard gun, and by certain modifications it may be applied to other styles of arms.

The principal feature of this invention consists in the application of a slide which is operated by the opening of the breech to cock the hammer, and if desired, to set the triggers. The slide is arranged within the moveable breech-block so that when the latter is moved to open the breech of the gun, the rear end sight of the slide strikes a fixed projection or obstruction on the frame and is moved forward so that its front end strikes the tumbler of the hammer, bringing it to half- or full-cock as desired.

Patent model gun for Freund patent no. 211728, "Breech-Loading Fire-Arm." It applies to improvements to the action of Ballard arms. (Smithsonian Institution collection)

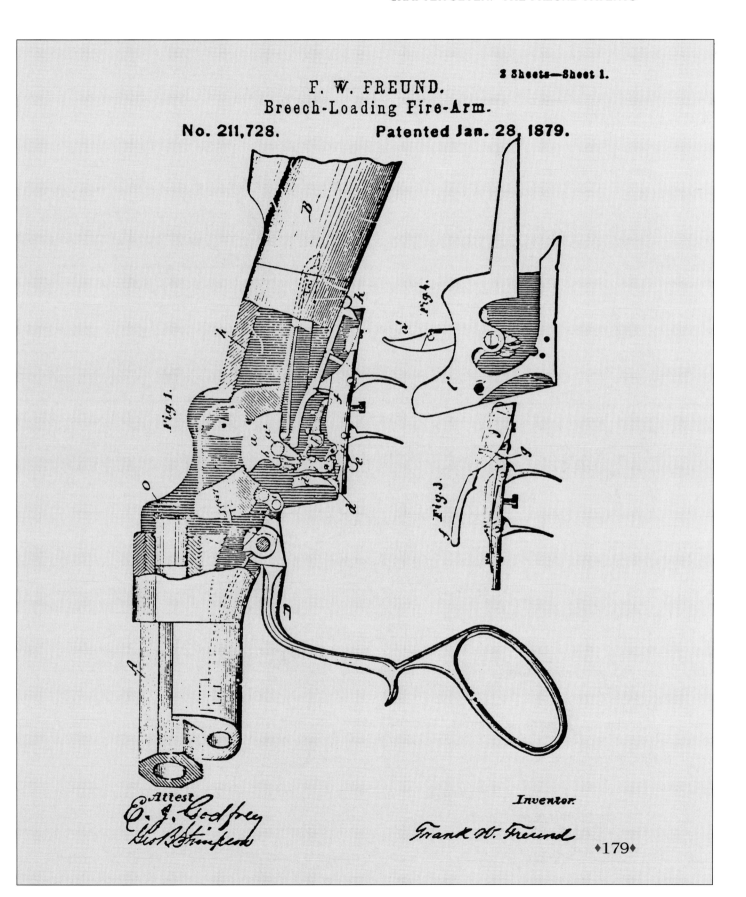

F. W. FREUND.
Breech-Loading Fire-Arm.

2 Sheets—Sheet 1.

No. 211,728.   Patented Jan. 28, 1879.

Attest

Inventor.

Frank W. Freund

♦179♦

2 Sheets—Sheet 2.

F. W. FREUND.
Breech-Loading Fire-Arm.

No. 211,728.   Patented Jan. 28, 1879.

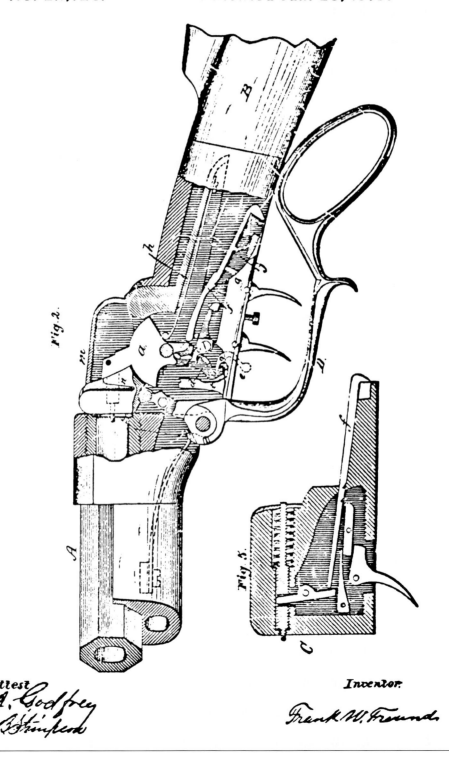

Attest
E. J. Godfrey
Geo. B. Stringer

Inventor:
Frank W. Freund

No. 216084

Frank W. Freund

Applied for October 30, 1876　　*Issued June 3, 1879*

IMPROVEMENT IN BREECH-LOADING FIRE-ARMS

The object of this invention is to produce an arm that shall be automatically cocked or half-cocked by the operation of the breechblock in the act of opening or closing the breech; and to construct the breechblock and the frame in which it operates in such a manner that when the breech is closed, the parts shall fit tightly together to prevent dirt or other foreign substances from entering. It consists of an intermediate lever placed between the breechblock in its downward movement to open the breech that will operate said lever and cause it to turn the tumbler upon its stem or arbor, and thereby cock or half-cock the hammer.

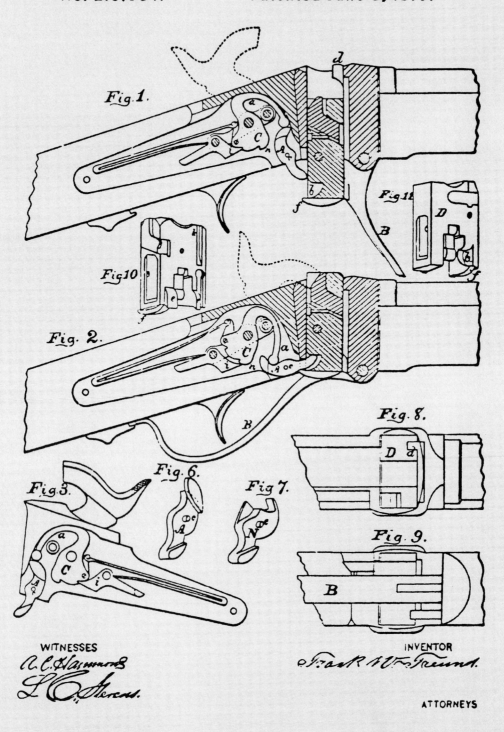

2 Sheets—Sheet 1.

F. W. FREUND.
Breech-Loading Fire-Arms.

No. 216,084.   Patented June 3, 1879.

Fig.1.

Fig.11.

Fig.10.

Fig. 2.

Fig.3.   Fig. 6.   Fig 7.

Fig. 8.

Fig. 9.

WITNESSES   INVENTOR

ATTORNEYS

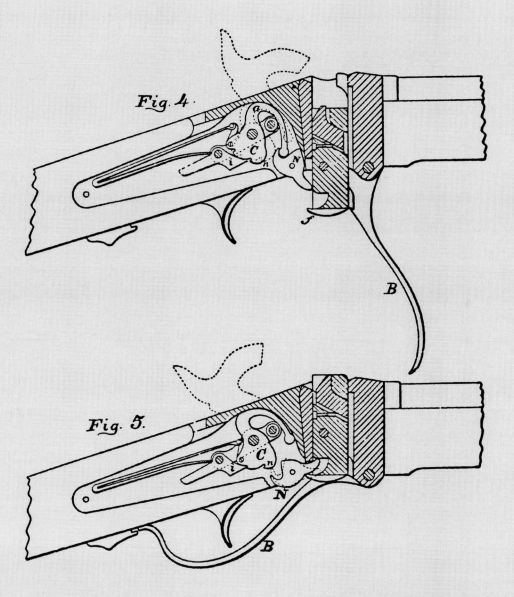

No. 229245

Frank W. and George Freund

Applied for March 9, 1880   *Issued June 29, 1880*

SIGHT FOR FIRE-ARMS

The object of this invention is to provide a rear sight for rifles, which enables the marksman to level the piece when aiming without the use of spirit levels or pendulums, and to aim at different elevations without raising or lowering the sight with respect to the barrel. To these ends, the invention consists of forming an angular, diamond-shape opening directly below the ordinary sight-notch.

The opening is of regular form so that when the barrel is level, the side angles shall be on the same horizontal plane, and the upper and lower angle on the same vertical plane.

This patent would not pertain itself to the diamond-shape opening below the ordinary sight notch, as a triangular form would be an obvious mistake. Some sights are very low, and it is impossible to make the diamond-shape opening, as there would be an obstruction in front when the sight is down. As the triangular is half of the diamond-shape, this is considered a modification of the full diamond-shape opening.

A horizontal line in the triangular-shape opening may have the sight notch indicating the center, when it is intended to take aim over the line.

(Model.)

F. W. & G. FREUND.
Sight for Fire Arms.

No. 229,245. **Patented June 29, 1880.**

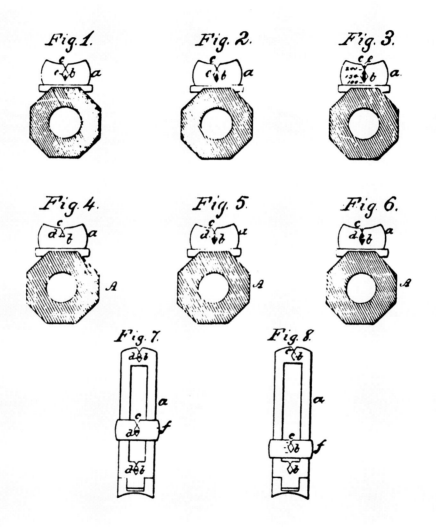

Fig. 1.
Fig. 2.
Fig. 3.
Fig. 4.
Fig. 5.
Fig. 6.
Fig. 7.
Fig. 8.
a
b
c
d
f
A

Witnesses.

H. E. Buechner.

Geo. Wilford.

Inventors.

Frank Wm Freund.

George Freund.

With regard to patent no. 229245, originally held jointly by Frank W. and George Freund, the "Sight for Firearms" became known as the "More Light" sight. Apparently after the Freund brothers had parted company, and George had moved to Durango, he solely applied for patents on this sight throughout the world. First was Canada, where he was granted patent no. 12673 by the government for this invention. George Freund then obtained additional patents for the "More Light" sight, as follow:

Germany	Patent no. 22,551
France	Patent no. 154,999
Belgium	Patent no. 63,109
England	Patent no. 2,233
Austria	Patent no. 27,946
Italy	Patent no. 15,811

In his prolific advertising, George Freund referred to this sight as "The only original 'More Light' Patent Gun Sights," and advised readers to "Address all communications to GEO. FREUND, Patentee, Colorado Armory Building, DURANGO, COLO., U.S.A."

No. 268090

Frank W. Freund

Applied for August 19, 1881   *Issued November 28, 1882*

SIGHT FOR FIRE-ARMS

This invention is an improvement on the sight shown in patent no. 229245, issued June 29, 1880.

It consists of applying to the "More Light" sight a moveable or sliding plate, by which the aperture below the sight notch can be opened, partially opened, or entirely closed, as desired.

Details of construction may be varied and are regarded as equivalents, and thus do not limit the precise construction shown in the drawings.

(No Model.)

F. W. FREUND.
SIGHT FOR FIRE ARMS.

No. 268,090.

Patented Nov. 28, 1882.

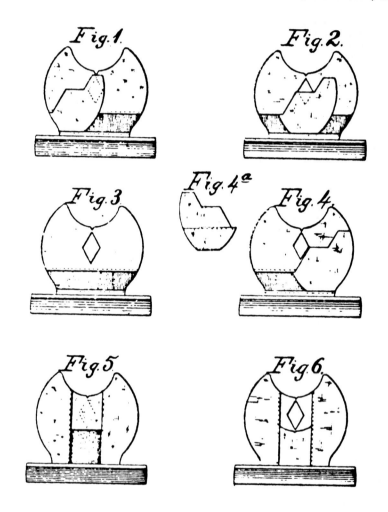

Fig. 1. Fig. 2. Fig. 3. Fig. 4ª. Fig. 4. Fig. 5. Fig. 6.

Witnesses.

Inventor.

No.273156

Robert B.F. Reed and George Freund

Applied for October 18, 1882   *Issued February 27, 1883*

SAFETY SHELL FOR BLASTING

This patent is for an improved safety shell for blasting purposes where high explosives are employed, such as Giant powder. The material is used in sticks or candles, and for firing the charge a cap is attached to the end of the stick or candle, the cap being on the end of the fuse.

The old method of attaching the cap was to bore a hole in the end of the stick or candle and insert the cap. The stick or candle had to be warmed to insert the cap, which was a dangerous procedure. The object of this invention is to remove these difficulties and to achieve safety in the use of Giant powder.

A safety shell of thin sheet-metal is formed to receive the cap or primer containing the fuminating material. The cap or primer can be attached permanently to the head of the shell or threaded to receive the cap or primer.

The construction of the shape is not limited to any particular shape or arrangement. The shells may be made of various lengths and diameters, so as to fit the sticks or candles of Giant powder or explosive.

All that is necessary to prepare the charge is to place the shell on the end of the stick or candle, then insert the cap with the fuse attached, which can be done with perfect safety.

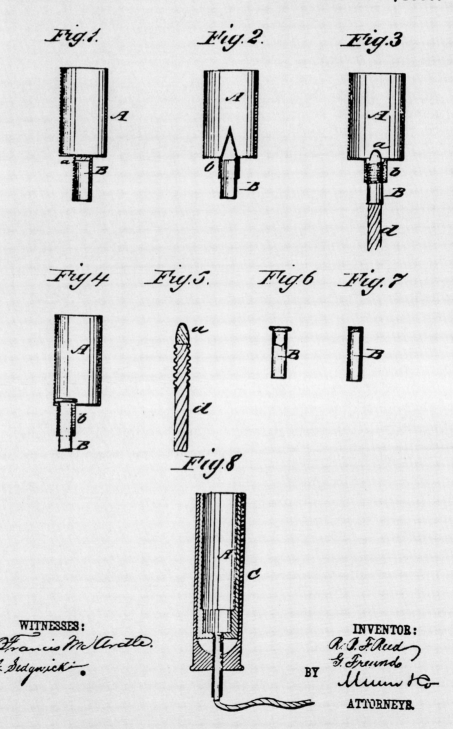

(Model.)

R. B. F. REED & G. FREUND.
SAFETY SHELL FOR BLASTING.

No. 273,156.   Patented Feb. 27, 1883.

Fig.1.   Fig.2.   Fig.3.

Fig.4.   Fig.5.   Fig.6   Fig.7.

Fig.8

WITNESSES: INVENTOR:
Francis McArdle.   R. B. F. Reed
C. Sedgwick.   G. Freund
                                        BY   Munn & Co
                                                ATTORNEYS.

No. 289768

George Freund

Applied for October 9, 1883   *Issued December 4, 1883*

CAP FOR FIRING EXPLOSIVE CHARGES

The object of this invention is to provide a new and improved cap for receiving the fulminate, that allows a full ten minutes for firing explosive caps in blasting rock, etc.

The invention consists of a screw-threaded tube adapted to receive the fulminate in its lower part, the lower end of the tube being closed by a cap. A cap connected with a conducting wire is screwed on the upper end of the tube, then the fuse is inserted in the tube.

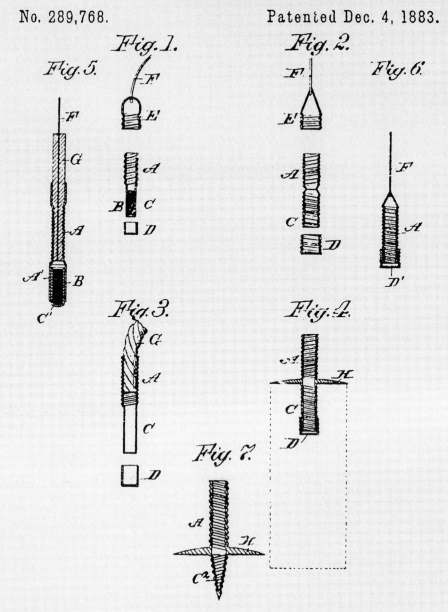

(No Model.)

G. FREUND.

CAP FOR FIRING EXPLOSIVE CHARGES.

No. 289,768.   Patented Dec. 4, 1883.

Fig. 1.   Fig. 2.

Fig. 5.   Fig. 6.

Fig. 3.   Fig. 4.

Fig. 7.

WITNESSES: INVENTOR:

G Freund

BY

ATTORNEYS.

♦192♦

No. 292642

George Freund

Applied for June 23, 1883   *Issued January 29, 1884*

SAFETY SHELL FOR BLASTING

This invention relates to safety shells of the type shown in patent no. 273156, granted to Reed and Freund on February 27, 1883.

This patent consists of certain novel forms of construction of a screw-threaded blasting shell formed with a screw-threaded nozzle in combination with a threaded firing cap, as shown and described.

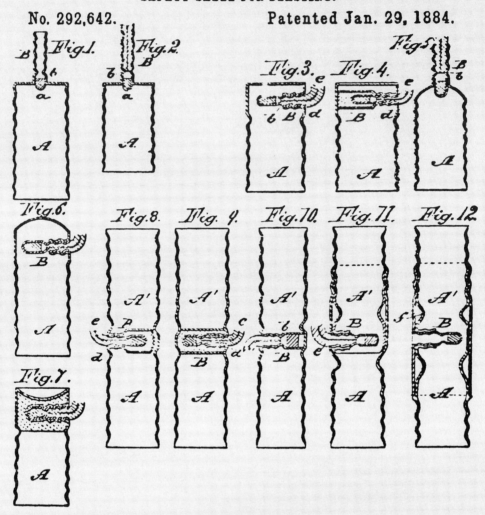

(No Model.)

2 Sheets—Sheet 1.

G. FREUND.

SAFETY SHELL FOR BLASTING.

No. 292,642.

Patented Jan. 29, 1884.

WITNESSES:

INVENTOR:

BY

ATTORNEYS.

(No Model.)   2 Sheets—Sheet 2.

G. FREUND.
SAFETY SHELL FOR BLASTING.

No. 292,642.   Patented Jan. 29, 1884.

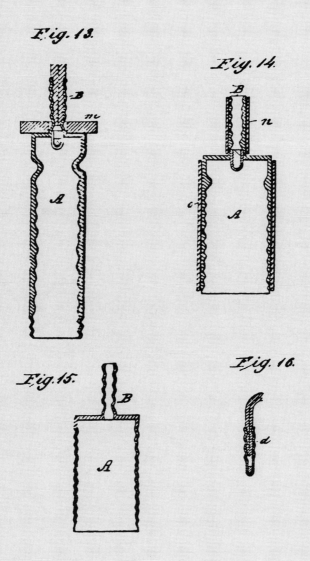

Fig. 13.

Fig. 14.

Fig. 15.

Fig. 16.

WITNESSES:

INVENTOR:

G. Freund

BY

ATTORNEYS.

No. 297375

George Freund

Applied for December 27, 1883   *Issued April 22, 1884*

POCKET KNIFE

This invention consists of a knife having a notch in the handle case, and a notch on the blade, the latter notch having a screw-thread formed on its bottom to press a screw-thread in the end of a blasting fuse placed in the notch in the handle.

It also consists of a blade pivoted to the case and forming a pair of pliers or nippers to be used to cut the fuse. The blade has an inner side hollowed out, to form a spoon for digging out a stick or candle of Giant powder to receive the fuse.

This invention provided the miner with a new and improved knife to facilitate the cutting and capping of a fuse.

(No Model.)

G. FREUND.
POCKET KNIFE.

No. 297,375.

Patented Apr. 22, 1884.

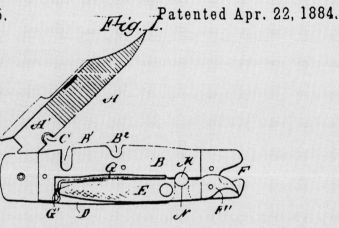

Fig. 1.

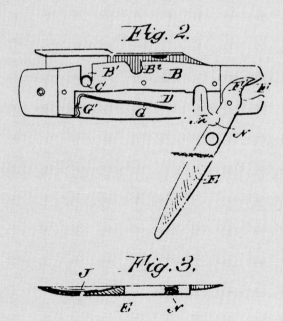

Fig. 2.

Fig. 3.

WITNESSES:

INVENTOR:

BY

ATTORNEYS.

No. 313414

George Freund

Applied for August 19, 1884   Issued March 3, 1885

KNIFE FOR MINERS

The object of this invention is to provide a new and improved knife for miners in facilitating the cutting and capping of fuses and opening cans. This is an improvement on the knife for which patent no. 297375 was issued on April 22, 1884.

This knife consists of the combination of a knife-casing, a can opener, and a cork-screw held in the blade for splitting or cutting the fuse.

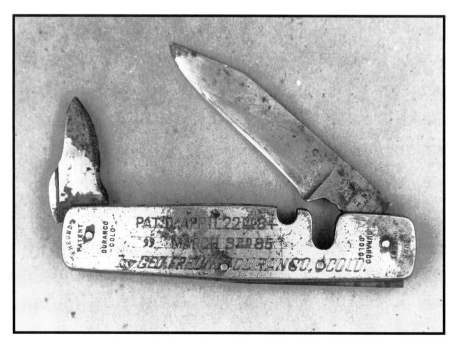

George Freund's "Knife for Miners," U.S. patent no. 313414. Actually an all-purpose tool for cutting and capping blasting fuses, and a can-opener for the miner's lunch, it is greatly sought-after among collectors today. (Author's collection)

(No Model.)

G. FREUND.
KNIFE FOR MINERS.

No. 313,414. Patented Mar. 3, 1885.

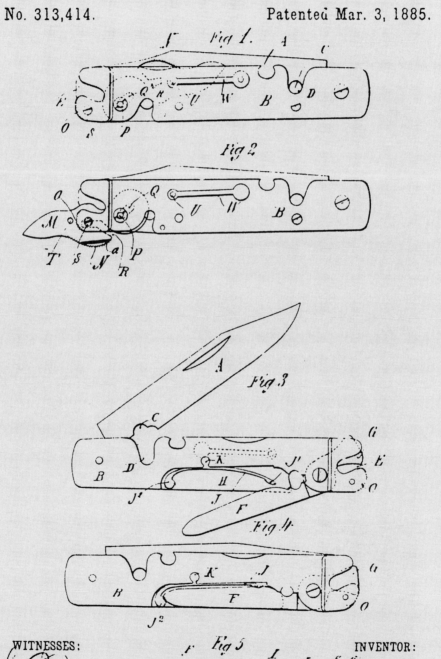

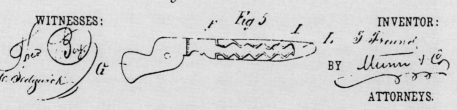

WITNESSES: INVENTOR:

ATTORNEYS.

No. 496051

Frank W. Freund

Applied for October 18, 1884   *Issued April 25, 1893*

GUN SIGHT

This invention relates to rear sights, and consists of a base rigidly attached to the barrel, and a series of interchangeable sight plates adapted to be attached to said base. The object is to provide a sight, or a series of sights, which may be changed at will.

It is oftentimes desirable to change the sights of firearms, especially rifles, to meet changes in the conditions under which the firearm is used. Therefore a construction is desirable which permits such changes to be made with as little delay and inconvenience as possible.

(No Model.)

F. W. FREUND.
GUN SIGHT.

No. 496,051. Patented Apr. 25, 1893.

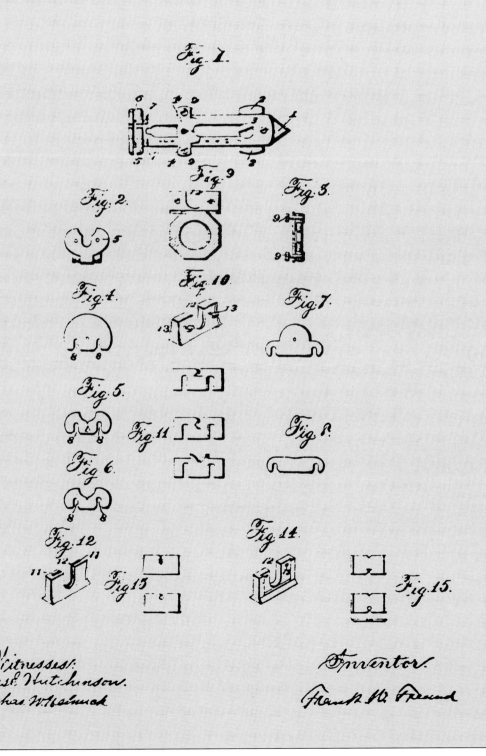

Witnesses:
Jas.F. Hutchinson.
Chas. W. Helmick

Inventor:
Frank W. Freund

No. D22406

George Freund

Applied for April 3, 1893   *Issued May 9, 1893*

DESIGN FOR A SPOON

The design consists of the surface ornamentation in the bowl of a spoon as shown in accompanying drawing, in which the figure is a front view of a spoon embodying George Freund's unique design.

The design represents a bird's-eye view featuring a river having a winding course, and crossed by several bridges. In the foreground appears a series of buildings of the general form of smelting works, together with a railroad with a train thereon, and a roundhouse. In the middle distance beyond the river, both to the left and right of the scene, appear additional smelting works with smoking chimneys, and in the background appear mountains, as illustrated.

DESIGN.

G. FREUND.
SPOON.

No. 22,406.   Patented May 9, 1893.

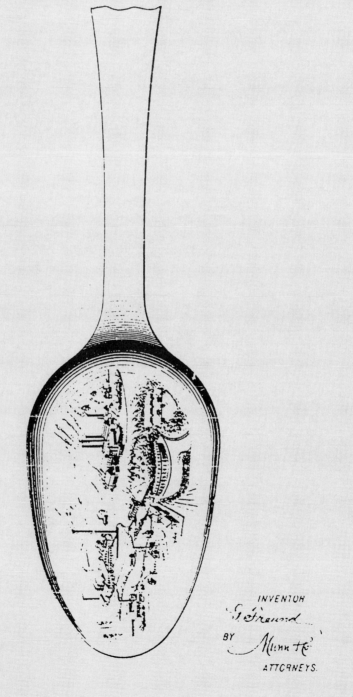

WITNESSES:

INVENTOR
G. Freund
BY
ATTORNEYS.

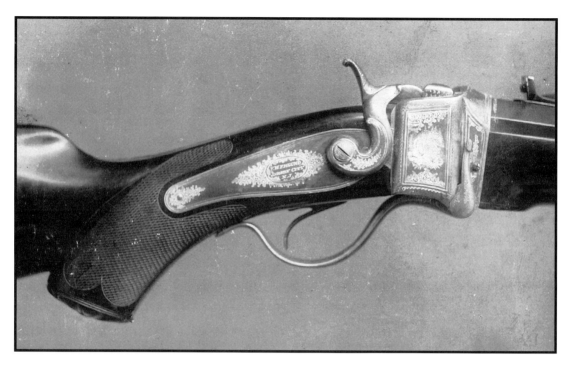

A period photograph of a Model 1877 Sharps rifle altered by Frank W. Freund while in Jersey City, New Jersey, for Theodore Roosevelt, right side.

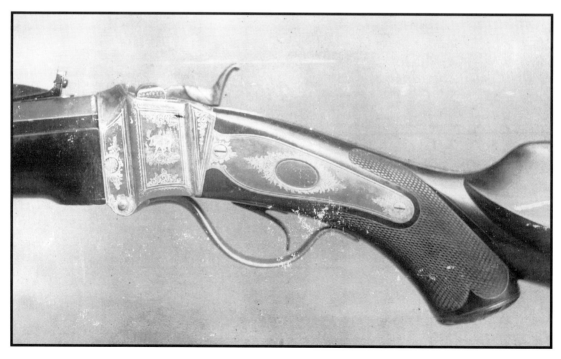

A period photograph of the left side of Theodore Roosevelt's Freund-altered Model 1877 Sharps rifle. Note the extremely fine-quality engraving and panel scene of a bull moose.

THE
FREUND
RIFLES

*A*ll indications point to the premise that the Freund brothers neither converted nor made any rifles in their early days in the West, while following the tracks of the Union Pacific railroad. At best, their shops along the right-of-way had to have been rather primitive, housed in tents or in hastily thrown-up clapboard buildings. During this period Frank appears to have been constantly on the move, scouting for new sites and visiting the several shops he always seemed to be operating at any given time.

Except for the stores at Cheyenne and Laramie, and the one at Salt Lake City, the other locations of Freund gun shops seemed to come and go after a period of one to three months, in the years spanning 1866 to 1870.

It was not until the Freund brothers settled in Denver in late 1870 that Frank had the time to start building rifles, or to begin any amount of experimenting with inventions that were patentable.

The first Freund-made rifle is a muzzle-loading percussion rifle of .48 caliber, marked "Freund & Bro., Denver, Colo." This is a heavy plains-type rifle with a fancy triggerguard and a crescent-style buttplate. It is easy to speculate that this gun may have been made by Frank Freund, for himself or George to shoot in the Denver *Deutcher Schuetzen* Club that was organized there in 1873[1] (*see* chapter twelve).

Most of the experimental work on single-shot rifles that was performed while the Freund brothers were in Denver was done on the Remington Rolling-Block rifle action. The patents applied for in Denver were for modifications or improvements to single-shot rifles, Remington Rolling-Block rifles, triggerguards, metallic cartridges, and sights.

These patents referred to double set-triggers, forked guard levers, cartridge extractors, changes in the cocking of the hammer, improvements in guard

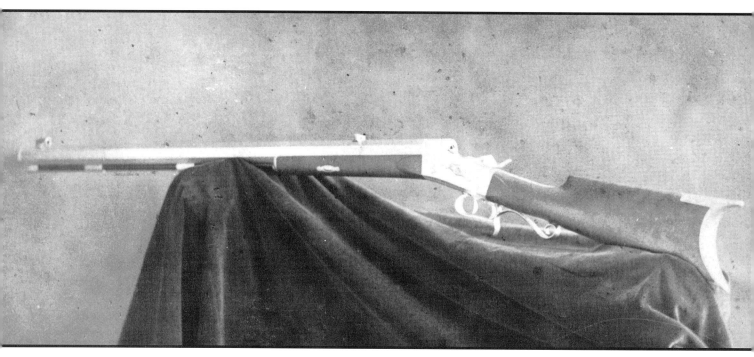

A Freund-modified Remington Rolling-Block rifle. (Freund Family Collection)

levers, and pistol grips for breech-loading rifles of the rolling-block type.

Aside from the patent models in the Smithsonian Institution collection, and the above percussion plains rifle, the only other surviving rifles that have Freund-Denver markings are two modified Remington Rolling-Block rifles, and one Springfield rifle.

A Springfield rifle that came from the Freund shop in Denver is mentioned in the book *Hunting at High Altitudes*, edited by George B. Grinnell. In it, Roger D. Williams describes his adventures in the Black Hills in 1875, with one of the "outlaw" or "sooner" parties. Traveling with the group was hunter and scout Moses Milner, better known as "California Joe." (This was not the Professor Walter P. Jenny expedition during the summer of 1875, for which California Joe also served as a guide.)

Williams was armed with a fine single-shot Springfield sporting rifle, which he called a "needle gun," a name often incorrectly applied to that action. The rifle was .45 caliber, had a heavy octagonal barrel, double set-triggers, and a curly maple pistol grip stock. It was chambered for a case holding 107 grains of powder, which was a rather heavy load for the Springfield action. In an encounter with a wounded deer, Williams, to his dismay, broke the stock of this

A Freund-modified Maynard rifle. Note the Freund operating lever, double set-triggers, and full-length scope. (Freund Family Collection)

fine Freund-Springfield rifle. He considered it the best hunting rifle he ever owned.[2]

There is another Freund-altered Springfield rifle in the collections of the Minnesota Historical Society (*see* chapter twelve). Freund's modifications included thinning the wrist to change the contours of the comb and the stock area behind the lock and lock screws, and the addition of a double set-trigger with a new triggerguard. The rifle is marked "Freund" on the triggerguard, and on the remaining piece of the original trigger-plate. It is unknown at which of the Freunds' shop locations this alteration was made.

Frank and George, no doubt, reworked the actions of a number of Remington Rolling-Block rifles, as several period photographs exist of them. One has a full-length scope, and others have cleaning rods and forearms with two key escutcheons. It is probable that other types of breech-loading rifles were worked on or improved while the Freund brothers were in Denver, but this author has not positively identified any examples of them.

The conversion work for which the Freund brothers are best known is the patented improvements to Model 1874 and Model 1877 Sharps rifles. The Sharps alterations may have been started while the Freund brothers worked in

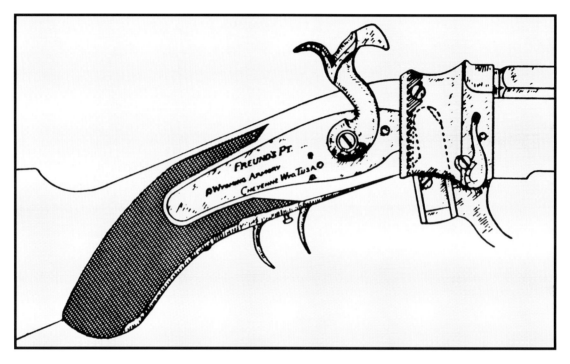

Sketch of Freund Improved Model 1877 Sharps rifle. Shaded section in receiver shows gib used to impart camming action to breech-block. (Courtesy John Barsotti, *Gun Digest*)

Denver, but the first patent for these improvements was applied for on June 1, 1876, about thirteen months after the Freunds had moved to Cheyenne, Wyoming Territory.

This patent as applied to the Model 1874 Sharps rifle closed the breech with an upward and forward motion, positively seating the cartridge into the chamber of the barrel. These modifications to the Sharps rifle started a correspondence between Frank W. Freund and the Sharps Rifle Company that continued for several years.

But the Sharps Rifle Company was not about to change their action to accept Freund's improvements. The company's opinion was that all Frank's "improvement" amounted to, was a way to force a cartridge into the chamber of a barrel that probably only needed cleaning for the cartridge to seat properly. However, what the factory representatives failed to recognize was that shooting conditions on the Western frontier were far different from those found on Eastern target ranges.

The Sharps Rifle Company was putting their sporting rifles into the hands of hunters and adventurers in the West who didn't always have the opportunity, or perhaps the time, to stop and clean their rifles after a few shots. Volatile weather, Indians, or a charging buffalo might not grant one the opportunity to have a clean chamber before loading the cartridge. Frank and his customers recognized this defect with the Sharps, and the need was apparent to them for a rifle that would chamber a cartridge every time without fail. The necessity of being able to extract the fired cartridge casing also was evident, and thus the need for Freund's double extractors.

Apparently the latter were not patentable, but Frank made reference to them in one of his first letters to the Sharps Rifle Company. Frank's efforts were aimed toward building a rifle that was readily acceptable to the cowboys, hunters, and sportsmen of the West.

The author has been able to gather information on 35 Model 1874s, and eight Model 1877s, for a total of 43 surviving Freund converted Sharps rifles for this study.

Barrel Length	Freund Sharps Model 1874	Barrel Length	Freund Sharps Model 1877
24 inches	2		
25 inches	1		
26 inches	10	26 inches	2
27 inches	1	27 inches	1
28 inches	10	28 inches	3
30 inches	8	30 inches	2
32 inches	3		

The average length of the barrels observed was just under 28 inches. The Freund brothers believed that the octagonal barrel was best for Western use. Of the 43 Sharps rifles surveyed above, 30 have octagonal barrels, 11 have half-round/half-octagon barrels, and only two have full-length round barrels.

There were four popular calibers during this period, with the .40 caliber and the .45 caliber being the most popular. Of the 43 surviving Freund-converted Sharps rifles, 24 were .40 caliber, one was .44 caliber, 17 were .45 caliber, and one was .50 caliber (a converted Model 1874 three-band military rifle).

Twenty-three of the surviving rifles have built-in trapdoors in their steel buttplates. These conceal a four- or five-piece cleaning rod in a recess drilled into the buttstock.

The Freund patented pistol grip was added to 18 of the surveyed Freund rifles. An additional twelve of the surviving specimens have built-in pistol grips, as were standard on some of the long-range and mid-range rifles. The other thirteen of the surviving Freund-modified Sharps have no evidence of a pistol grip, built-in or otherwise.

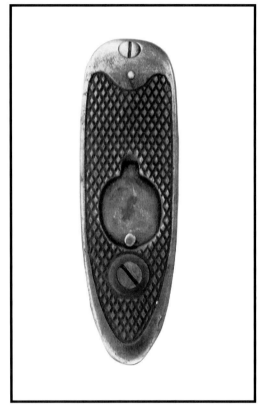

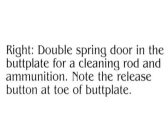

Left: Single sliding trap-door in the buttplate of a Freund-altered Sharps rifle, giving access to storage area for the cleaning rod.

Right: Double spring door in the buttplate for a cleaning rod and ammunition. Note the release button at toe of buttplate.

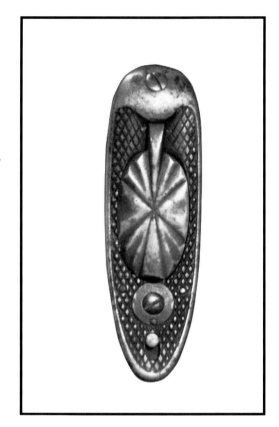

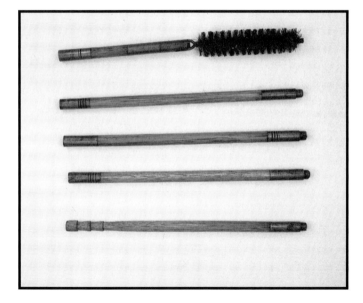

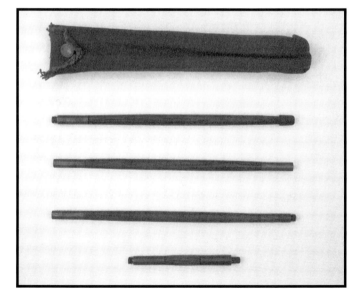

Above left: Five-piece wooden cleaning rod with brass ferrules. Above right: Four-piece wooden cleaning rod with brass ferrules, and canvas holder to keep rods from moving and rattling while stored in the rifle's buttstock.

Thirty-two of the 43 surviving specimens show engraving, some having just border engraving on the action and lockplate. Before 1880, the lockplates were engraved "Freund & Bro" or "Freund Bros". After that date, the guns were marked "Freund's Patent", or "F.W. Freund Pat." Their "Wyoming Armory, Cheyenne, Wyo., USA" inscription generally was included with the name.

Indians, wielding either a bow or tomahawk, were a favorite subject for embellishing the receivers of highly-engraved Freund guns. Animals, such as bears, buffalo, moose, and even flowers, were other favorites on the Sharps Model 1877 rifles converted by Freund.

"American Frontier", "Famous American Frontier", "Freund Improved", and initials also were engraved on the sides of their actions. "Boss Gun" was engraved behind the breech on four of the surviving specimens, with "Boss Gun" also engraved on the right side of one rifle's receiver.

No two Freund-engraved guns are ever exactly alike, one of the things that make them so interesting and important to gun collectors. Freund's custom rifles are things of beauty, and were very popular with the ranchers and sportsmen on the frontier. Highly engraved Freund-altered Model 1877 Sharps rifles are most sought-after by today's collectors.

Most of the Freunds' work was custom gunsmithing, and like their engraving, no two rifles which came from the "Wyoming Armory" shop were exactly alike in all details. Very little information is available on the prices charged for the Freund Improved Sharps; but from old letters, contemporary accounts, and various other sources it seems that thirty to forty dollars was the usual price for a plain job of remodeling a Sharps rifle. This included a Freund patent breechblock and work on the action, and even possibly Freund sights. But for a fancy Freund Improved Sharps, including engraving, alteration of the action, possibly a special stock, and the patented sights, the price ran as high as $100 or $125, and one letter mentions a price of $160 or more.[3]

Probably Frank Freund's experience on the buffalo range had something to do with the efforts he made to improve the already-famous sidehammer Sharps buffalo rifle. Much of his work at Cheyenne was in that direction.

Using ideas set forth in F.W. Freund's patents of 1876, 1877, and 1879, the Freund brothers remodeled the Sharps rifle so that the breechblock had a camming action, instead of a straight up-and-down movement, as it was opened and closed. The block itself was reshaped, and provided with a tapered steel plate at the rear which could be removed to clean or replace the firing pin. This plate, along with a gas-checked type of firing pin, was intended to eliminate the danger of powder gas escaping back into the shooter's face if a primer was punctured. The hammer also was reshaped considerably. The Sharps lock could

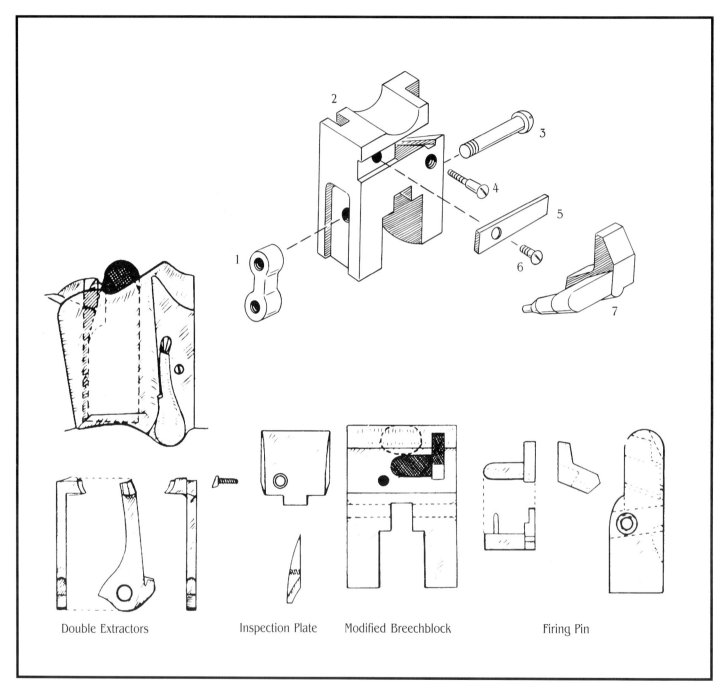

Double Extractors   Inspection Plate   Modified Breechblock   Firing Pin

Breechblock assembly of a standard Model 1874 Sharps rifle (top), as compared to the Freund-made assembly for the Sharps (bottom row). In the Sharps breechblock assembly, 1 is the lever toggle link, 2 the breechblock, 3 the upper toggle screw, 4 the firing pin screw, 5 the firing pin plate, 6 the firing pin plate screw, and 7 is the firing pin. Below it are shown the component parts of the P. Lorillard Sharps action, showing Freund's modifications, and how the camming movement occurs. The inspection plate permits quick removal of the Freund firing pin, and easier cleaning. The curved rear face of the breechblock gives it a camming motion, to give the extra power to seat a dirty or swollen cartridge.

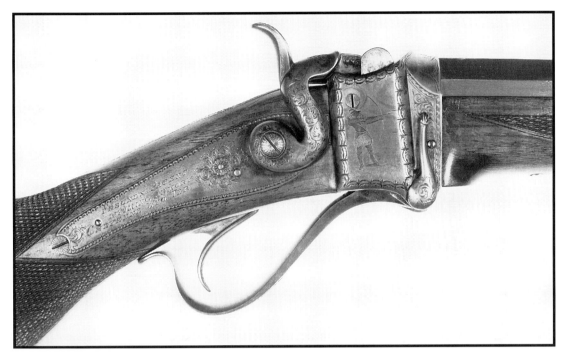

Right side of a beautiful Freund Improved Sharps Model 1877 rifle, finely engraved with a standing Indian with bow on the right side, and a bull buffalo on left side of receiver. Note "Freund's Patent/Wyoming Armory/Cheyenne Wyo. Ter." also engraved on right-side lockplate. (Dr. R.L. Moore, Jr. collection)

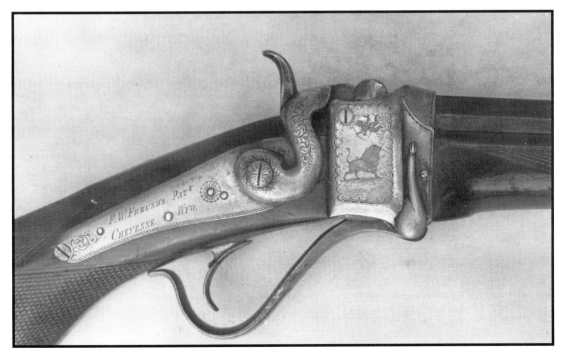

Freund-altered Model 1877 Sharps rifle. Note the characteristic Freund's bull buffalo engraved on right side of the receiver, but with an Indian being tossed overhead; "E.B.B." script initials on left. (Courtesy Dennis Brooks)

be altered, by means of a spur on the tumbler and a lever pivoted in the trigger plate, so that the hammer was automatically set on either half- or full-cock on opening and closing the breech. Freund also devised his own method of retracting the firing pin in the Sharps. Double extractors were another feature of the Freund Improved Sharps, as a second extractor was a big help in ejecting the large bottle-necked or long straight cases of the .40, .44, .45, and .50 caliber black powder cartridges in use at that time. Frank Freund's chief improvements to the Sharps action were covered by three patents, numbers 180567, 185911, and 216084.[4]

With regard to the Freund brothers' work on Sharps rifles, refer to the 1876 statement of professional market hunter J.A. Meline, regarding shooting accidents in the West, that is reproduced herein on page 55.

While Frank Freund was converting Sharps rifles in his Wyoming Armory, he also experimented with other makes of rifles. Some of this type of work might possibly have been done in Denver; however, picture postcards of these rifles appear to have come from Cheyenne.

It was a great blow to Frank when the factory whistle of the Sharps Rifle Company, located at Bridgeport, Connecticut, blew for the last time in October of 1881. Although Sharps rifle parts continued to be available for several years after the factory shut down, Frank saw the handwriting on the wall, and began

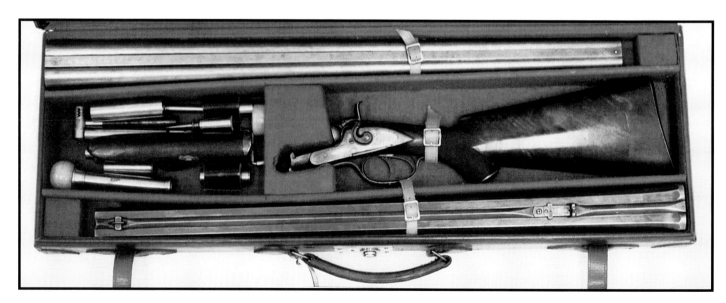

An English shotgun action fitted both with Freund double rifle barrels in caliber .45-90, and Freund shotgun barrels in 10 gauge. The leather carrying case is a restoration. (Author's collection)

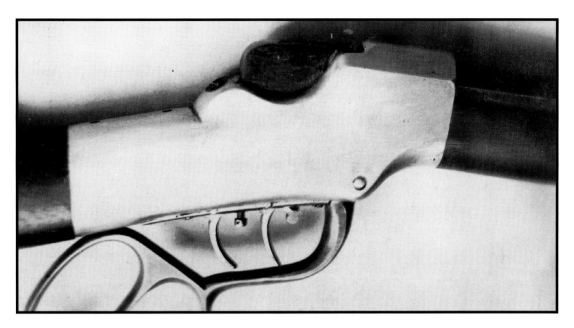

Close-up view of a rifle action — probably Ballard — modified by Freund. (Freund Family Collection)

to develop his own receiver designs. During the same period, Frank converted an English shotgun to a high grade double-rifle. Frank Freund also made a set of shotgun barrels in 10 gauge for this double rifle, that are marked as having been made at the Wyoming Armory.

Frank also designed and built a small, short rifle complete with his own action, which he called the "Wyoming Saddle Gun." Although various authors state that he made four or five Wyoming Saddle Guns, this author has been able to locate only one of them.

Freund's 1876 patents, numbers 184202 and 184203, covered two single-shot rifles entirely different from the Sharps, but with some resemblance to the Peabody-Martini. Both were designed so that they could be completely field-stripped without the use of tools, which was their most important feature. Freund did not develop the two patents any further, but he did retain his idea of making an action that could be taken apart without tools. Eventually it was incorporated into his design for a simplified and perfected single-shot rifle.

No doubt Frank Freund was working to produce a rifle that would be equally suitable, with modifications, for military and sporting use. The overwhelming defeat of Custer's 7th Cavalry at the Little Big Horn in Montana, on June 25th, 1876, aroused a storm of criticism and controversy over, among other things, the Springfield rifle and carbine, which were the standard U.S. Army

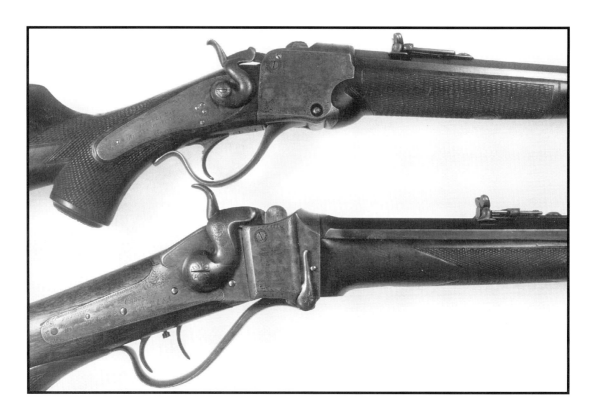

Comparison of a Freund "Wyoming Saddle Gun" (top, both pictures), and a Freund-altered Model 1874 Sharps rifle with receiver engraved "American Frontier" (bottom, both pictures), right and left sides. (Private collection)

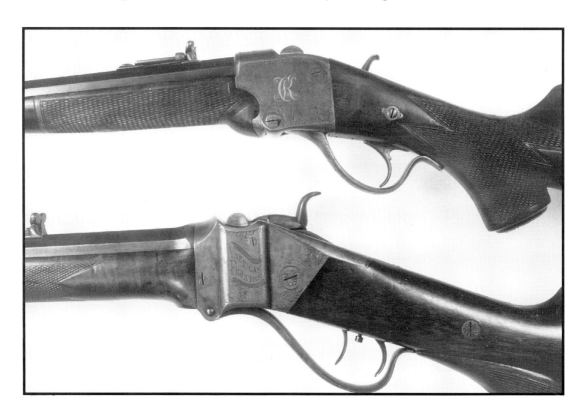

weapons at the time. A serious effort was made and continued to be made by some authorities to have the Springfield replaced with a different arm. That, of course, encouraged the efforts of other inventors and arms companies to design and produce a superior rifle that would be acceptable to the government if such a change were made. Military ordnance boards at that time were largely opposed to hammerless actions, and Freund probably knew that he would have to make a rifle with an outside hammer like the Sharps, or with some means of cocking the hammer concealed within the action.[5]

F.W. Freund worked alone producing his improved Sharps rifles and patented sights at the Wyoming Armory in the early 1880s. Fragmentary correspondence in his scrapbook shows that he increased his efforts to interest various firms in the commercial production of his rifle, but while a number of them acknowledged its merits, none undertook to manufacture it. The Sharps Rifle Company was no longer in business, and the single-shot rifle as a sporting arm was losing ground to the increasing popularity of the repeating rifle.

Freund's False-Hammer Cavalry Pattern carbine made for the U.S. military trials. (Freund Family Collection)

Among a group of old photographs of different F.W. Freund guns, are two showing a working model of a military carbine. The action is essentially the same as that of a fine Freund sporting rifle in the author's collection, known to collectors as the "False-Hammer Freund" for lack of a better name and because, strangely enough, it does have a false hammer.

Nothing is known about the history of the False-Hammer military model, except that it was made by Frank Freund before he moved back East in 1885. That it was made in Cheyenne would seem to be confirmed by a Cheyenne photographer's address on both pictures. It also is apparent that with this False-

Hammer action F.W. Freund had reached his goal of a perfected single-shot, falling-block rifle action. As the letter from Congressman M.E. Post quoted below suggests, this military model was made by F.W. Freund apparently in 1884 or 1885, to submit to the Army Ordnance Board for consideration.

The following captions appear on two photographs of the Freund False-Hammer military rifle:

> *Rifle of Cavalry pattern, 28 inch barrel, Half stock, Joint Cleaning Rods in the butt of Stock.*
>
> *The pattern and general style of this arm was suggested to me by the late Capt. Clark (of the 2nd Cavalry Reg't. and Lieut-General Sheridan's staff, U.S.A.)*
>
> *The Front Sight cannot be seen in the photograph on account of projection on each side, especially adapted for use on horseback.*
>
> *A corresponding drawing of this Rifle was sent to the U.S.A. Headquarters before the construction of this arm, as photographed.*

*****

> *Cavalry Pattern. Bands superseded. Lock and Breech mechanism detachable instantly without the use of instruments, and can be replaced as quickly.*

Both descriptions are pasted directly onto the pictures, and appear to have been clipped from a printed folder or brochure. Part of the printing has been erased for some unknown reason.

The officer whom Frank Freund credits with suggesting the general pattern of this rifle must have been Captain William Philo Clark, a young officer who served on the frontier for about ten years prior to his death in 1884. The Indians called him "White Hat"; as a lieutenant he played an important part at the surrender of the famous Chief Crazy Horse in 1877. Captain Clark was well-liked by the Indians, and he brought in the remnant of the Cheyennes under Little Wolf after their epic march of 1878-79. He was expert in sign language, and after his death the book he authored, *Indian Sign Language*, was published at Philadelphia, in 1885.

Exactly what did Captain Clark suggest to Frank Freund, regarding the design for his rifle? Was the false-hammer his idea to get around the Ordnance Board's usual objection to a hammerless action? The three letters which follow are from the Freund family scrapbook, and give some idea of the origin and plans for this military rifle.

House of Representatives U.S. Washington, D.C.
May 23, 1884
Dear Mr. Freund:

I saw Captain Clark yesterday and had a long talk with him in relation to your matters. He tells me that he has written too fully on the subject. I am convinced that the only thing to be done under the circumstances, is that you make a carbine in accordance with his suggestions, and if it stands the test, as I have no doubt it will, the War Department will recommend an appropriation for funds to make say five hundred for a field test. With such recommendation, there will be no trouble in securing the necessary appropriations.

Captain Clark tells me that some other arm will have to be substituted for the carbine now in use. I think it is a prize worth struggling for and would suggest that you make a big effort to secure same. I shall be glad to do all I can to assist you.

Yours very truly,
M.E. Post

*****

Temple Court, New York
April 14th, 1885
Messrs. Richards & Co.,
24 & 26 4th Ave. N.Y,

Dear Sirs:

I am in receipt of yours of the 9th inst., asking my opinion of the "Freund Rifle".

In reply, I beg leave to say, I have, myself, personally and most thoroughly examined the "Freund Rifle" and sights, and am, I believe, very well acquainted with all the other leading makes of fire-arms, and from my experience and knowledge of such matters (gained by four years' service in the field during the late rebellion) I feel no hesitation stating that the "Freund Rifle," military and sporting, and the Freund sights, are the best, most perfect and complete of any in existence.

As an illustration and comparison, the Freund Rifle and sights are as far superior to any gun of the present day, as the first percussion capped gun of the past was superior to the old smooth bore and flint lock.

The... leading points of merit and advantage of the Freund Rifle are:
1st. Its great simplicity, as all its parts can be instantaneously de-

tached and reassembled by anyone without the use of tools; because every part of the gun, lock, stock, sights, and barrel, seems to dove-tail and fit in its proper place without screws, clamps, or any other device so annoying to gunners. That fact, of itself, makes it a gun of especial value for military and frontier service, for no other gun in existence possesses that merit.

2nd. Its great safety and strength, on account of the solid breech block coming up squarely behind the cartridge renders it... safe, while at the same time its peculiar action in forcing the cartridge home, combined with the double automatic extractors, admits of greater rapidity of firing.

3rd. The great benefit derived through the double extractors in equalizing contraction and expansion.

Mechanically, it has no rival, and from every standpoint, it appears to be beyond competition.

Not only is it the best, but in my opinion, if properly managed, it can be produced ten (10) percent cheaper than any other first class gun on the market, and I will tell you why. Because it has but 31 pieces, all told—lock, stock, barrel, and sights; which is ten pieces less than any other gun of the same standard in the world. Read the following:

The total number of pieces in the different systems are as follows, not counting the sights as any sight may be used on either of the rifles:

Borchardt's Sharps [sic]......................64
Springfield, U.S. System 60
Sharp's [sic] *Michigan and other States* 60
Remington, State of New York and others..........54
Peabody-Martini, Turko-Russian 53
Martini-Henry, British System 53
Brown Standard Rifle 40
Freund's Rifle (including sights)......................31

Besides being less in number, the difficult pieces in Freund's system are far simpler and easier to make and clean than those in either of the others, being in the aggregate lighter whilst at the same time stronger.

In my opinion, if it is handled with only ordinary business ability, it will undoubtedly out-sell any other rifle now made, as greatly as it nows excells all other rifles in point of merit.

In conclusion, permit me to thank you for the value you... place on my opinion, and to say that I hope it may prove to be as of service to you.
Very Respectfully yours,
H.M. Munsell, Late Captain Co. C
99th Regiment Pa. Veteran Vol.[6]

Hartley & Graham Arms and Ammunition
Agents Union Metallic Cartidge Co.
Bridgeport, Conn., P.O. Box 1760

New York, May 23rd, 1885
Mr. F.W. Freund
Dear Sir:

We have examined your Patent Rifle (Single Shot) & consider it one of the best systems we have ever seen.

A good feature is the lock which can be taken off & apart without the aid of a screw driver and the Breech action is one of the strongest & perfectly safe for resistance of large charges of powder. For simplicity & durability we do not think it can be excelled.

Yours truly,
Hartley & Graham[7]

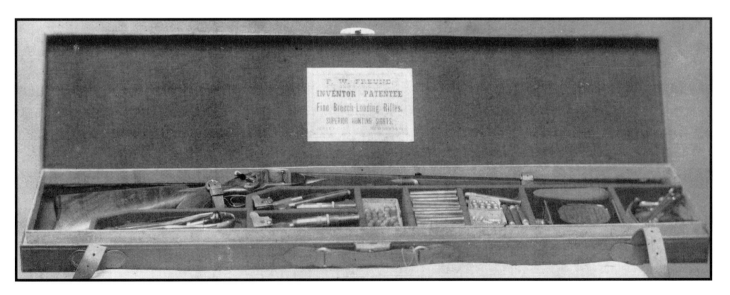

Freund's False-Hammer Sporting rifle shown in its original case. The photograph dates *circa* 1885. (Freund Family Collection)

Was the Freund False-Hammer rifle ever submitted to the Ordnance Board for trial? If it was, what happened to it, and why did it fail? What seems most logical is that this fine action may have been perfected too late; for in the 1880s, the Ordnance Department was earnestly seeking a suitable repeating rifle to replace the .45-70 Springfield. Single-shot rifles and large calibers were soon to be obsolete for military purposes.

While in Cheyenne, Frank was awarded U.S. patent number 183387 on October 17, 1876, for his "Improvement on Revolving Fire-Arms." This patent was to enable the operator in the field to disassemble a revolver without aid of tools. To the best of the author's knowledge, no known examples of guns having this patented improvement exist today.

The author has in his collection a fine False-Hammer rifle that was made after Frank Freund moved his family to Jersey City, New Jersey. It differs slightly from the military model, having a pistol-grip, but is essentially the same action.

The original photographs of this rifle show that at one time it had a full-length trunk case. Compartments in the case contained extra molds, a loading tool, a large quantity of extra 3¼" shells, paper-patched bullets, powder primers, and what appears to be either a spare lock mechanism or the parts for one.

While in New Jersey, Frank also converted two fine Model 1877 Sharps rifles for President Theodore Roosevelt. Both rifles are elaborately engraved, and numbered "1" and "2." There also were an extra barrel and forend provided with these rifles. Number "1" has an octagonal barrel, and number "2" has a round barrel. Theodore Roosevelt gave these rifles to members of the White House staff when he left the presidency. The two were separated for a number of years until they were reunited and sold to the daughter of Harold Schafer of Medora, North Dakota. She gave the rifles, along with the spare barrel and forend, to her father as a present. Today, the Roosevelt rifles are on display at the Theodore Roosevelt National Park at Medora, North Dakota.

According to the custom of the era of single-shot rifles, all rifles sold or passing through a shop were stamped with the dealer's name or mark, thus denoting the shop from which the gun originally came. This, then, was the source of many plain, unaltered Sharps or other-make rifles with the "FREUND BROS" or more commonly-seen "FREUND & BRO" stamps on their barrels. It also accounts for the large number of Sharps Model 1874 military rifles that were stamped and sold by the Freunds during this period.[8]

The author would be negligent if he did not warn the reader about possible forgeries. I know of at least four Sharps rifles that have been stamped and worked over to resemble Freund rifles. *Caveat Emptor*—let the buyer beware.

Perhaps even more prevalent than the spuriously-marked rifles are other go-withs marked "Freund", "Freund & Bro.", and "Freund & Bros." These take many forms; *i.e.* holsters, shell casings, cartridge boxes, and paper items. Not all of this material has been faked or copied—but, "be careful out there."

A hand-rendered label glued onto the lid of a pasteboard box of rifle cartridges, and marked "Freund's Best." The author knows of four examples; supposedly they were made by Frank W. Freund to be given away with the purchase of a newly-converted Freund rifle. Provenance unknown. (Author's collection)

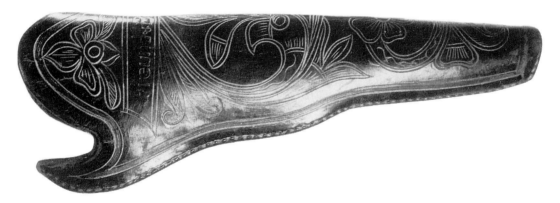

A revolver holster showing "Freund & Bro." stamped into its tooled design. It probably was made by E.L. Gallatin & Co., in Cheyenne, Wyoming Territory, *circa* 1870-1875, for distribution by the Freund Brothers. (Courtesy John E. Fox)

1. Information supplied by Mr. Frank Sellers.
2. Barsotti, Part I, 1957, pages 15-22.
3. *Ibid.*
4. *Ibid.*
5. Barsotti, Part II, 1958, pages 55-64.
6. *Ibid.*
7. *Ibid.*
8. Sellers, *op. cit.*, page 190.

This colorful poster featured the Freund Brothers' famous patent "More Light" sights, available at George Freund's Colorado Armory in Durango. (Courtesy Richard Labowski)

CHAPTER NINE

THE FREUND SIGHTS

The Freund Brothers, had they not been master gunsmiths, not held patents on their own rifle designs, and not converted rifles of other makes, still would have gained fame for their masterful work on gun sights.

Frank W. Freund alone held four United States patents on sights and a fifth U.S. patent in partnership with his brother George, for the well-known Freund "More Light" sight.

The first sight patent held by Frank Freund was number 160819 awarded on March 16, 1875, applied to the Remington Rolling-Block rifle. The author knows of only one way to describe this sight, and that is, "Outstanding!"

This sight design features a brass tail attached to the stepped sight support (elevator), the tail formed with a slot between its front end and the stepped support. To assist the user of the arm in adjusting to the sight, a scale is marked on it, with a long vertical slot cut through it for a small pointer. Previously, the main objection to this type of sight was the falling off and loss of the sight support. In the Freund sight, this was overcome by shaping the notches of the sight support to hold in two directions. Thus the sight and support cannot be easily moved upon one another.

The second sight patent held by Frank Freund, number 189721, was issued on April 17, 1877 and titled "Front Sights for Fire-Arms." Its object was to produce a sight that would be equally effective in dark, cloudy, and fair weather. The rear face is made in two contrasting colors, so the body of the sight is one color and the top side or edge is a line of opposing color. Thus are seen two contrasting colors, one of which forms a vertical line between two lines of the other color.

The third sight patent, number 229245, was issued on June 29, 1880 and held jointly by both Frank W. and George Freund. It enabled the marksman to

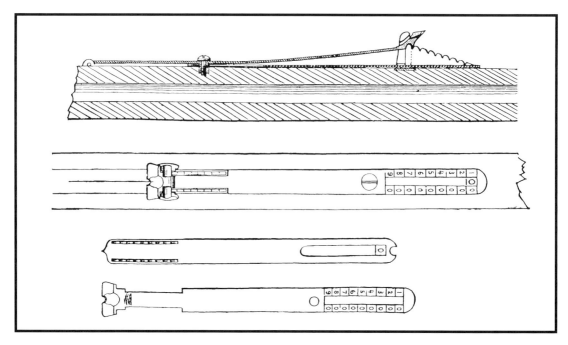

Detail drawings of Frank W. Freund's sight for the Remington Rolling-Block rifle. U.S. patent number 160819, awarded March 16, 1875.

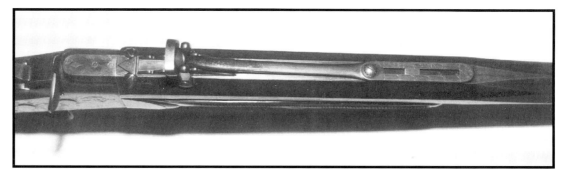

Close-up of a special Freund patent sight mounted on a Remington Rolling-Block rifle. (Courtesy Glenn Marsh)

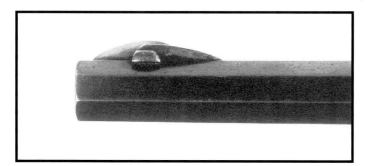 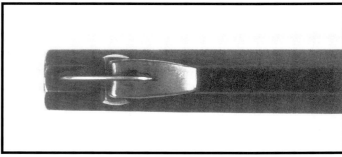

A special front sight installed on a Freund "American Frontier" rifle. This sight is different from those seen on other Freund rifles. (Courtesy Dave Carter)

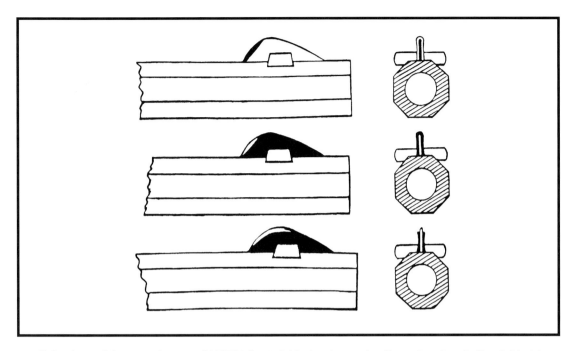

Detail drawings of the Freund patent #189721 front sight, showing contrasting colors for a better sight picture.

level the arm when aiming without the use of spirit levels, and to aim at different elevations without raising or lowering the sight with respect to the barrel. To those ends the invention consists of forming an angular, diamond-shaped opening directly below the ordinary sight notch. This sight later became known as the Freunds' "More Light" sight.

The fourth sight patent, number 268090, was issued to Frank Freund for an improvement on the previous patent. It consists of applying to the More Light sight a moveable or sliding plate by which the opening below the sight notch can be fully opened, partially opened, or fully closed. This sight has two screws to hold the plate, which can be adjusted as described above.

The fifth and last of the Frank W. Freund gun sight patents was applied for while Frank was still working in Cheyenne (on October 16, 1884), but not issued until April 25, 1893, long after the family had moved to New Jersey. This was patent number 496051, for a rear rifle sight. It consists of a base which is rigidly attached to the barrel, and a series of interchangeable sight plates that can be attached to the base, its object being to provide a sight or series of sights which can be changed at will. The base is provided with clamping screws to hold a sideplate, and a series of interchangeable sight plates which are notched to fit over the screws and clamp onto the base piece.

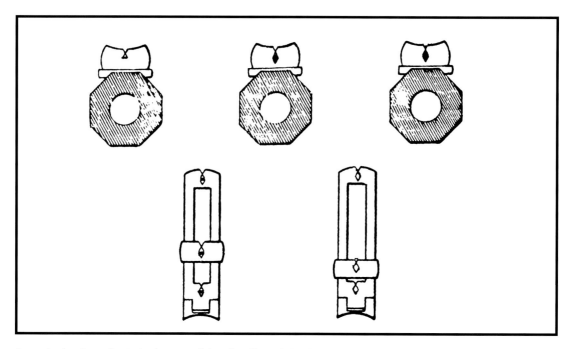

An early drawing of cuts in the rear sight of a rifle, originally intended to provide a leveling device. Later the design was developed into the Freunds' "More Light" sight.

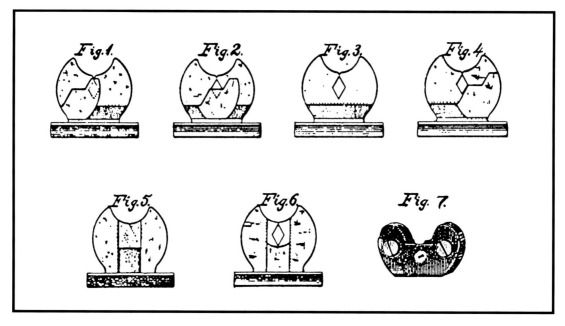

Period sketches of Freund rear sights, as seen from behind. *Fig. 1* shows the diamond-shaped aperture in the sight fully closed. *Fig. 2* shows it partially open. *Fig. 3* shows the moveable plate or slide removed from the sight. *Fig. 4* shows the plate with the aperture fully open. *Figs.* 5 and 6 show sights with a vertical slideway having slides to move up and down. *Fig. 7* shows a rear sight with a moveable plate attached by two screws.

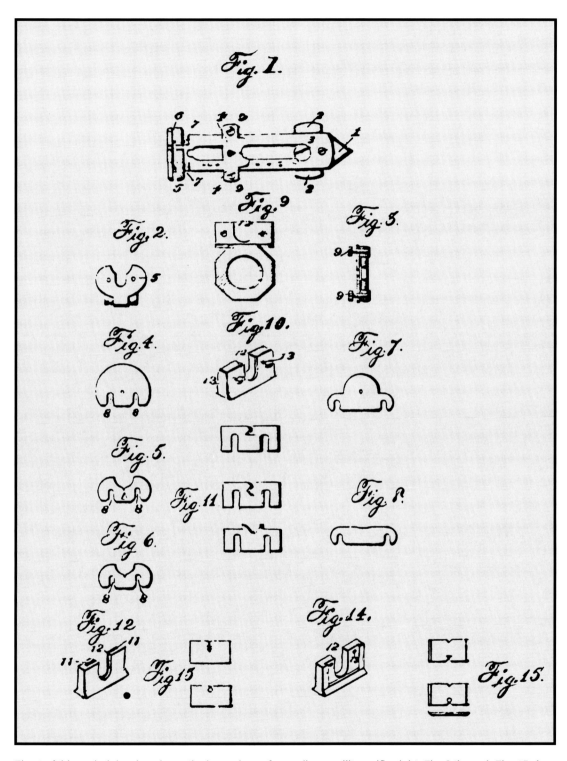

Fig. 1 of this period drawing shows the base piece of an ordinary military rifle sight. *Fig. 2* through *Fig. 15* show Freund's various plates and modifications to the sight.

The five Freund gun sight patents are described briefly above. But to understand their significance in the field as well as on the target range the following two articles, written in April of 1886 and published in *The American Field* magazine, better explain them as they relate to the hunter in the field.

FREUND RIFLE SIGHTS

Chicago, Ill.
April 13, 1886

EDITOR AMERICAN FIELD;—I have, of late, noticed a great many communications concerning the different styles of sights for hunting, in your valuable paper, and I would like to say a word about a new sight which has come under my observation. I had tried all the leading sights on the market, but had not been able to find a good open sight for hunting purposes for different colored objects and suitable for light or dark atmospheres. I found the buck-horn sight to blur too much, and that a perpendicular platina line in it was not much better, as it did not get a light from the right direction. Finally a friend called to my attention the adverisement in your paper, of Mr. F.W. Freund. I then sent for a circular to see if he had what I wanted, and, after reading it through, sent for a set which came in due season.

Both his front and rear sights have points of merit which cannot be equaled unless made under the same principle of contrasting colors. He makes three styles of beads—the target bead, coarse hunting bead, and fine hunting bead. All of them have contrasting colors of light and dark when held against a light or dark colored object. These beads slide in and out of a block on the rifle, and this is an important point, for it saves the trouble of knocking out the block when one wants to change a hunting bead for a target bead. The general construction of the bead is so fine and yet so strong that it makes it a very good sight for outdoor work. These beads have no projecting corners to catch and one can change the hunting sight to a target bead in an instant, and, as they are made tapering, they can be driven into the block very tightly. The hunting beads have round black shoulders, with a round piece of white metal projecting in the center and on an angle backwards of thirty degrees, so as to receive the light from above, and, as the perpendicular white line is just as high as its black shoulders, the elevation is always the same. The bead shows black against a dark object.

The rear sight has an adjustable white plate between two dark ones, which can be raised so as to show a white bar about the size of a fine needle just on the lower edge of the notch. This receives the light from above, which is desirable. This plate can be raised or lowered in an instant with the thumb and first finger of the left hand without taking the rifle from the shoulder or firing position. Consequently, if one is shooting in a bright light he can drop the white plate just 1/32 of an inch and he will have a black rear sight; but if shooting in a dark atmosphere or in the woods and he wants a rear sight which he can see plainly, this adjustable white plate, which can be raised in an instant, fills the bill the best of any rear sight I have ever seen. The rear sight also has a pear-shaped opening below the notch, which gives the hunter a chance to see game all around the front sight.

The sliding steps on this sight are placed on each side instead of being solid, thus giving the rear sight notch a chance to be placed much nearer the center of the bore of the rifle, for the hunter looks through the step and not over it, as he is obliged to do in all solid step slides. I think Mr. Freund's sights have also more combinations than others, and, I find, their advantages place them at the head of the list of hunting sights.

P.C. Bradley[1]

*****

THE FREUND PATENT RIFLE SIGHTS

Kingston, Pa.
April 24, 1886

EDITOR AMERICAN FIELD;—The subject of rifle sights is always interesting to the rifleman in exact proportion to his enthusiasm and interest in rifle matters in general. If he be of a critical temperament, and always looking for something better than the present sights upon his gun, he makes the subject a matter of study and frequent experiment. If easily suited he will find something better than the bungling excuse for sights commonly sent out by rifle factories; but, if he is a novice in fact, he usually thinks that anything is good enough that "comes from the city" and projects upon his rifle barrel. I have tried for about twenty years to be deserving of a place in the class first named; I have used every variety of sights and combinations known in the market or suggested by friends; I have frequently seen excellent guns condemned and laid away because they would not shoot well with poor sights and worse sighting, and, until recently, I never had the pleasure of examining a set of rifle sights the mechanical construction and make-up of which were beyond some hint of improvement which I could easily suggest.

A short time since, however, my attention was arrested by a very imperfect illustration of the patented rifle sights invented and manufactured by F.W. Freund of Greenville, N.J. I was well impressed with the novelty of their chief principle, and, though skeptical in some degree respecting their real character, I felt anxious to examine them in the real metal, and be permitted to test them at the target and in the woods. I therefore ordered a full set to be sent on approval consisting of one rear adjustable sliding silverplate sight, and four front sights or beads, which duly came to hand. The front sights are made from a fine quality of steel, spring tempered, thus enhancing their value by making them more elastic and less liable to become bent or broken. Each of the four front sights has a special purpose. All of them, however, have the general outline with a knife edge in shape, but the upper edge is so artistically rounded as to produce the effect of a perfect bead, while the remaining line of metal down to the base of the sight is worked down and polished so that the marksman is only aware of a beautifully formed bead upon a delicate stem. The hunting front sight is blackened all over by heat, except the white line which begins at its base and extends up into the black bead surmounting the sight. This is a little larger than those made for target work, thus enabling one to get a quick, coarse and distinct sight upon a moving object or one at rest. The second bead, for target use, is similar in shade and shape to the hunting pattern, but of smaller stem and bead. The sight is worthy of special notice to show the maker's somewhat philosophical drift in that it is practically a long range sight in view of the fact that its upper curve is made very short, in order to show less of the top when the rear sight is elevated. The third front sight in order is one of polished steel all over, save a diminutive black point in the face of the bead. This is a sight for very fine target work. The fourth sight presents a black line and bead with a silver band an eighth of an inch deep, perhaps, and located just back of the black face of the bead, the remaining color of the sight beyond the silver band being black. The silver band of this sight makes a break in the size of the diameter of the steel and enables the shooter to see more of the black center of the target.

Of course it is not possible to overcome the mischief of light and shade with unicolor sights. For example, a light colored sight becomes lost or absorbed in an object of similar color; so, also, is a dark sight invisible upon a dark object. The only way one can know when a black bead is upon a black bullseye is when he can see it, and it is quite impossible to determine what place is occupied by the bead upon the bullseye, whether the center, the outer rim, or neither. Black and white in combination necessarily produce the widest possible contrast, and it becomes an easy certainty to distinguish against the other.

Right here we begin to approach the superior character of Mr. Freund's rifle sights of contrasting colors. The projecting white line of the front sights is all that shows upon a shaded or dark surface, while the black shoulders, forming the bead of the same sight, become the indicator if held upon an object of light color. The simplicity and strength of these sights are excellent features of their construction.

Knowing it is possible to be fairly successful with the ordinary rifle sight under favorable conditions, I decided to try the new sights under nothing but difficulties to fully understand the effects of multi-colored objects, and changing lights upon them. In one instance I held the bright steel target sight upon a white object with a full glare of the dazzling sun upon it at one hundred yards, and found the object aimed at as clear and distinct as though the rays of light had been made for the occasion. The black point in the white face of this bead will always register its location upon a light target, while the white circle enclosing the black point can be seen upon the darker color with unmistakable certainty. Shooting over snow in bright sunlight, across the sun's rays, against them, and with their tantalizing effects playing "hide and seek" in the rear notch, were some of the positions occupied in testing these sights.

In passing these sights continuously along the line of objects of varying color, the effect is magical and amusing. Against a white or light surface the black pin head, only, is perceptible, except in the case of the silver cross band sight, which shows both stem and head. Against a dark or black object, only the white line is visible, except when using the polished steel sight, when the bright circle around the black point is seen. In one sense it is all chameleon like—"now you see it and now you don't."

In the make-up of our sight combining two colors we have also a provision against the deceptive effects of glare and gloom. The dark shoulders of the sight appear to neutralize or render inoperative the blaze of sun and rays of light, while the projecting white line of the same sight seems to illuminate or brighten the effects of darker light or shadow. Sunlight and

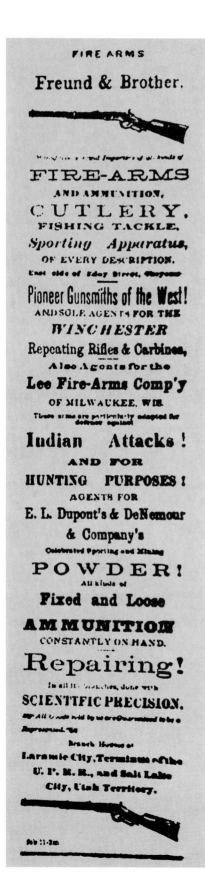

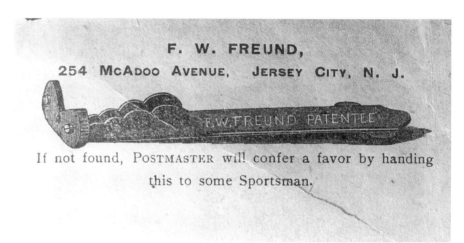

New Rifle Sights.

The kind they have been looking for.

Freund's Patent Front and Rear Sights.

The production of years of patient and thorough study by a practical riflemen, in the field and in the shop, in the East and West.

The satisfaction these sights give show that perfection has been attained in open sights for hunting, sporting, and target rifles. They have been thoroughly tested by expert riflemen, who unhesitatingly pronounce them superior to all other open sights. For particulars enclose stamp for illustrated circular. Address plainly.

F. W. FREUND,

Jersey City P. O.,   Greenville, N. J.

cloudy weather produce no ill effects upon these sights. In fact, they seem to be a prime necessity in overcoming mischievous atmospheric troubles. Where the true line of sight has been ascertained, the different sights can be used without removing the sight block from the barrel of the rifle. The sight block contains a dovetailed slot, into or out of which the sights are driven in changing them, the work requiring only a second or two of time.

It will be observed, upon a careful examination of the construction of these sights, that, besides the combination of colors which forms the contrasts already noted, the effect is also heightened by the delicate and intelligent shaping of the sights. The dark part of the sights will remain so, being out of the line of contact with anything that would be likely to rub them bright. The bright lines of the sight are constantly exposed to the effect of handling, hunting, and weather inclemencies, and will therefore remain bright. The superiority of steel sights over the ivory bead or any ivory or bone construction is very marked, their greater strength and unvarying color alone making this truth fully manifest.

The best description will fail to illustrate the merits of the Freund patent rear sight. Its greatest value begins where ordinary sights are a failure. In this rear sight, as well as by those in front, Mr. Freund has brought to his aid some useful principles of science and philosophy and embodied them in a mechanical way which successfully takes up and absorbs the difficulties of rifle practice arising from changing lights, now making it only necessary for the rifleman to look and see, with perfect confidence in his aim and the result of a trigger pull. The simple frame of the rear adjustable sliding plate sight of silver, or silver-plated, is much the same shape as the ordinary rear notch sight, except that this one has a pear-shaped aperture just under the notch, for the purpose of admitting all the rays of light as they come to a focus at the notch, instead of a portion of them, as in the case of solid rear sights. This improvement gives the eye clear images along the line of sight, instead of blurred ones, the latter always occurring when the lower semicircle of the rays of light fall upon the solid smooth front surface of the rear sight commonly in use. This pear-shaped opening was a very fine thought, as the light enters from all sides uninterrupted and unbroken, making a blurring or indistinctness quite out of the question. The rear plate of the rear sight is blued steel without a notch. The silver adjustable plate has a notch exactly opposite the one in the solid front plate of the rear sight. Here we have an adjustable line of silver between two blued ones of steel, preserving still the important principle of contrasting color, for use especially upon objects of light color. The silver edge of the sliding plate becomes visible upon slight pressure upward or invisible if pressed downward. This adjustable plate is of service in shooting when it is necessary to control the extremes of sunlight as the rays come over the shoulder and into the notch while aiming, as it projects just far enough above the dark notch to break the two bright rays of light.

The workmanship, fitting and finish of the sights, as well as their symmetry, is worthy of much praise, and may be partly due to a long period of apprenticeship required by the German masters before the mechanic is permitted to "paddle his own canoe." It was difficult for me to believe, before testing these sights under all circumstances, that such excellent results could be reached by means as simple, and think any gentleman who will make a study of the important principle involved in the construction of these sights will discover a deep meaning hidden in the different shapes, color, and location and size, as well as position, of the delicate beads which are so well wrought upon the knife-edge outline of the front sight. It seems necessary or advisable to have the full set of sights if one is engaged in all kinds of rifle shooting, as they are so well and sensibly adapted to different uses.

In conclusion, I am conscientous in the belief that F.W. Freund's patent rifle sights are the sights of the world today, and that their universal adoption by rifle clubs for military practice and by hunters is only a question of time and the proper opportunities for testing them.

W.T. Dodson[2]

Numerous other testimonials to Freund sights appear in Appendix I of this book, and may be found in chapter eleven (The Freund Catalogs), as well. Frank and George Freund both believed in promoting their products, and various examples of their advertising can be found in the shooting magazines and newspapers of the era in which they actively engaged in the gun business. Some examples of the Freunds' advertising are shown on the following pages.

Frank and George Freund did not limit themselves just to sights for Sharps rifles and the guns of other makers that they modified, but made or adapted sights for any type of firearm that their customers wanted including lever-action Winchesters and Colt revolvers. For example, in the author's collection is a Stevens "Bicycle Rifle" having Freund sights. Other gunsmiths of that era also installed Freund sights, while some customers ordered Freund sights by mail and did their own installations.

As referred to in chapter five, Theodore Roosevelt had Freund sights installed on several of his fine Model 1876 Winchesters. Examples of them can be found today in the Theodore Roosevelt National Park Visitors Center at Medora, North Dakota.

Frank Freund had infringement problems on his sights from one Walter Cooper of Bozeman, Montana, as evidenced by the following article:

GREENVILLE, N.J., October, 1885.

Editor of the Rifle: —

I desire to call the attention of riflemen to an article which lately appeared in the *American Field*, describing rifle sights made by Mr. Cooper, of Bozeman, and state that the bead sight described is a direct infringement upon my patent front sight for fire-arms as secured to me by letters-patent in this country under date of April 17, 1877, and numbered 189,721.

I have generally found the silver centre in the sight preferred to ivory on account of its being stronger and retaining its color better, but make the ivory beads set in a steel block; and I claim for this invention a dark bead with a white centre of any desirable material.

My sights have been thoroughly tested by many of the most intelligent and expert riflemen of this country, who acknowledge their superiority to all others, and having brought these sights to their state of perfection I wish to inform riflemen that the bead sight described on pages 348 and 349 of the *American Field*, of October 10, 1885, is a direct infringement upon my patent; and those desiring such sights, and with all the latest improvements, are cautioned not to purchase the same except from manufacturer and patentee, or responsible parties who have purchased of me. All of my sights have my name and the patent stamp.

F. W. FREUND,
912 BERGEN AVENUE, GREENVILLE, N.J.

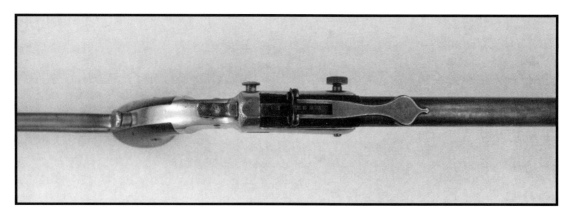

Above and opposite: Top and left side views of a Stevens "Bicycle Rifle" fitted with Freund sights. Note the pleasing, slim-waisted appearance of the sight base. Frank Freund was interested as much in the aesthetics of his products as in their function. (Author's collection)

Recently, the author received an advertising card for a tang sight made by Frank Freund, and mounted on a rifle. In showing it around, and talking with other collectors, no one else had seen a tang sight made by either of the Freund brothers. This was the case until I sent a copy of the picture to Dr. R.L. Moore, Jr. He phoned me after receiving the picture and asked if I would like to see this very tang sight. The next time I visited Dr. Moore he lent me the sight to study, and I am certain that this sight is the one shown in the photograph.

The upper adjusting screw was broken off in the closed position, so that the sight could not move. It was frozen in the down position, but gunsmith Tom Axtell restored the sight to its original condition. It is marked "Freund" in two places on the front of the sliding upper front (as shown in the original picture).

Dr. Moore purchased this sight from the well-known antique arms dealer and author Norm Flayderman a number of years ago, and it came with a "Freund"-marked front sight. Apparently at one time some controversy existed about a patent on this sight, as it came with two letters written by Frank Freund to an attorney (*see* chapter ten).

If one is inclined to use open sights today, Freund sights are found to be very practical and effective. They are still better for such shooting than virtually any other obtainable.

1. Bradley, P.C., "Freund Rifle Sights", in *The American Field*, April 13, 1886.
2. Dodson, W.T., "The Freund Patent Rifle Sights", in *The American Field*, April 24, 1886.

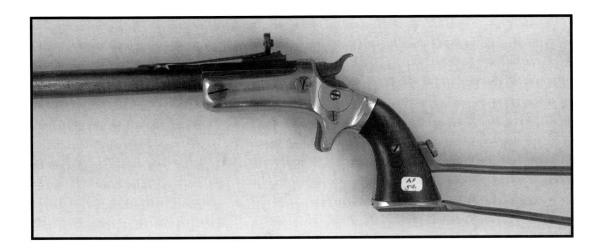

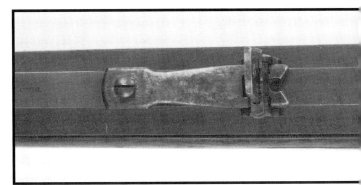

An R.S. Lawrence sight as modified by the Freund brothers. Note how the base has been narrowed behind of the mounting screw to give it a more pleasing appearance. (Author's collection)

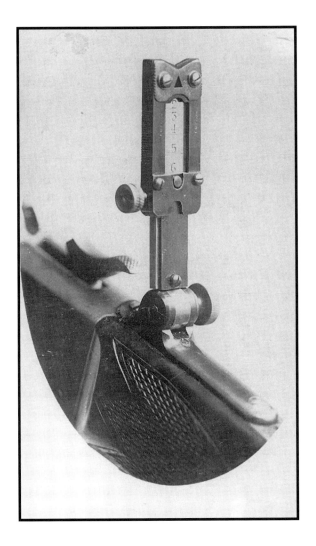

A tang sight made by F.W. Freund, mounted on a rifle with a very slim tang. The base of the sight appears to be a Winchester part, with the rest of the sight having been built by Frank W. Freund. (Freund Family Collection)

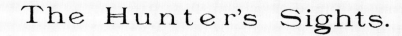

Close-up view of the "F.W. FREUND PATENTED" marking stamped into the base of a Freund-made blade front sight mounted on a rifle. (Author's collection)

"Freund's Patent" stamped atop a rifle barrel behind the rear sight, showing the date of the "More Light" sight patent jointly held by Frank and George Freund. (Author's collection)

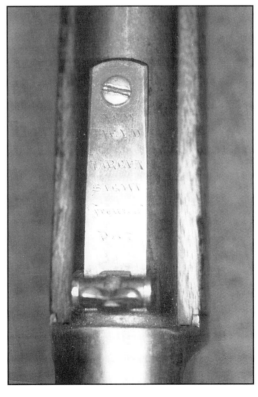

"New Field & Target Sight Freund's Pat.", reads the inscription hand-engraved by Frank W. Freund on the base of this rifle sight. (Private collection)

BURDETT, THOMPSON & LAW
Attorneys & Counsellors at Law
OFFICES, SUITE 501-4 OURAY BUILDING
Washington, D.C.

March 2, 1929

Miss Jeannette Freund,
Attorney at Law,
15 Broad Street, New York.

Dear Miss Freund: ⁓

　　　　　　We are pleased to inform you that the Bill H.R.
6705 for your mother's relief, has passed the Senate, and now goes to
the President for signature.　　The bill will no doubt be approved by
the President, and so become a law.

　　　We will mail this information to your home address, knowing what a
pleasure it will be to you to give your dear mother early information
of the good news.　　You will now know that much has been accomplished
since you were here on Sunday last.

　　　　　　　　Very truly yours,

　　　　　　　　Burdett Thompson & Law

1929 congratulatory letter to Jeanette Freund, by then herself an attorney, from the lawyers prosecuting her mother's infringement suit against the United States government regarding Frank W. Freund's 1875 detachable pistol-grip patent. This letter announces the successful settlement of that suit — forty-two years after it was originally filed. (Freund Family Collection)

CHAPTER TEN

THE
COURT
CASES

Freund vs. the United States

On October 19, 1875, the United States Patent Office issued to Frank W. Freund U.S. patent number 168834, for a "Detachable Pistol-Grip for Fire-Arms." On February 11, 1878, Frank, by proper instrument in writing, sold and transferred all rights, title, and interest therein to his wife, Clotilda Freund, the transfer being properly recorded in the United States Patent Office on March 22, 1878. Thus, Clotilda Freund became the sole holder and protector of this important firearms patent.

On June 21, 1887, Clotilda filed a petition to the court, asking judgement against the United States government for the sum of $14,293.00, for the unauthorized use of this patent. Between July 19, 1877 and November 9, 1887 the United States, at its National Armory at Springfield, Massachusetts, had manufactured 14,304 detachable pistol-grips for longarms. The reasonable value of the use of each grip was $1.00, that price having been set by the United States government itself.

With this, Clotilda Freund started the ball rolling on a claim against the United States government that was to continue for the next forty-two years. Clotilda was about thirty-three years old when the claim for the Freund pistol-grip was filed, and over seventy years of age when it finally was settled.

The next discovery on Clotilda's claim was a bill presented in the 59th Congress under the Tucker Act of March 3, 1887. The case was numbered 12807 Congressional, but never prosecuted because as presented it was not in the proper form.

Then, in the 61st Congress, Bill S.9436 for her relief was presented. On February 21, 1911, Clotilda Freund's claim was referred to the Court of the

Senate. It was delayed there until April 23, 1917, when the Court made their findings of fact, and on June 23, 1917, the Court submitted to the Senate a certified copy of its findings.

In each subsequent Congress, another bill was introduced for the relief of the Freund claim, resulting in its passage by the Senate in the 68th and 69th Congresses. However, it received no action by the House Committee on Claims until the approval of the pending House Bill No. 6705, on February 15, 1929.

Among the witnesses who testified in this claim on behalf of Clotilda Freund was General John C. Kelton, who stated that he had made a special study of rifle and target practices since 1876. Kelton testified regarding the detachable pistol-grip as follows:

> *The detachable pistol grip is a great advantage in that it gives a secure hold of the right hand upon the stock of the piece so as to press the rifle against the shoulder. By that means a freer use of the trigger finger is given, and the effect of the recoil on the shoulder of the firer is diminished.*
>
> *The handle of the rifle is of a conical shape; the greater circumstance of the cone is near the lock; and consequently, when there is no pistol-grip, in the effort to draw the gun against the shoulder, the hand necessarily slips; and to grasp it sufficiently tight to produce a pressure which will overcome the recoil, the trigger finger is so paralyzed as to prevent its free and proper use.*

Kelton's testimony followed the previous statement of a witness speaking of the detachable pistol-grip's use, and of its value to the military service:

> *I think it is of the utmost advantage to rifle practice in our service. I do not see how it is possible for the best firing to be done with the Springfield rifle without it.*

Evidence as to the per-grip value of the use of Freund's invention was introduced, and the Court found that the amount asked for by the claimant, $1.00 per grip, was reasonable. The claim was placed in the hands of Senator Brookhard as a sub-committee, he being an expert on the use of firearms. His conclusion was that the claim was a meritorious one, but that the Court had over-valued the amount the government should pay for the use of the invention. Brookhard recommended that 50 cents per grip be allowed for its unauthorized use by the Springfield Armory, thus reducing the claim from $14,304.00 to $7,152.00. In this form the claim dragged on for years, always passing the Senate, but never getting out of committee in the House. However, there was help on the way.

The youngest child of Frank W. and Clotilda Freund was their daughter, Jeanette, who had been born on December 4, 1897. Being the "baby" of the family, she lived with Clotilda long after the other Freund children had either married or passed away. Jeanette graduated from Jersey City High School with the goal of becoming a law stenographer. Her first job was with a young lawyer for $8.00 a week; in that position, she started reading law books. Jeanette Freund changed jobs several times, but she found that she enjoyed working for law firms as a stenographer to be the most satisfying.

Thus, she enrolled at New York University, and studied law at night for the next three years. In order to do so she needed a New York address, so Jeanette and Clotilda moved across the river to Brooklyn from Jersey City.

In July of 1925, Jeanette Freund received her L.L.B. degree from N.Y.U. She was working for the law firm of Stetson, Jennings, and Russell, later to become Davis, Polk, Wardwell, Gardner, and Reed, and there she began her legal career. Her name was put on the directory in the building at 15 Broad Street, the only woman's name so included at the time.[1]

This close-up view illustrates the obvious design similarities between the "pirated" Springfield Armory detachable rifle pistol-grip, shown at the top of the photo, and Frank W. Freund's 1875 patented pistol-grip as fitted to a Model 1874 Sharps rifle, shown at bottom. (Top rifle courtesy Tom Lewis; lower rifle, author's collection)

Records of the Freund family contain voluminous correspondence initiated by Jeanette Freund on behalf of her mother, concerning Clotilda Freund's claim against the United States government.

One of the chief roadblocks to clearing the Freund bill through the House concerned the amount to be paid the attorneys. At that time, the House took a strong position against any great percentage of a settlement going to the lawyers. (We should be so lucky today!) When it was settled that the attorney fees would be $2,000.00, the bill passed the House on February 15, 1929. After years of waiting, and filing of the proper papers to the satisfaction of the Court of Claims, Clotilda Freund received her payment of $7,152.00 from the government—less the $2,000.00 paid to the attorneys.[2]

Frank W. Freund vs. Nelson King and the Sharps Rifle Company

The legal case previously discussed, *Freund vs. the United States Government*, was documented by volumes of correspondence collected by their daughter, Jeanette, who participated in the action. Unfortunately, very little documentation has turned up regarding the patent infringement case of Frank W. Freund *vs.* Nelson King and the Sharps Rifle Company.

It is the opinion of this author that whatever documentation existed probably was lost in a fire at Angie Freund Keegan's house in New Jersey, during the time that Clotilda Freund lived there with the older daughter and her family. A few references to this infringement case are to be found in letters written to Clotilda Freund from W.E. Simonds, a Hartford, Connecticut patent attorney,[3] and a reference or two in the Sharps Rifle Company letter books.[4]

This patent infringement case appears to have been over a self-cocking device as applied to the Sharps rifle. On December 3, 1875, Frank W. Freund had applied for a patent on an "Improvement in a Breech-Loading Fire-Arm." In fact, Freund's patent application provided for several significant improvements:

1. To provide a gas check.
2. To half-cock the hammer.
3. To move the firing pin in a straight line.
4. To prevent firing when the hammer is bearing on the firing pin.

The patent was not actually granted to Frank W. Freund until January 3, 1877, *thirteen months* after it was applied for. It is U.S. patent number 185911.

On March 4, 1876, Nelson King had applied for a patent for a breechloading firearm, its purpose also being to provide a self-cocking device for the Sharps rifle. However, King's patent was granted just *two months* later, on March 4, 1876, as patent number 177852.

This author and some other historians believe Frank W. Freund suspected that there may have been something wrong with the handling of his patent request. While Freund applied earlier, and patiently waited for a patent on his self-cocking device, King secured his patent improvement almost a year sooner.

Virtually all of the correspondence between patent attorney W.E. Simonds and the Sharps Rifle Company occurred during 1875 and 1876. It is possible that a hearing may have been held during that time by the Patent Office, to decide which of the two requests for the patent was the strongest.

Although the words "patent infringement" are used in some of the correspondence, on page 317 of Letter Book 9 of the Sharps Rifle Company a letter to patent attorney Simonds states:

> *We have concluded not to put another dollar into the King patent. We should be pleased to have Mr. Freund make us an offer for withdrawing or for any claim we may have in the matter.*
>
> C.H. Pond, September 23, 1876.[5]

It also was at about this time that Nelson King was discharged from the Sharps Rifle Company over an unrelated matter. Thereafter, it would seem that there was little interest on the part of the company over anything to do with Nelson King.[6]

All of the correspondence between E.G. Westcott, W.E. Simonds, Nelson King, and F.W. Freund suggests strongly that Frank W. Freund had filed an interference against the King patent, apparently over the half-cock mechanism. Nevertheless, it is evident that the U.S. Patent Office ruled in favor of King.[7]

The author has come into the possession of two letters and a postscript written by Frank W. Freund, on December 31, 1897 and January 9, 1898. The letters were written from East Boston, Massachusetts. Why Frank was in that city at the time, we do not know. He may have been working there, or perhaps was taking a short vacation. It appears from the letters that he had retained George D. Seymour, an attorney, to defend him in a suit, and also to help him sell a sight patent. Accompanying the two letters is an advertising card showing a tang sight, likely the sight referred to in the letters, although there is no record of a patent to Frank Freund for this type of sight.

Although Frank Freund spoke English, his letter-writing left a lot to be desired, particularly in terms of their construction and spelling. The letters are reproduced on the following pages just as written.

East Boston, Mass.
Dec. 31st, 1897
Geo. D. Seymour, Esq.

Dear Sir

Your favor of 30th inst. has been received, I thank you for having recommended the purchasing of my Sight Patent, and I think you have given a good advice, only I think it would be better to do that before you wish me to testify as it may look as this has something to do with the case.

You have in contention which it has not. It is a separate matter and of another invention, and the price ask for is easily the cost of cash outlay of the patent.

I prefer the Neinsh Dealer to have the Pat as they are where they first made the sight the rest only followed up.

The Neinsh Co can make it they all can do the same but that can put a stop to it, and I want to get it off my mind and I shall never talk of another patent and as long I live.

I am perfectly cured on that. By the way I saw one of your clients the Burnam Co. of Chicopee Falls went over board so you may get a little rest from that quarter.

I hope you had no losses sustained on that account and I don't plan anyone to do only cash.

Besides when it can be done as well.

Should you come to Boston I shall be glad to see you any time. Wishing you a Happy New Year.

I remain yours
very respectfully
F.W. Freund [8]

1. Freund Family Collection.
2. *Ibid.*
3. Letter in the collection of Dr. Richard J. Labowski.
4. In the collection of Dr. R.L. Moore, Jr., owner of the original Sharps Rifle Company records.
5. *Ibid.*
6. Sellers, *op. cit.*
7. Dr. R.L. Moore, Jr.
8. Freund Family Collection.
9. *Ibid.*
10. *Ibid.*

E. Boston, Mass.
Jan. 9, 1898
Geo. D. Seymour, Esq.

Dear Sir

I omitted to ask you to be so kind and let me know about a week before you intend to come here or also in case you changed plan on the matter. There are some little matters to send for that are at my Jersey City home which I think may be used to advantage in your case referring to them in connection.

I also include you a photo of a folding grip sight. Also mail notes the 1893 Pat. to some extent it makes a very accurate and convenient rear sight open or peep sight quickly catching bead an object aimed at.

It shows the bead off to good advantage and used by sportsmen that are used to open middle sights but eyesight failed they are using this now and claim they are doing as well as ever.

I also saved a slip of paper showing impression of center plate with aperture. The easy manner in which it can be in the present construction making a complete middle sight.

You may submit this to Mr. Benett should you have occasion to see him again and let me know the result.

Yours very truly

F.W. Freund [9]

P.S. I almost forgotten in regard to mention in regards to case referred to, friendly toward the Neinsh Co. as I am in that matter may disclose some points which may have been overlooked and may be a little surprise—so you need not hold back on anything for fear that I could go back on south is just and honorable, but I don't care to get fooled. I am like a burned child that don't like fire.

I have the points but need your assistance to frame it properly in which case, of course, we will have to talk the matter over before taking deposition as not to get matters mixed. However, it seems very simple and clear to me as far as you stated the case to me. [10]

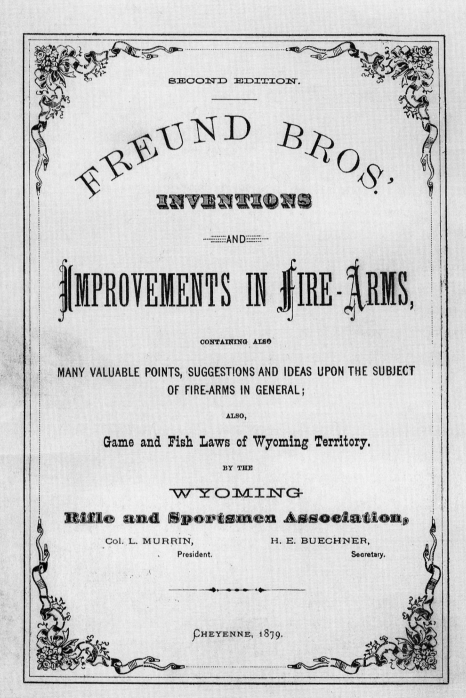

SECOND EDITION.

FREUND BROS.'

INVENTIONS

AND

IMPROVEMENTS IN FIRE-ARMS,

CONTAINING ALSO

MANY VALUABLE POINTS, SUGGESTIONS AND IDEAS UPON THE SUBJECT
OF FIRE-ARMS IN GENERAL;

ALSO,

Game and Fish Laws of Wyoming Territory.

BY THE

WYOMING

Rifle and Sportsmen Association,

Col. L. MURRIN,
President.

H. E. BUECHNER,
Secretary.

CHEYENNE, 1879.

CHAPTER ELEVEN

THE
FREUND
CATALOGS

ortunately for today's arms collectors and historians, the Freund brothers inherited the German gunmakers' practice of profusely marking their products. Probably, too, they simply enjoyed seeing their names in printed advertising, or stamped on the barrels of their rifles and almost everything else they sold, repaired, or worked on.

This is especially true of their self-promotional broadsides and catalogs. The author has in his collection copies of four catalogs that were issued by the Freund Brothers, and perhaps there are still others out there which have not as yet come to light.

The first of these catalogs was issued at Cheyenne, in 1879. It is a small, gray booklet, measuring $5\frac{5}{8}$ inches by $7\frac{3}{4}$ inches. It was reprinted by collector Bob Borcherdt about 1980. Interestingly, this catalog is titled "Second Edition." Is there a first edition out there somewhere? It would have been issued after 1876, and before 1879.

Perhaps the most interesting by-products to have come to light during the research of the Freund catalogs are two scrapbooks put together by someone in the Freund family, probably Frank's wife Clotilda, or one of their daughters. Both survived in the family collection until recently.

The first scrapbook contains early pictures of the Freund family in Europe, several early pictures of Frank and George and their children, photographs of early Freund guns, and locations of some early Freund gun shops. It was assembled by pasting the pictures onto a red canvas-type page that deteriorated over the years; thus the pictures had to be removed from the scrapbook. Fortunately, the photographs themselves still exist in the Freund Family Collection, and a microfilm of the scrapbook also exists.

The second scrapbook consists of a number of letters of testimony about the unsurpassed features of the Freund rifles and Freund sights. The interesting thing about these testimonial letters is that the majority were written within a span of about three days. It appears that Frank went up one side of the main street in Cheyenne, and then down the other, soliciting them within that very short period of time. This scrapbook came down through Frank's daughter, Angie, who gave it to her son, Jack Keegan. Keegan's second wife sold it in an auction on the West Coast within the past several years, after his death. Fortunately for collectors, a researcher at the University of Wyoming in Laramie was able to secure the loan of this scrapbook from Jack Keegan, around 1960, and made a microfilm copy of it for the university library's collections.

The cover of the Freund Brothers' 1879 catalog done in Cheyenne is reproduced on page 250, and the rest of it appears on pages 253 through 272. Note that on the back cover of this catalog, F.A. Dammann is named as the successor to Freund & Bro.; this in turn is stamped over by "C. FREUND, SUCCESSOR." Perhaps the first edition of this catalog was a "Freund & Bro." catalog, and the second edition carried the "F.A. Dammann" imprint.

The second Freund catalog came out subsequent to the introduction of Freund's Wyoming Saddle Gun, after 1880 when George had left for Durango but before Frank moved to the East. It is stamped over on the inside of the front cover as well as on an inside page, "F.W. FREUND, 54 STEVENS AVE., JERSEY CITY, N.J." This catalog also measures approximately $5\frac{5}{8}$ inches by $7\frac{3}{4}$ inches. It has an orange cover, and consists of sixteen pages plus the covers (*see* pages 273 through 282).

The third Freund catalog is more of a monograph on the Freund sights. It consists of four pages measuring approximately 5 by 9 inches, and lists F.W. Freund's address as Jersey City, New Jersey (*see* pages 283 through 286).

The fourth Freund catalog was issued by George Freund from Durango, Colorado. It measures approximately $4\frac{3}{4}$ inches by 8 inches, and is printed in color. It has 42 pages, plus covers, and is quite an imposing catalog. It is obvious that by then George had evolved into quite an entrepreneur, as while the first thirteen pages are devoted to George's Colorado Armory, the remaining 29 pages (with two exceptions) are taken up by advertisments for other businesses in the community. Every type of legitimate enterprise in the frontier town was represented: hotels, jewelers, hardware, clothing, and harness suppliers, as well as others. George Freund called this catalog number 389, which the author interprets as having been issued in March of 1889 (*see* pages 287 through 308).

FREUND BROS.' INVENTIONS

AND

IMPROVEMENTS IN FIRE-ARMS,

CONTAINING ALSO

MANY VALUABLE POINTS, SUGGESTIONS AND IDEAS UPON THE SUBJECT OF
FIRE-ARMS IN GENERAL,

ALSO,

Game and Fish Laws of Wyoming Territory.

BY THE

WYOMING RIFLE AND SPORTSMEN ASSOCIATION,

COL. L. MURRIN,
President.

H. E. BUECHNER,
Secretary.

Publishers, New York.

FREUND BROS.,
GUN MAKERS,

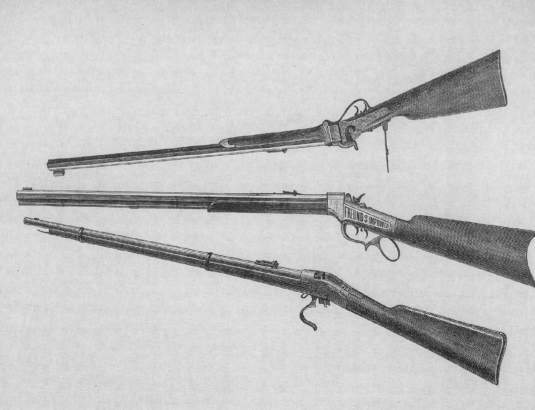

CHEYENNE, WYOMING TERRITORY

THE IMPORTANCE OF FIRE-ARMS

IN GENERAL.

THE importance of fire-arms is now well understood throughout the civilized world, and the more efficient the weapon the better it is appreciated by the people. In olden times warfare was conducted and carried on between hostile nations by means of the spear, the club, the rude sword and the battering ram, fire-arms were unknown.

The cannon, the rifle, and the revolver were unheard of, and as yet not invented, and when Rome was in the height and zenith of her glory, her soldiers, when the strife of battle came, relied more upon their physical strength and endurance to enable them to win victories over the enemies of the "mistress of the world" than they did upon their skill and prowess. For centuries the world was ignorant of that which is now patent to all, namely, that there is a more efficient and yet less barbarous method of carrying on war, than the system which was in vogue among the ancients. The discovery and invention of fire-arms has proved a blessing rather than a curse to the human family, and the whole course of events in modern times tends to prove and establish this fact.

Fire-arms in America.

It can be truthfully said that in all things that pertain to the invention, improvement, and manufacture of fire-arms, America to-day takes the lead. This is evidenced by the fact that the European nations send their purchasing agents to our ports whenever the storm cloud of war lowers in the eastern horizon. It was more on account of the superior efficiency of American fire-arms that the Turks were enabled to hold at bay so long the mighty hosts of Russia in the late war than from any other cause or reason.

The American improved fire-arms was a more potent agent in the cause of Turkey than the "Banner of the Prophet."

Fire-arms in the Far West.

While America has of late years become as it were one vast arsenal, yet for obvious reasons there are portions of our country where much more attention is paid to the subject of fire-arms than in other localities of our broad domain. We allude more particularly to the Far West. In this region nearly every man is a gunner, and the boys as well as, in a vast number of cases, the women, well understand the use of the shot gun and rifle, and many of them are experts in their use.

There are, of course, good and valid reasons for this, which are well understood west of the Missouri River. In the first place it behooves the dwellers upon our exposed frontiers to be at all times prepared to repel the incursions of the ever-treacherous red men, for the ranchmen, the stock growers, and the freighter, can never tell at what hour of the day or night his enemy will come, and, as a general thing, when the danger arises he is compelled to rely mainly on his own skill and efforts to protect his wife, family, home and property from the savage foe.

Often, too, the pioneer is compelled to rely on his trusty rifle to procure food for a time for himself and those who are dependent on him for their support. There are many other reasons which might be mentioned as tending to establish the fact that the people of the Far West are unusually proficient in the use of fire-arms, but it is sufficient to remark, in conclusion, upon this point, that what men are necessarily compelled to learn for their own safety and self preservation, they generally learn well and thoroughly, and, when once learned, it is not easily forgotten.

The Kind of Fire-arms Needed.

There being so much need then of fire-arms in the Far West, and for a right understanding of their use, it follows that people naturally seek an efficient weapon. An inferior one is worse than none, more especially in this part of the country, where distances are so deceiving, and where oftentimes the life of the wielder of the weapon depends upon the accuracy of a single shot. Hence the people here are now, and have been for some time, anxiously inquiring after the most efficient "shooting iron" that can be obtained.

Nor is this all. They are also turning their attention to the in-

vention, improvement and manufacture of fire-arms. Taking it all in all it can be truthfully said that nowhere in America is there more attention paid to this subject than here in the territories, and nowhere are the people, as a general thing, more competent to judge of the merits or demerits of all kinds of fire-arms now in use in this country than those who dwell here upon these plains, and within the shadow of the "Rockies."

The Freund Brothers, of Cheyenne, Wy. Ter.

Among those who have turned their attention to this subject, and who have made it a specialty to improve by invention and otherwise the efficiency of American fire-arms, the Freund Brothers, of Cheyenne, Wyoming Territory, stand among the first and foremost. Their wonderful inventions, of which mention will be made further along, bid fair to revolutionize and overturn the entire system of the manufacture and use of nearly all kinds of fire-arms, in America at least.

The Experience and Business of the Freund Bros.

The two Freunds are Germans by birth, and while mere boys and yet residents of the Old Country they turned their attention to fire-arms, and before coming to America they had been engaged in some of the largest and best manufactories of Europe, including Paris and London, where they of course thoroughly learned the business in which they have ever since been engaged. Coming to America in 1863, they three years later located in the Far West, where they have industriously and profitably pursued the business of remodeling, readjusting and improving fire-arms, and in addition thereto have been doing a large general business as wholesale and retail dealers in all kinds of fire-arms and ammunition. The headquarters of the firm are now at Cheyenne, Wyoming Territory. They have always made it a specialty to observe and study the defects of every species of fire-arm that has been in use in the Far West for the past twelve years, and hence their skill, experience and judgment in relation to all such matters has been and is now well known throughout the West, and are duly appreciated by all who take an interest in these matters. About a year ago the firm sold out their general business to Mr. F. A. Dammann, so as to be able to devote their entire attention to that which they have for so long made a specialty. Of the present relations that exist between the Freund Bros. and Mr. Dammann more definite mention is made under the head of "Miscellaneous Matters."

What the Freund Bros. Say.

In presenting to the public our new improvement on Sharps rifle, we take occasion to thank our customers and friends for their kind and unprejudiced examination and criticism of this new invention. The favorable consideration with which this arm has met from all who have taken sufficient interest in it to examine and test its merits is due to the fact that a sportsman, or any one skilled in the use of fire-arms, recognizes at once in the perfection of its movements and the singular adaptation of its several parts to the accomplishment of the design intended, all the requirements of a perfect gun.

The [report of Sharps Rifle Company for 1878, contains the following:

"In computing percentages made in the great International Match of 1876, with forty competitors in the field, using rifles of six different makers, including all the *crack* British muzzle-loaders, it was found that Sharps headed the list with 877, the next highest scoring only 867." (See official report in *Rod and Gun*, September 30, 1876.)

By reference to a challenge by Mr. Rigby (one of the firm of William and John Rigby, the most celebrated riflemakers of Great Britain) to the Sharps Rifle Company, we discover Mr. Rigby's opinion of the American breech-loader as it now stands. (See *Spirit of the Times*, March 9th, 1878.)

Breech-Loaders vs. Muzzle-Loaders,

Read the following communication from Mr. JOHN RIGBY, the original challenger, to the Sharps Rifle Company:

"DUBLIN, February 12, 1878.

"To THE SHARPS RIFLE COMPANY:
"*Sir*—Your letter received, &c. * * * You appear to misconceive, somewhat, the real purport of my proposal quoted by you from a letter written to the *Volunteer Service Gazette*, in London. It was made for the purpose of demonstrating that the American

breech-loading match rifle, as used in the Centennial and other matches at long range, does possess the advantages of rapid and simple manipulation which properly belongs to breech-loaders, and that it is, in this respect, superior to our muzzle-loaders. I did not propose to enter into a contest of muzzle-loaders against breech-loaders in general, but against Creedmoor rifles loaded and treated in the special manner which was found at the Centennial matches to gain the best results."

We coincide with Mr. Rigby's sentiments in this matter, for we claim that the shooting qualities of a gun depend upon the perfection of the barrel and the skill of the marksman. An old-fashioned flint-lock muzzle-loader, possessing a good barrel, placed in the hands of a good marksman, would have attained the same result at Creedmoor as that of the finest rifle used at the match. We claim for our *action* only speed and safety to the operator—the accuracy depends entirely upon the perfection of the barrel and the skill of the marksman—that is, for Creedmoor or target practice. The highest and most satisfactory results are attainable only by a combination of the best qualities of a barrel and facilities for rapidity of action. This rapidity of action pertains exclusively to the breech-loader, and stamps it at once the superior of the muzzle-loader, or indeed any other gun, other things being equal, such as using the cartridges from the belt without being expressly prepared and dipped in oil, as at target practice. In such practice the guns competing should be interchanged among the members of the team to arrive at a fair and impartial estimate of their respective merits. Not only so, but target practice does not furnish sudden emergencies which call for quick and decisive action, but allows a given prescribed time for every necessary motion of the marksman—loading, shooting, and cleaning his gun. An Indian or a deer would have no such delicate consideration for his convenience. The one would shoot him down while he was leisurely making his preparations, and the other would scamper off undisturbed by even the *report* of the so-carefully-handled weapon. To the length of range and accuracy of a gun must be added rapidity of action, in order to make it available in every contingency, and where continuous firing is required it must possess appliances for the instant forcing home of the cartridge. We claim, and can demonstrate to any one interested, that our new breech-block is the safest made, and that it accomplishes the forcing home of the cartridge with perfect ease after a great number of shots have been fired. The Sharps rifle without the Freund improvement cannot lay claim to these merits. This improvement has been applied to the new improved Sharps rifle, model of 1878, 1* 2*, and when so applied to either the old or the new model, we feel confident that the claim which the Sharps Company make that their gun is the "best in the world" can not be controverted. Although our invention has been introduced but a short time, orders from all quarters are constantly pouring in, and it has been commended and recommended by all military men, hunters and sportsmen who have used it. The following highly flattering encomiums and testimonials which we have received from time to time explain themselves.

HIGH AUTHORITY.

The Freund improvement on the Sharps rifle, of which so much has been said in praise in the *Sun*, is thus highly and flatteringly recommended by Gen. Sheridan :

HEADQUARTERS MILITARY DIVISION OF THE MISSOURI,
Chicago, January 23, 1878.

Mr. FRANK W. FREUND, Cheyenne, Wyoming Territory :

Dear Sir,—The Sharps rifle, in which at my request you inserted your patent breech-block, pleases me very much, and the more I see of it the more strongly am I impressed with the practical ideas you have developed in your invention. My friends in the city, who have often used Sharps rifle in hunting large game, are quite as well satisfied with the improvement as I am, and the army officers to whom I have shown it all think well of it. I trust that it will eventually prove a source of profit to you. With thanks, I am,

Very truly yours,
P. H. SHERIDAN,
Lieutenant-General.

GEN. CROOK'S ENDORSEMENT.

The following letter from Gen. Geo. Crook speaks in the highest terms of the improvement attached to his rifle by Freund Bros. The General's rifle is of the Sharps make, and here is what he says in relation thereto :

HEADQUARTERS DEP'T PLATTE,
COM'DG GENERAL'S OFFICE,
Omaha. Neb., April 30, 1878.

FREUND BROS., Cheyenne, Wyoming :

Gentlemen.—I have received the Sharps rifle left in your hands for repairs and alteration. The double ejector spring and the rounding off of the inner edge of the breech-block are improvements which only need to be seen by practical men to commend themselves for general use. The beauty and perfection of workmanship in this, as in other repairs I have had made at your establishment, are entitled to the highest praise.

Very respectfully your obedient servant,
GEORGE CROOK,
Brigadier-General.

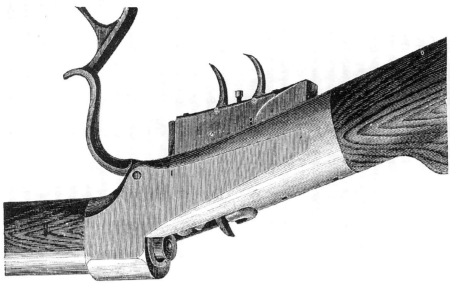

BALLARD RIFLE,

WITH FREUNDS' IMPROVED BREECH MECHANISM.

The above cut represents a Ballard rifle with our new improved breech action. The shooting qualities of the barrel and the general finish of the Ballard rifle are inferior to none ; but, unfortunately, it has, in our opinion, a very serious defect in its breech movement, possessing no facilities for forcing the cartridge to its proper position in the barrel, and lacking the strength in the resisting parts so essential to a reliable service gun. By inserting our new improved breech movement, we have overcome both these defects. Our breech movement forms a direct resistance at a right angle to the line of recoil, against heavy charges. Without our movement, the Ballard gun cannot offer the necessary resistance to heavy charges of powder. For this reason, they have not as yet made a 40 calibre gun to shoot 90 grains of powder; the strain is so great. Consequently, they have made cartridges to suit the gun. But, to meet the demands of the public the gun must be made to suit the cartridges in market. By applying our movement to the Ballard gun, this is accomplished, and any cartridges in the market may be used with perfect safety. We make this movement either with single or double extractors, as ordered, and when applied to the Ballard rifle, it becomes equal to any in the market. The Ballard rifle possesses an advantage over the Sharps in use, in the fact of its having a cleaning-rod attached; to attach it to the Sharps rifle would entail additional expense. This gun is splendidly sighted for hunting or target practice, and has double triggers. By the insertion of our new movement, we make a direct resistance in the rear, and there is neither spring nor strain, and there is less recoil.

In conclusion, we say that we guarantee every rifle which we alter to our new breech movement, and if any rifle fails to conform in every respect to our representations, we will take it back and pay all charges of transportation, besides refunding the purchase money. We make the insertion of our new breech movement a specialty; but are prepared to do any work, whether new or repairs, in the gun-making line.

In sending orders, be particular to describe fully the size, length, and weight of barrel, the calibre, single or double trigger, specify whether the pistol grip is desired, and whether plain, open or peep sight is preferred, that no error may occur.

Orders may be addressed to F. A. Dammann, who has our agency, or to us.

FREUND BROS.

CHEYENNE, Wy. Ter.

Some Questions Answered.

The question is often asked, "Why do not the leading manufacturers of the United States, or some of them, adopt the Freund im-

provement?" The question is easily answered. It is natural for men everywhere to prefer their own work and inventions to that of others, and this is one of the reasons why a company engaged in the manufacture of improved fire-arms feel adverse to giving countenance to an invention not their own, because by so doing, their own inventions and discoveries may be thrown in the shade. Then again, the proprietors and owners of some of the largest fire-arm manufactories in this country are men who are not themselves really acquainted with the business. They put subordinates in charge who are supposed to be entirely competent and capable, and are in many cases, of course, efficient enough in most respects, but they take too limited a view of some things which are of the greatest importance to them and to the people at large. While many of these superintendents who are in the employ of proprietors that have no real or practical knowledge of the business, have the power and authority (in some cases) to adopt improvements which will add to the efficiency of the fire-arm, they all can give advice and make suggestions to their employers in regard to these matters, which would no doubt in many instances be acted upon. They fail to do this, however, because they entertain the erroneous opinion that it is not a part of their duty to secure for their firm that which is most efficient, but they act upon the idea that their own peculiar inventions and arms must be held up to the world as being the best, and that nothing which some other firm has invented and discovered should be countenanced in the slightest degree. Thus it is that between the jealousies which the majority of men are prone to, and the lack of broad and correct views as to what the true interests of the firm or manufactory are, the Freund improvement has thus far been shut out. No complaint is made by the Freunds on account of this, for they well know and understand that it all comes about because a wrong view is taken of the matter by the parties previously referred to; yet they contend, however, that when the merits of their inventions come to be well understood, they *will* be adopted and used in preference to less serviceable and imperfect inventions that are now in use. The Freunds do not assail or decry the business of anybody else. They contend that their improvement is the best, they set forth to the world the facts which support the claim, and they say to all, "If you do not coincide with us in our

A SHARPS RIFLE (*top view*),

With Freunds' Improved Breech Mechanism Adjusted.

This style of breech-block having resistance of equal hight on each side with Freunds' new firing-pin, half-cocking slide and cover, see engraving, page 14.

For adjusting the improvement, $35.00. Work first-class in all respects.

---

belief, and insist that we are wrong, give us the proofs and arguments which disprove what we claim!"

Other Questions Answered.

The Sharps Rifle.

"Why is the Sharps rifle breech-block better than that of any other gun?" is a question we have frequently heard asked. In answering this question, we might as well refer to both the old and the new Sharps rifle, and state the reasons which we have for our faith in that efficient fire-arm.

As to the breech-block of the old Sharps rifle manufactured in 1848, it moves vertically in line of resistance and is the same style of breech-block adopted by Krupp in Germany, in the manufacture of the celebrated Krupp cannon. In times when the gun had a larger calibre and less powder was used than now, it done very good service. But since the calibre has been cut down from 52 to 40, bottle-neck shells are substituted for the old style of cartridge and the charge of powder, increased to more than double the former quantity, the breech-block of the old Sharps rifle has become doubly efficient.

Freund's new breech movement which is now being adjusted to the old rifle by the firm of Freund Bros., has the vertical slide, and the facility for forcing the cartridge home, and the strongest extracting power of any breech-block mechanism ever used in any gun in this country.

The following testimonials on this point will be read with interest:

DEADWOOD, D. T., Aug. 15, 1874.

Dear Sir,—Things in and about Deadwood are quite lively now, and the mines are paying well. The principal feature of excitement here, during the past few days, is the arrival in town of George Freund from Cheyenne, and I had an opportunity to see the latest improved Sharps rifle, of which we have heard so much lately. The gun, however, is fully up to all the representations we have heard concerning it, and, to my mind, is the "boss" gun for this section of the country. The breech of the improved Sharp is made so as to afford ample lever power for forcing the cartridge into the chamber, thus doing away with the principal objection to the Sharp. Much fault has been found here against the Sharp on account of the difficulty experienced in getting the cartridge in the chamber when the arm has been fired a few times without cleaning, and especially the forty-calibre rifle, which I use, and have any amount of difficulty in introducing the cartridge, even when hunting game. Sometimes I want to fire, when hunting, several times without cleaning, but from three to four times is the best I can do with the old model Sharps. Every one here thinks the Sharps with Freunds improvement is the most perfect arm they have ever seen. The gun has to be made by hand now but Mr. Freund informs me that he will soon complete arrangements for having them altered by machinery. This being the case, I think nearly every one will have their Sharps changed to the new system; I shall, at least. They are using the plain bullet now in this rifle and do away with the patching, which I think is a good thing for use on the frontier. Well, I presume you have heard enough about guns this time, and as there is not much news to write, will close.

A SPORTSMAN.

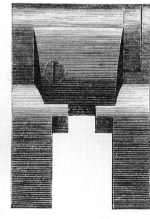

There are of course two kinds of Sharps rifles, the old and the new. While the old style is universally preferred to the new, at least in the Far West, the new rifle is perhaps a better looking gun. It is claimed by nearly all who understand its merits and demerits, that nothing has been gained by the production of the new gun, and it is entirely too impracticable for use in the West. The people of this region demand something serviceable and efficient, and care but very little for that which is fancy and stylish.

The breech mechanism of the new gun cannot be removed as easily as that of the old, in fact it requires considerable skill and a small tool chest to accomplish the feat in any satisfactory manner, while that of the old Sharps rifle can be removed, cleaned, oiled and put together again in a moment's time without instruments.

VIEWS OF BREECH BLOCK, SLIDE, PARALLEL MOVING FIRING PIN AND COVER.

The Best Kind of a Gun for Service on the Plains.

Another question which is frequently asked, is, "What is the best kind of a gun for use and service on the Plains?" The answer to this question is a simple one and quickly made. The long-range gun is by all means the best.

The hunter who goes in quest of game is generally compelled to bring down the deer or the antelope at a distance of several hundred yards, if he secures the game at all, and for that reason a short-range gun, or one on which he is compelled to elevate or depress the sight is of little or no value. On the other hand, the gun which will throw a ball instantaneously and accurately a half mile or such a matter, and hit the exact object aimed at, can be relied upon, and is the one preferred by the hunter and sportsman on the Plains. Then, too, in an encounter with the Indians, it is of vital importance that he who is thus unfortunately situated should have a long-range gun, for if this is the case, he may be able from a long distance to defeat and drive off his assailants without harm to himself. While but few of the Indians who go on the "war path" in the Far West, are armed with long-range rifles, yet nevertheless they are all supplied with very effective short-range weapons, and they are deadly shots. The great majority of the men shot and killed in the Black Hills country since its first settlement, have been shot directly through the heart. Yes, we say emphatically to all our friends, both far and near, that the long-range gun is the best and safest one for service on the Plains and in the mountains.

The Sharps rifle with the Freund improvement attached and adjusted is exactly the weapon that is needed for the rough service on the Plains, and this fact is now generally conceded by all of the experienced hunters and sportsmen who have tried and tested them, often when their lives depended upon a single shot. We, however, forbear any further remarks upon this point, and call especial attention to the following letter, published in the Cheyenne Daily *Sun*, July 18th, 1878, from Mr. C. C. Miller, the Union Pacific agent at Table Rock, Wyoming, a gentleman who has had constant experience with all kinds of fire-arms in the Far West for a number of years.

WHY IT IS THE BEST.

THE VIEWS OF C. C. MILLER UPON THE FREUND IMPROVEMENT.

The following letter was received Tuesday by Freund Brothers from Mr. C. C. Miller, the Union Pacific agent at Table Rock station, who, as it will be seen, has had considerable experience with the Sharps rifle as it comes from the factory, but recently got one improved by Freund Brothers:

TABLE ROCK, July 15, 1878.

FREUND BROTHERS, CHEYENNE:

Dear Sirs,—The Sharps rifle with your new improved breech mechanism is as perfect a fire-arm as could be produced. I take considerable pride in owning a good rifle, but I had begun to get discouraged in trying to find one. I have owned and tested all of the modern breech-loaders now in use. They all shoot well, but in the breech mechanism they are far from being perfect. I discarded all of them in preference to the factory-made Sharps sporting rifle—weight 12 pounds, 40-calibre, shooting 90 grains of powder. The one I now own makes the third one of the same kind. I sold the two former, and was also going to get rid of this one, on account of the caps in the shells bursting nearly every time the rifle was fired, and letting the gas come back through the hole of the firing bolt into my eyes, nearly blinding me. Another objection to them that used to aggravate me badly was when I went hunting, and got up within fifty yards of four or five hundred antelope, all in a bunch, maybe the first shot fired would be one of the old shells that had been reloaded three or four times, and in trying to extract the empty shell after firing (there being an extractor on only one side), I would pull the head of the shell off, leaving the remainder sticking in the barrel; that would end the shooting right there; would have to come home, and spend about half a day with a ramrod and hooks, trying to get the rest of the shell out. This would happen, too, nearly every time I went hunting; but with yours, if I had only one extractor, it could not happen, as the extractor falls back twice the distance of the Sharps, and takes a better hold of the shell; further, you have no cut in the breech-block for extractor, while in the Sharps there is an opening for dust to accumulate, and makes the extractor in such a case useless. Another thing that would also occur very often was, when trying to insert an old shell in the rifle it would get stuck just before getting it in far enough to clear the breech-block, and it would be impossible for me to get it either in any further, or out. This also ends the shooting, unless you have a ramrod along to punch the cartridge back out again, as it is very dangerous to try and force it in any further by pounding, etc. It is impossible for any of these things to happen with your improved breech-block.

Another fault with my Sharps was that after firing them and extracting the empty shell, the hammer had to be drawn back to a half-cock for safety before inserting another cartridge and raising the lever. With your improvement, the simple and easy way in which the hammer is drawn back to a half-cock, either in throwing the lever down or up, making the same perfectly safe against premature discharges in loading or carrying the arm; the improved firing bolt and making the firing hole so much smaller, thereby making it impossible for a cap to ever burst, or for the firing hole to ever get clogged up with brass off the caps, is perfect.

The rifle I have now was open to all the objections I have stated, as you well know, until I heard of your recent improvement, when I determined to try it, and I am very glad I did so, as you have now made it the kind of a rifle I have long been trying to obtain, and I would not use any other. I have fired the rifle about two hundred times since I got it back from you, six weeks ago. The most of the shells used have been reloaded as many as twenty-five times each, but the gun works exactly as well with them as new shells that have never been used.

For sportsmen, hunters, and men in an Indian country, where it is an imperative necessity that a rifle should never fail them, and where, if it does, they are liable to lose their lives, your improvement is invaluable to them, as it can never fail to do all that is claimed for it.

I remain,

Yours very respectfully,

C. C. MILLER,
Agent U. P. R., Table Rock.

Something of Importance which the Public Should Know.

There are at the present time so many different kinds of fire-arms manufactured and sold all over the broad extent of our country, that unless a person is interested in the matter of efficient fire-arms, he is apt to get "a little mixed," if we may be allowed to use this oft-repeated Western phrase, in regard to making a judicious selection if he desires to purchase. We have already seen that the long-range rifle that throws a bullet with irresistible force

is the best one for service; but as there are a great many old hunters and experienced marksmen who do not pretend to be posted as to the scientific principles upon which the gun is constructed, and by which this quality is attained, we propose right here to give to the public a fact or two by means of which the more effective, powerful, and at the same time long-range rifle can always be selected. Take, for instance (to illustrate our point), a twelve-pound Napoleon brass cannon. Its extreme accurate range is 1,728 yards. The ball, of course, weighs twelve pounds, and the charge of powder necessary to be used will be a trifle over two pounds. While the extreme (accurate) range is 1,728 yards, yet the shot can be thrown about 2,400 yards, less than a mile and a half. Now, then take the 6-pounder Rodman gun, which does not require more than two-thirds the amount of powder, and you can throw the ball over three miles, and the extreme accurate range will be at least two and one-third miles. Why this difference? To be sure, the twelve-pound ball is just as heavy again as the 6-pounder, but then the large gun takes the most powder, and one will just about balance with the other, so far as the means to attain the end are relatively concerned, and yet there is a difference of nearly two miles in the extent of the range in favor of the smaller calibre gun. This fact is well known to all who, like the writer, have served for a couple of years in the light artillery, and every one knows that the difference in favor of the 6-pounder is accounted for by reason of its having a smaller calibre. As regards long and short range rifles, the same principle holds good, and the same results follow as in the case of the Napoleon and Rodman cannons. The larger the calibre the less power there is to force the bullet out, or rather the smaller the bullet the more concentrated the power that emanates from the explosion of the powder, and hence the greater the force with which the bullet speeds on its mission.

The true test then to be observed, if you are seeking for a long-range rifle, is to pick the one with a small calibre (other things being equal) and then you will be sure to get what you are after, and remember always that in firing a gun the greater the force behind the bullet the greater the force in front, and the 40-calibre Sharps rifle with the Freund Improvement, you will find exactly fills the bill.

How The Freund Improvement Works, and its value.

We propose now to refer again to the Freund Improvement, and lay before the readers of this pamphlet such statements in reference to this valuable invention and its value, as will enable them to have a just appreciation of its merits.

Upon this point, however, we propose to call attention to the following testimonials, as they set forth fully the merits of this invention in language so plain, direct and pointed, we feel that no better statement of the matter could be made, than that which is contained in these articles.

The first is from the Cheyenne *Daily Sun* of October 28, 1877, and while it refers more particularly to the gun of General Sheridan, to which the Freunds adjusted their improvement, yet, nevertheless, the whole subject is handled in a satisfactory manner, and the entire ground is covered.

The second is an affidavit of J. A. Meline and certificate of the Acting Commissioner of Patents, which documents were also published in the same paper, June 2d, 1878.

GENERAL P. H. SHERIDAN'S GUN.

WITH FREUNDS' IMPROVEMENTS.

We stepped into Freund Bros'. gun store to inspect this much talked of gun, and found it to be a splendid piece of mechanism, not only in the material, but in the beauty and simplicity of the improvements, which are the following: First, for forcing the cartridge into the chamber of the gun by the closing of the breech; second, it "half-cocks" the hammer by opening or closing the breech; third, it has two extractors working simultaneously; fourth, the construction of the firing pin; fifth, the construction of all these movements with the guard lever. The object of having the breech with the improved movement is to prevent any accident by the premature discharge of the cartridge by forcing it into the chamber with the breech-block or clubbing it in with a stick or hammer when the gun becomes foul from frequent discharges. The improved movement will force the cartridge into the barrel by the simple pressure on the lever, even if the cartridge does extend out quite a distance, so that the gun can be fired an unlimited number of times without cleaning.

The half-cock improvement is also done with the movement-lever, which, when it opens and closes, sets the hammer on a safety notch, and in no case can the hammer set on the firing pin, as in the old style of rifle, which is liable to a premature discharge by a sudden jar on the ground or by any other cause that may give a sudden jar or blow on the hammer; but prevents all accidents to which other guns are subject between the act of loading and firing, and obviates the necessity of the operator half-cocking the gun, where in a case of emergency, such as an attack by Indians, quick work is wanted.

For military purposes this movement cannot be surpassed, as in a battle men will become excited; some will put their guns at "half-cock," some at "full-cock," and most men will not cock them at all, and in either of the two last-mentioned cases the arms in the hands of an excited man becomes dangerous either to himself or those about him, as it is impossible for an officer to watch the guns of his men continually to see that they put them on the "half-cock" before loading. We will further illustrate: An officer gives the order to a company of soldiers, either of infantry or cavalry, to "load," so as to be prepared to use their guns at any time after the order is given. It would require a great deal of valuable time of each officer to inspect the arms of the company, to see that each man has his gun on half-cock, to prevent accident; but by the above improvement on the Sharps rifle, the operation of loading, puts the gun on half-cock, and so it remains until it is again cocked for the purpose of firing. It also has two strong and simple extractors, which work simultaneously. They take hold of the cartridge on each side, and by the guard lever power (which is the greatest possible power) extracts the cartridge. These extractors are placed on each side of the chamber and fit in close, and no openings are left, as in the old gun, so that no dirt or dust can enter the barrel of the gun by the breech.

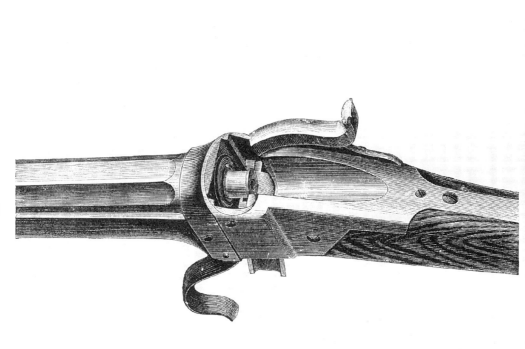

Representation of top view of Sharps Rifle Model of 1874.

WITH FREUNDS' IMPROVED BREECH MECHANISM,

which illustrates how the cartridge is held in its place, and is forced home by the closing of the breech, which also half cocks the piece, and has the cover shown in cut on page 14.

Price, ordinary finish, for adjusting as in cut on page 13, $25.00.

20

The firing pin has a gas cup that prevents the force of the gas escaping from the Sharp rifle, in case of a breakage of the cap in the cartridge.

All the above improvements are combined with the guard lever of the Sharp rifle, which is constructed with a link giving the whole movement a power that cannot be surpassed on a scale so small as a gun breech. The resisting power of the Sharps breech, against a large charge of powder, is greater than any other gun now made, and the above entire mechanism attached to a Sharps 40-calibre rifle will surpass any military gun in the world. All the work on this gun displays a perfect knowledge of the art of gun making, and cannot be excelled for the ground essentials of strength, effectiveness, simplicity and beauty.

The Freund Brothers inform us that they will have this magnificent rifle on exhibition at their armory for several days. Of course they will be glad to welcome sportsmen and others interested, and exemplify the advantages of the arm that has already received the highest possible compliment—that of being adopted by such eminent authority as the Lieutenant-General of the United States Army.

CONVINCING EVIDENCE.

AS TO THE VALUE OF THE FREUND IMPROVEMENT.

The convenience of the improvement made by Freund Bros. upon the Sharps rifle is generally understood, but its value as a protection against accidents is best appreciated by those who have been in critical situations, or had personal knowledge of injuries inflicted by the same rifle without the improvement. For the information of readers of the *Sun* we give below a sworn statement of a well-known hunter, J. A. Meline, which is directly to the point:

DEPARTMENT OF THE INTERIOR, } U. S. Patent Office.

To all persons to whom these Presents shall come, Greeting:

This is to certify that the annexed is a true copy from files of this office of the affidavit of J. A. Meline, in the matter of the letters patent granted Frank W. Freund, dated August 1, 1876, No. 180,567, for improvement in Breech-Loading Fire Arms.

In testimony whereof, I, W. H. Doolittle, Acting Commissioner of Patents, have caused the seal of the Patent Office to be hereunto affixed this 21st day of May, in the year of our Lord one thousand eight hundred and seventy-eight, and of the Independence of the United States the one hundred and second.

W. H. DOOLITTLE,
Acting Commissioner.

TERRITORY OF WYOMING, } ss.
County of Laramie,

To whom it may Concern:

J. A. Meline, of said county and Territory, being by me duly sworn according to law, deposes and says that he is a hunter by profession, and has been engaged in such business for twenty-two years; that he is well acquainted with the rifle known as Sharps Breech-Loading Rifle, and has used it exclusively for five years; that he has had them made to order for him at the factory, and always has considered it the best rifle in use, finding but one objection to it—that there is found a difficulty in forcing in the cartridge after the rifle has been fired a few times, and by reason of this trouble, or difficulty, he has known many accidents to occur, such difficulty being more observable where the rifle has a small bore, and a long cartridge used.

This deponent further says that he has personal knowledge of the following accidents as having occurred from the said defect in the rifle:

While attacked by the Cheyenne Indians in the Fall of 1874, on the north fork of Smoky Hill River, one Charles Brown, a hunter, lost his life by his inability to force a cartridge into his rifle—which was one of Sharps breech-loading rifles—soon enough. The Indians succeeded in taking his life before he could put his rifle into a condition to fire, the rifle being found after his death with a portion of the cartridge exhibited out of the gun, showing the fact of this difficulty in placing it in.

Also, that one McLaughlin, a hunter of experience, was severely injured in the hand and lost one finger by reason of the explosion of a cartridge while he was in the act of forcing it in, requiring the aid of a stick.

That one Henry Campbell and another James Campbell, both hunters, were injured in the hand and eye, while in a similar act as above mentioned, and that he has knowledge of many others injured thereby, and that he has experienced considerable trouble with the rifle in that respect.

This deponent further says that he has lately been shown an improvement in the said Sharps Breech-Loading Rifle, made by F. W. Freund, of Cheyenne, Wyoming, it having an improvement of the breech block which avoids the danger of all such accidents, and that he admired the said improvement very much, for which said improvement this deponent is informed that the said inventor, F. W. Freund, has applied for a patent; and this deponent says that it is his opinion that no man using rifles will want any other kind after once seeing this improvement; that great credit is due the inventor, which all hunters will appreciate.

J. A. MELINE.

Subscribed and sworn to before me, this }
11th day of April, A. D. 1876.

E. P. JOHNSON,
Notary Public.

[SEAL]

21

A Sharps Rifle Model of 1874,

with Freunds' Improved Breech Mechanism and Patent Front Sight, together with Freunds' Patent Detachable Pistol Grip; also, Key and Cleaning-out attachment. Hand-made butt. Stock weight from 10 to 16 pounds.—Price for work of this kind (first-class), from $100 to $150, according to finish required, including full set of reloading apparatus.

An Explanation.

Many persons complain of the Hammerless gun, many others inquire as to what the objections are to this kind of weapon, and hence we have thought it best to state here briefly what the objections are. Several objections are made, but the main reason why the new Sharps hammerless gun is practically of little value, is that after firing a few shots it is almost an impossibility to force the cartridge into the chamber. The objection, of course, applies to the old Sharps rifle unless it has the Freund Improvement attached. On account of the breech frame being so narrow, the defect in the new rifle is the greater of the two. Again, as we have already remarked, the breech mechanism of the hammerless gun cannot be removed for the purpose of oiling and cleaning without a great deal of trouble and with the aid of tools and implements; added to these objections is the further one, that of uncertainty as to whether the gun is cocked or not, which fact, as it is easy to understand, would be liable at times to lead to serious results.

We refer our readers to the following letter of Mr. S. Bock, Veterinary Surgeon of the 5th Cavalry regiment, who has had much experience, knows what he wants, and does not consider himself fully armed and prepared for all emergencies unless he is armed with a Sharps rifle with Freunds' Improvement attached.

ANOTHER ENDORSEMENT.

The firm of Freund Bros. received the following letter yesterday from a gentleman who has had a Sharps rifle in their hands:

HEADQUARTERS 5TH U. S. CAVALRY, }
Fort D. A. Russell, Wy., May 16. }

MESSRS. FREUND BROS. Cheyenne, Wy.:

Gentlemen,—The Sharps rifle, model of 1878, altered by you, is a perfect success. Your new breech-block works to perfection. I fired sixty-five shots in ten minutes without cleaning, the cartridges being more or less dirty, nor did any shell stick in loading or being expelled, the

that just as soon as the patch becomes marred, torn or defaced, the exact accuracy of the ball when fired is gone, and instead of the patch being the means by which the accuracy of the shot is attained, it has an adverse effect and largely contributes to destroy it. For fancy firing, when the opportunity exists to clean out the gun, etc., every time a shot is fired, the patch ball would perhaps do better service; but for rough experience on the plains and in the mountains the patchless ball is far preferable.

Shells.

For the very same reasons stated elsewhere as demonstrating the fact that the small calibre gun must necessarily be a longer range weapon than the large calibre, we also hold that the long bottle-neck shell is the best one, into which the ball can be fixed by the manufacturer of and dealer in ammunition.

On account of the contraction at the neck of the shell, or rather at the end into which the ball is fixed; the force of the explosion is thereby concentrated into a smaller compass, which necessarily throws the ball from the gun with more force and power than would otherwise be the case.

Gun Sights—The New Freund Invention.

In regard to gun sights, we have this to say: for the reason that distances in the Far West are so deceiving (on account of the high elevation and thin atmosphere) nearly all of the Eastern-made guns have to be re-sighted before they can be used. The globe and peep sights both have to be taken off as being impractical.

Some time ago the Freund Brothers invented and have since been manufacturing and attaching to long-range fire-arms, a new sight, which is the most perfect of anything in that line ever invented or adjusted to a gun, and all those who have tested this new invention agree with what we have stated.

The sight is made of fine steel and is spring tempered so as to give it strength and elasticity. It *may* bend but it will not break with ordinary usage, neither will it stay bent.

It has a white centre line and dark shoulders, so that under all circumstances the gun will show plain in taking aim, and is now being brought into general use throughout the West.

---

old trouble of forcing the shell in with the thumb being done away with. Trouble was always had on account of the cavity in the breech being so narrow that it made it difficult to force all of the cartridge into the barrel, especially if the shells were not perfectly clean.

Yours respectfully,

S. BOCK,

V. S. 5th U. S. Cavalry.

The following, however, more fully explains the rapidity with which firing can be done with the Freund Improvement attached to the Sharps rifle. The extract is from the Cheyenne *Sun* of May 2d, 1878.

Sharps Creedmoor Rifle Model of 1874,

WITH FREUNDS' NEW BREECH MECHANISM.

Applied for $25 to $35, in addition to the factory price of the gun.

FAST SHOOTING.

Yesterday Mr. F. W. Freund tested, at Fort D. A. Russell, a Sharps hammerless rifle of the model of 1875, which he changed to Freund Bros.' new improved breech movement. He fired fifty-four shots in five minutes, taking the gun from his shoulder, loading and sighting it at every shot. He used Government cartridges, which were taken without care or selection, and the exhibition was satisfactory in every respect, being admired by all present. This gun is made with safety trigger, which has to be moved at every shot in order to release the proper trigger. The breech movement is changed on the same principle as in the old Sharps to force the cartridge into the barrel, which it is almost impossible to do by the pressure of the thumb after the gun has been fired a few times, but goes easily into place by Freunds' improvement.

About Ammunition.

We desire to say just a word in reference to the different kinds of ammunition now in use, or rather we will refer to that which we consider the best for use in long-range firing. The most practical ball for frontier purposes, is the patchless ball, that is taking into consideration the way which most of our hunters and sportsmen carry their ammunition, to wit, in a belt around the waist; they are certainly entitled to the preference.

The reason why the patch ball is defective, or practically so, is

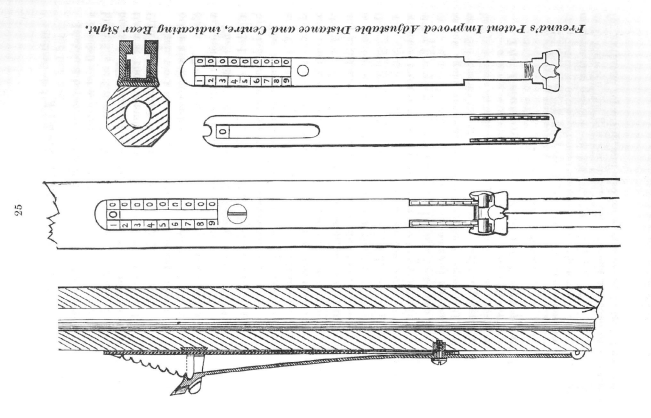

Freund's Patent Improved Adjustable Distance and Centre, indicating Rear Sight.

The workings of this sight are somewhat as follows : the gun is, in the first place, generally sighted so as to shoot with a very fine bead a distance of one hundred yards, and when the patent sight is used with the white centre line, the rear sight does not need to be

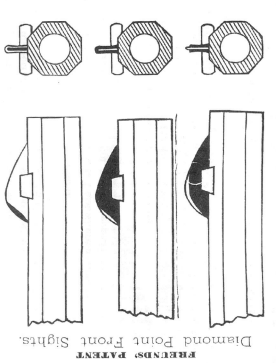

FREUNDS' PATENT

Diamond Point Front Sights.

rammed up so as to point out the distance of three or four hundred yards. The sight can be taken coarser even clear down to the barrel. A great saving of time is thus effected.

Rear Sight.

They have also invented and are now making a style of rear sight which is usually called the "Spring Sporting Rifle Sight," which is the quickest and readiest adjustable gun sight ever put upon a rifle. It is so arranged that it indicates the different distances by means of notches, and has a guide preventing the bending or breaking of the sight, and which also keeps it in its seat. These sights are made of the buckhorn shape, or any other style desired.

The readers of this pamphlet are referred to the following cut of the sight, which will give them a correct idea of this ingenious invention.

26

Miscellaneous Memoranda.

It should be borne in mind by all that the Freund Brothers make a specialty of repairing and remodeling the latest and most valuable kinds of fire-arms. No matter how valuable the fire-arm, or how much it has cost the owner, it can be repaired or remodeled by them at their stand in Cheyenne and come out a better weapon than when it was left with them for repair.

Their work in this respect, and of this particular kind, as well as other kinds of work done and performed by them, is always first-class. They do no other kind, and keep themselves aloof as much as possible from any other class of work.

Their skill and efficiency as repairers of valuable fire-arms is alluded to in complimentary terms by Gen. Crook, as will be seen by his letter in another part of this pamphlet.

The Freund Brothers were educated for their work in Europe, and they probably to-day have no rivals that can compete with them in skill and efficiency, and not only this, but they have a full and complete assortment of scientific instruments by which alone the work of repairing and manufacturing fire-arms can be tested by any satisfactory criterion. They are able to convince any one who is familiar with their class of business that with these instruments they can measure and test any work in their line scientifically and with complete accuracy. Hence, when the question is asked as to whether the Freunds can take a four or five hundred dollar gun, of either European or American manufacture, that has a broken stock, or an injured barrel, or is in any wise damaged, and make it as good as new again, you may safely answer the question in the affirmative, and the reasons for their so doing are clearly apparent from what has just been stated.

The Relations of the Freund Brothers with F. A. Dammann.

Mr. F. A. Dammann, the gentleman who succeeded to the general business of the Freund Bros., is also thoroughly posted in all matters that pertain to his line of business. The relations that exist between this gentleman and the Freunds are such that orders for work and repairing, and also for anything else in their line, may

be sent to F. A. Dammann, and if so, they will reach their destination and receive attention just as promptly as though they had been sent direct to the Freunds. On the other hand, orders to F. A. Dammann for arms, ammunition, and everything else in his line of business, can be sent in the first instance to the Freund Bros. It is at the Dammann establishment that all work done by the Freunds is tested.

Mr. Dammann keeps on hand a full and complete assortment of guns, pistols, revolvers, rifles, ammunition, cutlery, etc., etc., and his place of business is at the old stand of the Freunds, on Ferguson street, Cheyenne.

Further Testimonials and Evidence in Regard to the Freund Improvements, etc.

The following from the Packmaster at Gen. Bradley's Camp at Sun Dance Hill, Wyoming, speaks for itself. Mr. Delaney has had a large and varied experience in the Far West, having participated in nearly all of the Indian wars that have raged in the territories for the past twenty years.

SUNDANCE HILL, JUNE 12TH, 1878.
GEN'L BRADLEY'S CAMP,
EXPEDITION DEADWOOD, DAKOTA.

Messrs. FREUND BROS., Cheyenne, Wyoming:

Gentlemen,—I have been for a long time engaged on the frontier, and for many years I used a SHARPS RIFLE, but always found considerable difficulty, while out, in forcing the cartridge into the barrel of the rifle, after firing it a few times, and thereby permitting the chamber to become foul, and cartridges that I reloaded without a swedge, I could not use at all.

I have had my rifle changed to your new improved breech movement, and had it with me and in constant use all through the Indian trouble in the Big Horn country, and find that I have entire confidence in my rifle now. I, like all other persons that are engaged in out-door life, depend on my rifle a great deal of my time for subsistence, and often for life. I hope it will not be long ere the Sharps Rifle, with the Freunds improvement, will be the only gun called for and in use on the frontier.

Yours respectfully,

WILLIAM DELANEY,
Packmaster of Bradley's Command.

Mr. Morris Appel, an old and experienced marksman, a well-known freighter and a reliable gentleman in all respects, also gives us his opinion.

FORT LARAMIE, Wy. Ter., June 1st, 1878.

Messrs. FREUND BROS., Cheyenne, Wy.:

Gentlemen.—I have been for many years engaged in the freighting business, and while on the road am at all times handling arms. The first time that I really felt lost, with a good gun in my hands, was in 1876. I had a good Sharps rifle, but had used up all the factory cartridges, and after reloading them I attempted to load my rifle, and found it impossible, as I could not force them in. I am not a mechanic, but still I know that there was a defect in *that* gun. Since having my rifle changed to your system, it gives me perfect satisfaction in every particular, and I shall be pleased to make my ideas of the gun known to all freighters and hunters of my acquaintance, and recommend it for their use.

Very respectfully yours,

M. APPEL.

Mr. W. C. Irvine, one of Wyoming's "cattle kings" and a gentleman of large experience, speaks as follows:

Messrs. Freund Bros., Cheyenne, Wy. Ter. :

Ogallalla, Neb., January 21st, 1878.

Dear Sirs,—For the last five years I have used the Sharps rifle, shooting the 40 cal. bottle-neck shell, and must say that it is the strongest shooting gun, and therefore the best gun for the plains.

The objections I had against this style of gun you have entirely done away with, by applying your breech mechanism thereto.

The last gun with your improvement which I bought of you, is a gun of which it can truly be said, that it is the best gun known up to this date, and I feel secure with this rifle in hand against any danger, even in the roughest handling of the same, and therefore highly recommend the same to those that pass through countries where a reliable gun is their best friend.

Yours very respectfully,

W. C. IRVINE.

Mr. A. Schoomaker, an old frontiersman, also says a word as follows:

Messrs. Freund Bros., Cheyenne, Wy. Ter. :

Evanston, Wy. Ter., July 28th, 1878.

Dear Sirs,—The 40 cal. long-range rifle I received of you is the strongest shooting gun I ever saw. We gave it a splendid trial on July 4th, when several of us shootists went up to Echo, and were surprised over the accurate shooting of the gun, and the truly wonderful working of your breech mechanism. We tried the gun over 1,600 yards, and found that it carried up accurately, and all present came to the conclusion that the rifle with your improvement was a great achievement.

The wedge motion of the breech-block gives your rifle at least 100 per cent. advantage over the Sharps perpendicular motion in sending the cartridge home. Your extractors fall back further, thereby clearing the shell entirely from the chamber, and are entirely covered, so as to exclude all sand or dirt.

We made a splendid score at a bull's-eye. 10 inches square, 300 yards off hands with your gun, three of us, shooting five shots each, making nine bull's-eyes out of fifteen shots, at 1,600 yards, with peep sight. We made some excellent shots; the most of them would have killed a deer, as they struck in a space the size of one.

The Editors *Evanston Age,* also the *Argus* and *Ogden Freeman,* of Utah, were there. These gentlemen, and all others present, expressed themselves in the most favorable terms, and agreed that your gun is the most perfect of all they ever saw used on trial shooting.

Yours, &c.,

A. SCHOOMAKER.

The following we consider convincing evidence of the strength and durability of the Freund's mechanism when applied to the Sharps rifle. This evidence is from F. A. Dammann, the successor to the Freunds in the general business.

TO ALL WHOM IT MAY CONCERN.

[*Daily Sun, June 20, 1878.*]

To illustrate the perfection of the Freund Bros. mechanism, when applied to the "Sharps rifle," the following trial was made yesterday, June 19th:

To find out whether there would be any possible chance for a cap to burst, so as to throw the gas rearward, (which has been such a serious defect in the Sharps rifle), and also to show the resisting strength and perfect operation of the improved rifle, a 40-calibre Sharps rifle was loaded with 150 grains powder and three bullets, Sharps make, swedged and patched, each weighing 370 grains, making 1,110 grains of lead, and fired. No defect in the mechanism was shown, the cap did not break, and not a particle of gas escaped rearward. After this trial, without cleaning the gun barrel or breech, ten shots were fired in less than a minute, and conveniently taking aim from the shoulder, with 90 grain cartridges, and the result was that the last as well as the first cartridge was inserted and extracted without any difficulty, and I am convinced that if a thousand shots should be fired out of the rifle without cleaning the same, there would be no more difficulty encountered. Therefore I would say to those that wish their rifles changed, that every rifle will undergo the same test, and in their presence if desired.

The Freund Brothers have explained to me the merits of their improved firing-pin, which was patented January 2d, 1877, and have convinced me that it is an impossibility for a cap to break, and it also has an additional cover on the breech-block, closing the same gas tight rearward. A rifle with this improved breech mechanism is in my opinion superior to any other gun, and I therefore would recommend it to all who desire a perfect and reliable arm.

F. A. DAMMANN,

Wyoming Armory.

Cheyenne, W. T., June 19, 1878.

SHARPS HAMMERLESS RIFLE MODEL OF 1878,

with Freunds' Breech Block (top view),

showing the cartridge projecting. Price for altering to this system, $25.00.

The following is from the Deadwood *Times,* the leading paper of the Black Hills, and our readers will see from this what is thought of the Freunds inventions, etc., in "the Land of Gold:"

Mr. Geo. Freund, of the firm of Freund & Bro., Cheyenne, has arrived in the city with a full and complete sample of arms, ammunition, etc. Mr. Freund is at Stebbins. Wood & Post's bank, where he may be found ready to take orders for anything in the gun line. The past reputation of Freund & Bro. throughout this section of the country is a sufficient guarantee that everything coming from their hands will be first class in every respect. Mr. Geo. Freund has with him the latest improved Sharps rifle, which has commanded the admiration and excited the curiosity of all throughout the West. Military men and hunters alike pronounce it the best gun in the world for army and frontier service. The daily *Tribune* of Denver and the Cheyenne *Sun* concur in the same opinion. The improved Sharps does away with the defects of the old Sharps breech, while every important feature of the old model is preserved. The gun as constructed under Freund's improvements, offered all the requisite facilities for forcing the cartridge in the chamber when projecting rearward some distance, and the hammer is carried to half-cock by either opening or closing the breech. In addition to this, Freund has two other improvements which are attached to this rifle: The combination front sight, showing dark on light objects and light on dark, and the detachable pistol grip.

After thoroughly testing the matter of using a plain or patchable bullet, Freund & Bro. have adopted with great success the plain and unpatched explosive bullets in the use of their new gun. This does away with the patch, which hunters and others have always found difficult to preserve intact. To prevent the patching from getting crumpled or worn by carrying, has always been a source of trouble and care to the hunter and others. In using the unpatched ball, no trouble of this kind is experienced, while the quality of shooting is just as good. It is impossible to give a minute and detailed description of the arm, in question, owing to our limited space, but let it suffice to say that the gun, in our opinion, is far superior to any other gun in use for hard service, and all will find it worth their while to call on Mr. Geo. Freund, who will be pleased to show the arm and give the necessary explanations.

The following from the Denver *Tribune*, published at Denver, Colorado, shows what is thought of the Freund Improvement in the Centennial State:

In a recent issue of the Denver *Tribune* we find the following notice of an improved Sharps rifle, the work of Mr. F. W. Freund, of this city:

"On calling upon Mr. J. P. Lower at his gun store yesterday, a *Tribune* reporter was shown the latest improved Sharps rifle, which all who have seen it have pronounced the most perfect arm that it has ever been their pleasure to examine. The improvements are the inventions of Mr. F. W. Freund, formerly in the gun business here, but now at Cheyenne. To say that the improvements are very valuable would be simply to reiterate the conclusions of the many who were present and examined the arm with the reporter. It certainly deserves at our hands more than a passing notice.

"The rifle as changed under these improvements, does away with all the objectionable features of the old Sharps breech system, leaving only the good, the perfect resisting qualities and movements undisturbed. This is done, too, without the addition or use of a single extra piece. The improvement on the breech action so modifies the old model Sharps as to let the breech block swing back when out of the direct line of resistance, affording ample space to admit the use of double extractors, permitting the cartridge to project rearward some distance, which on the closing of the breech is gradually and with ease forced into the chamber. This is accomplished by the same movements of the lever as under the old system, and the cartridge, too, being pressed over the breech-block is held in position by an eccentric, so that the gun, held in any position, will not cause the cartridge to drop out of the gun before the breech is closed.

"Another improvement consists of a very simple change in the construction of the firing pin, so that by either opening or closing the breech the hammer is carried to half-cock or into the safety notch, thus preventing the frequent premature discharge of the arm occasioned by the closing of the breech with hammer resting on the firing pin.

"Lastly we come to two more patents. The detachable pistol grip and the combination front sight. The pistol-grip is made so that it can be put on any kind of an arm without the least difficulty and gives to the gun a more complete and symmetrical appearance, and to the operator better facilities for taking an accurate aim. The combination front sight is made so as to form nearly a perfect pin head with a white centre and dark shoulders, which shows light on dark objects and dark on light objects. To go into the minute details and point out the many defects of the Sharps rifle would be both tedious to the reader and a waste of time and space. For every one who has ever used a Sharps rifle, knows wherein its chief defects lie, and will consequently appreciate above all others the improvement which affords the requisite facilities for forcing the cartridge into the chamber without even marring the strength of the old Sharps breech system, for which it has a world-wide reputation."

Here is further testimony on the subject from the Cheyenne *Daily Sun* of August 23d, 1877, and as will be seen by a perusal of the Freunds' same, that Gen. Sheridan was struck at the first sight of Freunds' improvement with its efficiency and value:

GENERAL PHIL. SHERIDAN'S RIFLE.

We called upon Messrs. Freund & Bro. at their armory yesterday, and were shown General Phil. Sheridan's fine rifle which he carried with him during his late trip through the Big-

Horn country with General Crook. The arm in question is one of the Sharps Rifle Company's make, and is gotten up in the finest and most tasty manner possible, and without question is as fine a weapon as that company is capable of constructing.

General Crook is also the possessor of an arm of the same manufacture, which is equally as fine as the Lieutenant General's. General Crook, however, had Freund & Bro. attach their improved breech system to his rifle prior to his trip with Sheridan through the Big Horn regions. General Sheridan, after seeing General Crook's gun with the improvements thoroughly tested under all circumstances, fully recognized their value, and as a further evidence of his implicit belief in the new breech system, has left his fine Sharps rifle with Freund & Bro. to have them attach their new breech and half-cocking systems.

The Generals both unite in the opinion that Freund's new system of breech is an invaluable improvement to the Sharps rifle, and obviates the objections which have always been raised against the arm whenever laid before military men for examination and adoption into the army. What is true, however, as regards the value of the improvements for army service, they are equally valuable for hunters and frontiersmen, who, like the Generals, are loud in their praises of the improvements. In fact the want of the hunter, frontiersman, and army are identical. They want a strong, effective and durable weapon. The Sharps, with the Freund improvements, supplies the want, and we cannot but commend the good judgment of Generals Sheridan and Crook in seeing and recognizing the value of the improvements in question. The gun must on its merits take the place of the old and defective style now in use by the army, for in our opinion it is the only perfect gun in the world.

We conclude our list of testimonials with the following from Harry Yount, one of the most renowned, experienced and daring Indian fighters and scouts that ever drew a bead on an Indian or roamed over the plains of the Far West, not excepting Buffalo Bill. Harry Yount has for years been out with scouting and surveying parties, and no man in the Far West knows better the use or value of an efficient rifle than he:

THE SHARPS RIFLE—FREUND'S IMPROVEMENT.

WHAT AN OLD HUNTER SAYS ABOUT IT.

I am a hunter by profession, if such an occupation can be dignified with such a title. As it requires large and varied experience, coupled with no small degree of skill, acquired at the expense of arduous and perilous labor, to entitle one really to such a distinction, I think I may classify my occupation among the profession.

I have hunted on "the plains" and in the Rocky Mountains for seventeen years. The objects of pursuit have varied with the seasons through all the gradations of game, from the "jack-rabbit" to the grizzly bear and buffalo. I hunt not only for sport, but also for the purpose of pecuniary gain; so that I am interested in all the latest improvements in gun manufacture, and endeavor to keep posted in regard to the most important changes. Knowing the requirements of a good weapon, I can easily distinguish between the catch-penny patent (styled an "improvement") and an alteration or addition which accomplishes a valuable purpose. Of course, I am speaking more particularly with regard to the demands of the frontiersman and mountaineer. What would answer admirably in "the States" for shooting quail, geese and ducks, would be poorly adapted to an encounter with a mountain lion or a grizzly bear. I have hunted and killed not only these, but also antelope, deer, elk, buffalo and the "big horn," or mountain sheep, which last are so numerous in some parts of our country that I have ranged over as to give geographical name to an extended and remarkable range of mountains—the Big Horn mountains. I have hunted from Montana to Arizona, through all the intervening territory. Nor do I confine myself at all to Spring and Summer, but am engaged as assiduously in the depth of Winter, killing game and curing large quantities of meat, which I haul to market in the Spring.

Many men call themselves hunters who go out only in fine weather, in the Fall, for "sport." Of course, such are easily deceived by the *appearance* of a gun, not requiring, as I do, to know the points of excellence or weakness. My gun is my best friend, serving as my means of subsistence and being my most reliable dependence in the moment of danger; so that, as another one loves his horse or his dog, I love my gun, and when I shoot a good one I know it as well as the manufacturer. I have tried quite a variety in my experience—the muzzle-loader, the fine English breech-loader, the Winchester, needle-gun and the Remington; but the best, most complete and reliable gun I ever used is Freund's improvement on the Sharps rifle—in short, it is the "chief" for the Rocky Mountain region, and all hunters should have one at whatever cost. I consider Freund's breech the finishing touch on the Sharps gun, forcing in the cartridge with the greatet facility, and, at the same time, half-cocking the piece. With its double-extractor, and its perfect adjustment and finish, excluding all dirt, there is nothing more to be desired. Being closed in the rear, and operating without any obstruction and with unerring precision, it obviates the necessity for cleaning until an almost unlimited number of discharges have been made. The needle-gun was a vast improvement over the now discarded muzzle-loader, but the weakness and imperfection of the breech mechanism have rendered them unreliable, and even dangerous to the operator. For the purpose of consuming the old stock of muzzle-loading barrels, it was an excellent device, but since heavier charges are demanded they have become unserviceable.

HARRY YOUNT.

Why the Freund Brothers do not Undertake the Manufacture of the Gun Entire.

The question is sometimes asked "Why do not the Freund Brothers manufacture the gun entire, inasmuch as they are making so many and such valuable improvements to arms that are already in use?"

We propose not only to answer this question, but we propose to go a little further and state to the readers of this pamphlet what the Freunds can and possibly will do in the near future in reference to the manufacture of the most complete fire-arm that the world has ever seen. In answering the question as to why the Freunds do not engage in the manufacture of the entire fire-arm, all that need be said here is that it takes a large capital to carry on a business of that kind, and the firm have not at present the means to engage in an enterprise of such magnitude. In regular manufactures, of course, everything is made by machinery, and that too, of the costliest kind, such as requires a very large outlay of capital to procure. Hence it will be easily understood upon a moment's reflection why the firm does not engage in the manufacture of fire-arms. They make their own improvements, of course, and do it all by hand, but to undertake to construct the entire piece in this manner would, in any view that might be taken of the matter, be absolutely impracticable, especially as the prices are now, for they could not compete with machine manufactories.

They claim that they could, with proper facilities, manufacture a more complete and effective fire-arm than has ever yet been produced in America, without infringing in the slightest degree upon any patent that has ever been issued from the United States Patent Office. They claim that they can manufacture a rifle that would combine all the good qualities of the Ballard and the Sharps rifle (old and new model), and yet without any of the defects or imper-

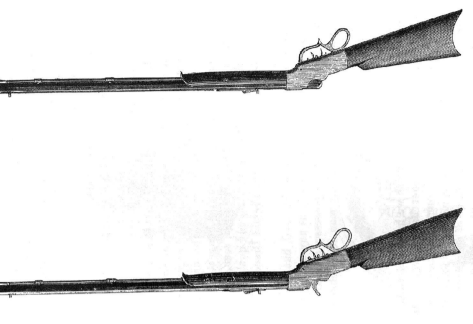

fections of either, and this too without any infringement upon the patents that have ever been issued to any of the inventors or the manufacturers of those fire-arms. They already have patents of their own, covering all of the different parts of the proposed fire-arm, from which they could make up a gun such as we have spoken of. They would not in any event be compelled to rely upon anybody

FREUNDS' New Interchangeable Rifle.

else, or to tread on any other man's reservation or exclusive rights. The facts are that not a single invention has come out for years past emanating from any other source that they have not improved upon and obtained a patent for, and they now have valuable improvements on the principal fire-arms in use. In fact, they have already invented this superior gun, and before this pamphlet comes before the public they will have obtained their patent for the same. Cuts of this new and superior rifle appear on page 33.

This rifle, as before intimated, combines all of the good points and elements of the old and new Sharps rifle, and also of the Ballard. It can, at the option of the owner, be changed from an open to a concealed hammer rifle, from a half-cock to a full-cocked fire-arm, and reverse in both cases, and in either case a double or single trigger can be used. In fact, this rifle is an interchangeable, interconvertible fire-arm, such as has never been invented, seen, or heard of before, and in addition to its other good parts and qualities it combines in its mechanism and make up the Freund Breech Block Improvement.

This gun was, of course, made entirely by hand, but it could be manufactured by machinery just as cheaply as any of those which are now in use. This rifle would be to its possessor several different kinds of fire-arms combined in one. Every man could change it to suit himself and his own fancy.

Conclusion.

We have now concluded the task which we have undertaken, and we hope and trust that our labors have not been in vain. It has been our aim throughout to present to our readers some matters and suggestions upon the invention, manufacture, improvement, and right use of fire-arms, such as would be of service to them in the future, and also to direct the attention of the public to the source from whence the best inventions and the most valuable improvements in fire-arms have emanated.

We submit the facts and suggestions which we have made, and caused to be set forth in the foregoing pages, to the candid consideration of the public at large, confidently believing that their verdict will be in our favor, and that the stamp of approval will be imprinted upon the pages of this pamphlet as being the embodiment of undeniable truths and a candid statements of facts.

35

The silver medal, of which the foregoing cut is a correct representation, was offered by the Colorado Industrial Society in 1873 as a prize for the best gun made within the borders of the State (then Territory). The Freund Bros. won the medal. This medal is a beautiful and valuable one, and is a fitting tribute to the skill [and mechanical ingenuity of this well known firm, and it is the only one ever offered or awarded as a prize for gun making in that State up to the present time.

WYOMING GAME AND FISH LAW.

An Act Entitled an Act for the Protection of Game and Fish in the Territory of Wyoming.

Be it Enacted by the Council and House of Representatives of the Territory of Wyoming:

SECTION 1. It shall be unlawful for any person or persons to kill or offer for sale any elk, deer, antelope, buffalo or mountain sheep, or young of either kind, between the first day of March and the fifteenth day of August, in each and every year; *Provided,* That it shall not be unlawful for any person or persons to kill enough of the animals aforesaid to supply their own immediate wants.

[margin: Time when unlawful to kill or sell the elk. etc. Proviso.]

SEC. 2. It shall be unlawful for any person or persons to kill, trap, or ensnare any pheasant, quail, prairie hen or prairie chicken, or sage hen, in this Territory, from the first day of March until the fifteenth day of August in each and every year.

[margin: Time when unlawful to kill quail, etc.]

SEC. 3. It shall be unlawful for any person or persons, at any time during the period when game may be killed, to kill or take a greater amount than can be disposed of to advantage or profit.

[margin: Unnecessary killing forbidden.]

SEC. 4. It shall be unlawful for any person or persons to take, ensnare or trap trout in the waters and streams of this Territory otherwise than by hook and line or such other mode by which fish can be singly obtained.

[margin: Manner of taking trout]

SEC. 5. Any person who shall violate any section or sections of this act, shall forfeit and pay the sum of fifty dollars, one-fourth to go to the informer and the remainder to the public schools within the county where the offense was committed, to be recovered in any action of debt, in the name of the complainant and the people of Wyoming Territory.

[margin: Forfeiture of fifty dollars.]

SEC. 6. Non-payment of the fine and penalty of this act shall subject the offender to sixty days imprisonment in the common jail, or place of imprisonment belonging to the county in which the provisions of this act have been violated.

[margin: Non-payment.]

SEC. 7. This act shall take effect and be in force from and after its passage.

Approved first December, 1869.

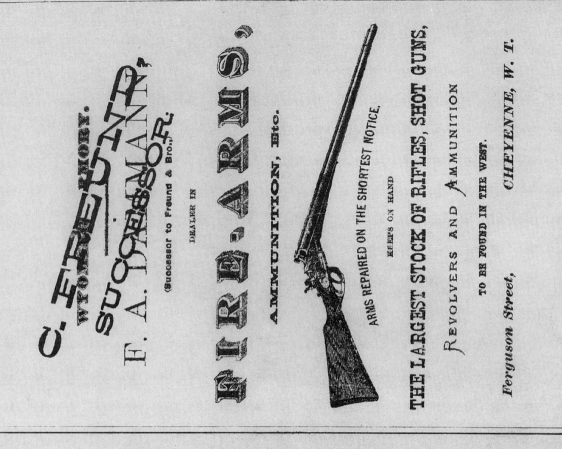
An Act for the Protection, and to Prevent the Destruction of Wild Game in the Territory of Wyoming.

Be it enacted by the Council and House of Representatives of the Territory of Wyoming :

Time in which the killing of elk, deer, etc., is unlawful. SECTION 1. It shall be unlawful for any person or persons to kill or destroy any elk, deer, mountain sheep, or antelope within the limits of this Territory, between the fifteenth day of January and the fifteenth day of July henceforth.

Said animals to be killed for food only. SEC. 2. And it shall be unlawful for any person or persons to kill or destroy any animal named in the foregoing section, at any time, for any purpose, or under any pretext whatsoever, except for food, and then only when necessary for human subsistence, governed in amount and quantity by the reasonable necessities of the person or persons killing such animal.

Waste of any part prohibited. SEC. 3. And it shall be unlawful for any person or persons killing any animal mentioned in the first section of this act, to let or cause any portion of such animal to go to waste and destruction except the entrails, head and shanks ; otherwise, every portion of such animal, when killed (unless diseased and unfit for food), must be used and converted into necessary food for human subsistence.

Speculation in such animals forbidden. SEC. 4. That no such animals as are named in this act shall be killed and used for speculative purposes, beyond a reasonable home market supply. for food purposes only.

Penalty for violation of provisions of this act. SEC. 5. That a violation of any of the provisions of this act shall be punished by a fine of not less than fifty dollars, nor more than five hundred dollars, or by imprisonment in the county jail (belonging to the county in which the provisions of this act have been violated) not less than thirty days nor more than six months ; or by both such fine and imprisonment, in the discretion of such court.

Disposition of fines collected under this act. SEC. 6. The informer of any and all violations of this act shall be entitled to receive one-half of any and all fines collected under this act. The other one-half of all such fines, after deducting the costs of the prosecution, shall be paid into the county school fund, and be used for school purposes only. And in case said informer shall not demand one-half of said fine, then the whole of the said fine so received (deducting costs), shall be paid into said school fund.

Conflicting acts repealed. SEC. 7. All acts and parts of acts in conflict with this act are hereby repealed.

In force. SEC. 8. This act shall take effect and be in force from and after its passage.

Approved December 7th, 1875.

And Frontiersmen of the Far West, all say the same.

Military Men of Highest Rank, Long Range Shootists, Miners, Stockman, Tourists

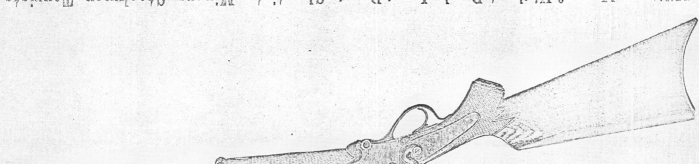

Acknowledged to be the Best in the World.

The F. W. Freund's Patent Rifle

Awarded Silver Medal

to

F. W. Freund,

for the

Best Rifle,

at Denver, Colorado.

THE IMPORTANCE OF FIRE-ARMS

IN GENERAL.

The importance of fire-arms is now well understood throughout the civilized world, and the more efficient the weapon the better it is appreciated by the people. In olden times warfare was conducted and carried on between hostile nations by means of the spear, the club, the rude sword and the battering ram; fire-arms were unknown.

The cannon, the rifle, and the revolver were unheard of, and as yet not invented, and when Rome was in the height and zenith of her glory, her soldiers, when the strife of battle came, relied more upon their physical strength and endurance to enable them to win victories over the enemies of the "mistress of the world" than they did upon their skill and prowess. For centuries the world was ignorant of that which is now patent to all, namely, that there is a more efficient and yet less barbarous method of carrying on war, than the system which was in vogue among the ancients. The discovery and invention of fire-arms has proved a blessing rather than a curse to the human family, and the whole course of events in modern times tends to prove and establish this fact.

Fire-Arms in America.

It can be truthfully said that in all things that pertain to the invention, improvement, and manufacture of fire-arms, America to-day takes the lead. This is evidenced by the fact that the European nations send their purchasing agents to our ports whenever the storm cloud of war lowers in the Eastern horizon. It was more on account of the superior efficiency of American fire-arms that the Turks were enabled to hold at bay so long the mighty hosts of Russia in the late war, than from any other cause or reason.

The American improved fire-arms was a more potent agent in the cause of Turkey than the "Banner of the Prophet."

Fire-Arms in the Far West.

While America has of late years become as it were one vast arsenal, yet for obvious reasons there are portions of our country where much more

TO THE PUBLIC.

Attention is directed to the fact that for the improvements on articles as represented and shown in this pamphlet, Letters Patent of the United States were granted solely and exclusively at the times and dates mentioned to F. W. Freund. Also, that the Wyoming Armory is the only place where these improvements are made, and that no right, privilege or permit has been granted or given to any person or persons to manufacture such improvements; that any one representing themselves as patentees, or in any way imitating such improvements are impostors, and not entitled to the patronage or recognition of the public.

F. W. FREUND,

Wyoming Armory, Cheyenne, W. T.

84 STEVENS AVE.

IMPROVEMENT OF FIRE ARMS,

Produced by Practical Observation of the Defects of the Different Styles of Breech Loading Rifles, used on the Western Frontier for the last 18 years, up to the present day.

Sectional view, position and construction of Breech when open, and ready to force the cartridge home by closing.

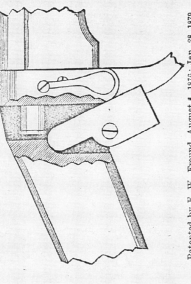

Patented by F. W. Freund, August 4, 1876; Jan. 28, 1879.

Showing Breech closed with square backing against the charge.

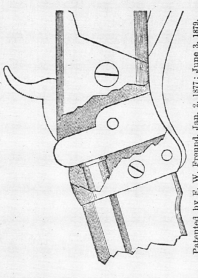

Patented by F. W. Freund, Jan. 2, 1877; June 3, 1879.

Freund's Detachable Pistol Grip.

Adapted to the U. S. Officer Springfield Rifle, and applied by the National Armory.

Patented by F. W. Freund, April 20, 1875; and April 20, 1877; also Oct. 19, 1875.

The Grip can be applied to any Gun having no Grip and can be sent by Mail.

attention is paid to the subject of fire-arms than in other localities of our broad domain. We allude more particularly to the Far West. In this region nearly every man is a gunner, and the boys as well as, in a vast number of cases, the women, well understand the use of the shot gun and rifle, and many of them are experts in their use.

There are, of course, good and valid reasons for this, which are well understood west of the Missouri River. In the first place it behooves the dwellers upon our exposed frontiers to be at all times prepared to repel the incursions of the ever-treacherous red men, for the ranchmen, the stock growers, and the freighter, can never tell at what hour of the day or night his enemy will come, and, as a general thing, when the danger arises he is compelled to rely mainly on his own skill and efforts to protect his wife, family, home and property from the savage foe.

Often too, the pioneer is compelled to rely on his trusty rifle to procure food for a time for himself and those who are dependent on him for their support. There are many other reasons which might be mentioned as tending to establish the fact that the people of the Far West are unusually proficient in the use of fire-arms, but it is sufficient to remark, in conclusion, upon this point, that what men are necessarily compelled to learn for their own safety and self-preservation, they generally learn well and thoroughly, and, when once learned, it is not easily forgotten.

Nor is this all. They are also turning their attention to the invention, improvement and manufacture of fire-arms. Taking it all in all it can be truthfully said that nowhere in America is there more attention paid to this subject than here in the territories, and nowhere are the people, as a general thing, more competent to judge of the merits or demerits of all kinds of fire-arms now in use in this country than those who dwell here upon these plains, and within the shadow of the "Rockies."

Among those who have turned their attention to this subject, and who have made it a specialty to improve by invention and otherwise the efficiency of American fire-arms, Mr. F. W. Freund, of Cheyenne, Wyoming Territory, stands among the first and foremost. His wonderful inventions, of which mention will be made further along, bid fair to revolutionize and overturn the entire system of the manufacture and use of nearly all kinds of fire-arms.

In presenting to the public my new improvement rifle, I take occasion to thank my customers and friends for their kind and unprejudiced examination and criticism of this new invention. The favorable consideration with which this arm has met from all who have taken sufficient interest in it to examine and test its merits, is due to the fact that a sportsman, or any one skilled in the use of fire-arms, recognizes at once in the perfection of its movements and the singular adaptation of its several parts to the accomplishment of the design intended, all the requirements of a perfect gun.

4

THE F. W. FREUND'S PATENT RIFLE.

Exhibited in the Wyoming Section of the Exposition, at Denver, Colorado, 1882, in Finished and Rough state; and American Institute, New York, 1883.

Six years in constant rough use all over the West. Acknowledged to be the most perfect Long Range Breech Loader by the highest Military and Civil Authorities, Experts, Hunters, Sportsmen, Miners throughout the country, and the pet of the Stockmen. The Rifles are always made to order, and adapted to any desirable charge of powder up to 150 grains.

In addition to the standard Rifle, manufactured at the Armory, other styles of Rifles, single or double barrel Rifles, or Rifle and Shot combined, or interchangeable, may be made to order to any style or action to suit the customer, and some of the valuable improvements in fire arms, ammunition, sights, etc., given below, patented by Mr. F. W. Freund, may also be applied thereto.

Improvement in Breech Loading Fire Arms.........Oct. 13, 1873.
    "   "   "   "   "   April 20, 1873.
    "   "   "   "   two patents...Nov. 7, 1876.
    "   " in Revolving Fire Arms.............Oct. 17, 1876.
    "   " in Cartridges....................March 16, 1876.
    "   " in Primers, for Cartridges........Nov. 29, 1876.
    "   " in Rear Sights, for Fire Arms.....March 16, 1875.

THE PERFECTION OF SIGHTS.

A Rear Sight for everybody, changeable to different styles of Sights, as well as to Dark or Light, by the new adjustment. Patented to F. W. Freund, Cheyenne, Wyoming Territory, Nov. 28, 1882.

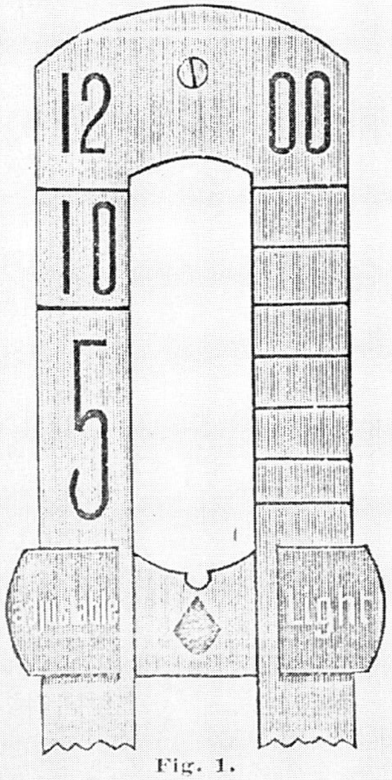

The New and only perfect Field and Target Sights in the world.

Commence supplying with 1883.

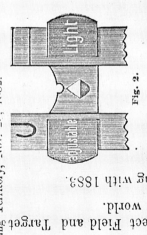

Fig. 1.

Fig. 2.

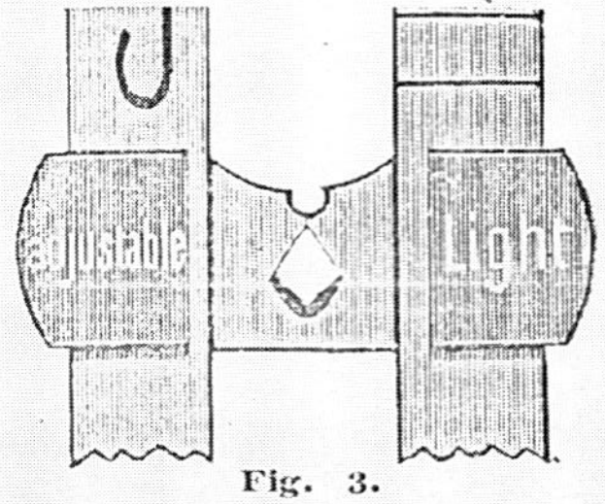

Fig. 3.

THE PERFECTION.

THE NEW REAR SIGHT, ADJUSTABLE TO DARK OR LIGHT, BELOW THE SIGHT NOTCH.

The annexed cut represents the rear sight closed, and when used by its operator it may be changed to such style as to suit himself for his own use and according to the Light.

This Sight has lately been introduced, and proves to fill a long felt vacancy. It is now put on all the leading rifles with the greatest success. It excels all other kinds of sights, for hunting, sporting and frontier use, beyond all comparison. Some of the best experts wonder why this sight had not been invented long ere this. Gun makers and sportsmen have felt the necessity of different open sights, but all their shaping and constructing amounted to the same old sights. In ordering sights without having the guns, it would be well to send the old sights of the guns as guides, giving a description of the gun, length of barrel, caliber and amount of powder the gun shoots. Any person can put these sights on themselves, providing they are for standard Rifles, and follow directions.

One of the main advantages of the rear sight can be adjusted to light or dark as may be required, below the notch (and this has not before been explained in print), is that the eye has a clear track. In the old sight the center of the eye's rays occurred in the notch. The lower half of the rays were interrupted at the near sight below the notch, because the rays could not penetrate the solid matter there ; of course the result was that the eye did not receive a clear image, but the notch was always blurred. With Freund's rear sight, however, the rays form a perfect focus in the eye, because of the open space above and below the notch. Light comes in on all sides ; the rays are therefore perfect, and there can consequently be no blurring whatever. Again, the bead is under perfect control ; should it chance to fall below the notch it can readily be seen through the angular shaped opening and at once adjusted in the notch.

They can be sent by mail to all parts of the country. Price, $10 per set. $5.00 extra for Wind Gauge, only made when ordered. Extracts of letters from persons who have used these sights are given herein, in which the famous patent front sight and the angular opening in rear sight have been used.

P. S.—When using the above sights be sure to direct the aim so that the TOP of the BEAD will appear over the top sight notch, and NOT THROUGH THE ANGULAR OPENING.

THE SIGHTS FOR ALL

THE ADVANCED INVENTION IN GUN SIGHTS

OF THE NINETEENTH CENTURY.

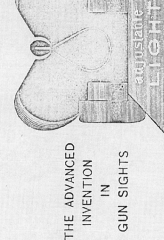

Fig. 4.

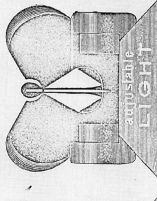

Fig. 5.

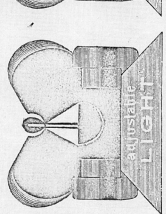

Fig. 6.

Since the introduction of what is known as the

FREUND'S PATENT DIAMOND SHAPE REAR SIGHT,

which heretofore has been made of but one sized opening below the Sight-notch, I have found it necessary, because of the difference in strength and weakness of eyesight of different persons, to modify and change the angular opening below the Sight-notch, so as to adapt it to all the differing eye-sights of sportsmen. This change in the size of angular opening is effected by means of an adjustable sliding plate, so arranged as to make

the opening of a size suitable to any eye-sight, and can be made in an instant of time, and without use of any instrument.

This Sight is now constructed under my new and improved plan so as to be changed from the Sight without any opening to a sight with full angular opening below sight notch, and of either triangle or diamond shape, as will be seen by reference to above cuts.

Figure 1 represents the graduated leaf sight closed. Figure 2, the graduated leaf sight adjusted so as to form a triangular opening below sight notch. Figure 3, the same sight adjusted to make a full diamond opening. All of these changes are made by simply moving the adjustable sliding plate.

In the lower cuts, which give a view of both front and rear sights when taking aim, the value and working of the sliding adjustable plate is seen. Figure 4 represents angular opening in rear sight entirely closed. Figure 5, same sight with triangular opening, the sliding plate being slightly moved so as to make opening of any size and dimension. Figure 6, the same rear sight with sliding plate shifted to make full diamond opening under sight notch.

The value of this improved adjustable slide will be readily understood and appreciated by all, for the reason that one sight is made to conform to the needs and requirements of the differing degrees of eye-sight, as also to the changes of light and darkness. On a clear, bright day a much smaller opening of sight will suffice than on a dark or cloudy day.

I have now a sight which I can highly recommend to all persons, as meeting a long felt want, and its simplicity, together with easy manner of adjustment, render it desirable for everybody; and patented November 28, 1882, by F. W. Freund.

Rear View of Patent Front Sight.   Side View of Patent Front Sight.

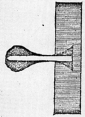

This sight is made of fine steel, and is spring tempered so as to give it strength and elasticity. It may bend but it will not break with ordinary usage, nor will it stay bent.

It has a white center line and dark shoulders, so that under all circumstances the sight will show plain in taking aim, and is now being brought into general use throughout the West.

The workings of this sight are somewhat as follows: The gun is, in the first place, generally sighted so as to shoot with a very fine bead a distance of one hundred yards, and when the patent sight is used with the white center line, the rear sight does not need to be rammed up so as to point out the distance of three or four hundred yards. The sight can be taken coarser even clear down to the barrel. A great saving of time is thus effected. It combines four different styles of sights in one, and is patented by F. W. Freund, April 17th, 1877.

The patentee and inventor and manufacturer of Freunds' Patent Improved Rifle desires it to be understood that this is no "fancy" gun, and that buyers pay only for genuine merit. The improvements are entirely made by hand by the most skilled mechanics. The breech and other parts are wrought out of solid steel forgings, care being taken to manipulate them so as to retain their excellent quality throughout the process of manufacture. At the same time it will be observed that the manufacturer has succeeded admirably in retaining the utmost strength, with excellence of design and beauty of finish. In order to furnish a cheaper rifle, the manufacturer uses most of a Sharp's rifle, introducing all the features of his improvements, such as the breech mechanism, firing pin, extractors and patent sights, etc., as described below; but in Freund's own make all parts are new and all made by hand. The work is thoroughly finished, all the parts fitting together and adjusting perfectly. Buyers of these rifles do not pay for useless ornamentation; their money is for hard, honest, skillful and necessary labor. On account of the extensive demand the manufacturer has endeavored to have these guns made economically in the East, but in no case was he able to do so, as the cost was higher than in Wyoming, while the workmanship was not satisfactory to the patentee.

During recent years there has been extensive rivalry among makers of magazine and other breech loading rifles, but the palm has been awarded, by competent persons, to the long range breech-loader, and why? Because it has a heavy breech and can accommodate extra heavy charges of powder; it is always operative, being simple and strong in its mechanism, and it never fails to work. This is especially true of Freunds' Patent Improved Rifle, which is absolutely without equal in the world. No other long range rifle combines the advantages of strength, beauty and lightness of this rifle. Expert hunters and tourists, most of whom have tried various guns, prefer Freunds' Improved Rifle, because they find that this long range rifle is also the best at short range, being even more certain in hitting the object desired. There have been efforts made to imitate legitimate long range guns by "express" rifles. These imitations, while they shoot heavy charges of powder, absolutely fail to carry the necessary weight of lead; indeed they employ hollow bullets. But in Freunds' rifle the weight of powder supports weight of lead, so that when the missile arrives at its destination it is heavy enough to do the work required, which is not the case with "express" bullets. As a matter of fact the makers of "express" rifles dare not use heavy bullets in small bores, with a large amount of powder, because it would become unsafe in the breech to do so.

In ordering goods, sufficient money should accompany the order to cover Express charges both ways; also, attention is called to give name and place plain. Often, we receive letters, but no one can make out the name, and we cannot be held responsible for any delay. Send your orders at an early date to avoid delay.

In Freunds' Rifle there is no clubbing home of cartridges, after the gun has been fired, which has led to so many serious accidents, and often proved very unfortunate in critical moments. No rearward escape of gas to injure the eyes. No premature discharge by loading or carrying the rifle, and no hammerless, no working loose of breech parts, and no flaying out of breech block.

The Rifle is extremely simple, and its parts very durably constructed. The breech mechanism can be taken apart and put together in a moment's time, without the use of instruments or tools. The breech block has a rocking motion, thereby forcing the cartridge home by the closing of the breech. It has double extractors. The lock is thrown back at rest or half-cock by the positive motion of opening or closing the breech. The operation of the breech mechanism is done by the guard lever, giving the greatest power, and places the hand to the trigger by the closing of the breech. The breech block has its resistance in a right angle with the line of the bore, and direct in rear of the charge. The inside of the barrel can be seen and cleaned from the rear. The firing bolt and breech block are so constructed as to prevent breaking of caps and rearward escape of gas, and exclude all dirt, thereby preventing obstructions and misfire. Single or double triggers can be used. The guns are made, usually, 40 caliber, 90 grains, and 45 caliber, up to 150 grains of powder. It has a pistol grip and a cleaning rod of hickory wood, with wiper and brush in butt of stock, and the New Adjustable Patent Sights. The Rifle is made of the very best material and in the most careful manner, is severely and thoroughly tested, and the price, including all the above, is one hundred and sixty ($160) dollars, but are made up to $300, according to extra work, engravings, fine cases, fixtures, etc.

In ordering Rifles the payment of $25.00 should accompany the order from parties not known. Satisfaction guaranteed in every respect. Extracts of letters given herein.

11

EXTRACTS FROM LETTERS.

[What Gen. Ph. Sheridan says.]

HEADQUARTERS MILITARY DIVISION OF THE MISSOURI,
Chicago, January 23, 1878.

The Sharp's rifle, in which, at my request, you inserted your patent breech-block, pleases me very much, and the more I see of it the more strongly am I impressed with the practical ideas you have developed in your invention.

[What Gen. Geo. Crook says.]

HEADQUARTERS DEP'T PLATTE, COM'DG GENERAL'S OFFICE,
Omaha, Neb. April 30, 1878.

I have received the Sharp's rifle left in your hands for repairs and alteration. The double ejector spring and the rounding of the inner edge of the breech-block are improvements which only need to be seen by practical men to commend themselves for general use. The beauty and perfection of workmanship in this, as in other repairs I have had made at your establishment, are entitled to the highest praise.

[From a letter from Mr. Zimmermann, a gun maker of long and practical experience in Europe and America.]

DODGE CITY, KAN., March 2, 1880.

From the first breech-loading needle gun of Prussian manufacture to the very last of the new patterns at to-day, your improved Sharp's towers above them all, where the points of value are settled. An arm, to be effective these days, must possess the highest grade of shooting qualities, combining therewith these three other imperative qualities, to-wit: rapidity of action, durability and safety. The improvement of yours on the Sharp's covers all these requirements. It combines all these points, which thousands of gunsmiths have tried to harmonize and failed. The arm pleases every one who sees it, and attracts the undivided attention of stockmen, officers of the army and hunters.

Maj. Smith, of Fort Dodge, is well pleased with the rifle you sent him. All the officers at the Fort examined the arm, and were gratified with it in every particular. The breech mechanism was, they said, well adapted to hard service. They praised the fine workmanship upon the arm, and expressed themselves astonished that the Sharp's Rifle Company did not adopt the improvement at once, instead of making the hammerless gun now turned out by them, which, in everybody's opinion in this section, is as worthless as it is hammerless.

[From an old Buffalo Hunter—an old frontiersman.]

While attacked by the Cheyenne Indians, in the fall of 1874, on the north fork of Smoky Hill River, one Charles Brown, a hunter, lost his life by his inability to force a cartridge into his rifle—which was one of Sharp's breech-loading rifles—soon enough. The Indians succeeded in taking his life before he could put his rifle into a condition to fire, the rifle being found after his death with a portion of the cartridge exhibited out of the gun, showing the fact of his difficulty in placing it in.

Also, that one McLaughlin, a hunter of experience, was severely injured in the hand and lost one finger by reason of the explosion of a cartridge while he was in the act of forcing it in, requiring the aid of a stick.

That one Henry Campbell and another James Campbell, both hunters, were injured in the hand and eye while in a similar act as above mentioned, and that he has knowledge of many others injured thereby, and that he has experienced considerable trouble with the rifle in that respect.

This deponent further says that he has lately been shown an improvement in the said Sharp's breech-loading rifle, made by F. W. Freund, of Cheyenne, Wyoming, it having an improvement of the breech block which avoids the danger of all such accidents, and that he admired the said improvement very much, for which said improvement this deponent is informed that the said inventor, F. W. Freund, has applied for a patent; and this deponent says that it is his opinion that no man using rifles will want any other kind after once seeing this improvement; that great credit is due the inventor, which all hunters will appreciate.

Subscribed and sworn to before me this 11th day of April, A. D. 1876.

[SEAL.]

E. P. JOHNSON, Notary Public.

J. A. MELINE.

[What Mr. John P. Lower says, in his letter March 3, 1880.]

I consider the model of 1874—Sharp's rifle with your improved breech action—as being the most perfect and reliable arm to be found anywhere.

New York, Oct. 17, 1879.

[From Messrs. Geo. Breed & Hogarth, exporters.]

The rifle arrived safely yesterday. We must say the improvement you have added to it increases its original value more than two-fold, and as our correspondent in Africa is an old hunter, he will at once appreciate your skillful and practical alteration; and we wish you the success which you so justly deserve.

[From Mr. John P. Lower, gun dealer, and member of Denver Gun Club.]

March 5th, 1880.

It was surprising what accuracy was obtained in shooting at a target off hand at 500 yards. W. J. Foy, the president of our club, who is second to none at long range shooting, tried it, and was well pleased with its performance.

Gen. W. E. Strong, of Chicago, president of the Peshtigo Lumber Company. That gentleman writes as follows, regarding the sights:

Chicago, Oct. 22, 1880.

I have just returned from a hunting expedition on the head waters of the Peshtigo River, in Northwestern Wisconsin, and during the trip I tested thoroughly your patent sights recently put on my Winchester rifle, model of 1873. In my opinion your sights are by long odds the best ever invented for hunting purposes. From my earliest boyhood I have been fond of a rifle, and have used

THE WYOMING SADDLE GUN.

40 Caliber; weight 5 lbs; shoots 70 grains of powder. Exhibited in the Wyoming Section of the Exposition at Denver, Colorado, 1882, in finished and rough state; American Institute, New York City, 1883, and to be on exhibition in the great World's Exhibition, London, England, in 1884. Patented. Designed and Manufactured at the Wyoming Armory by F. W. Freund, Cheyenne, Wyo. Ter.

The annexed cut represents F. W. Freund's Pat. Improved Rifle made entirely by hand. It has a low hammer and small breech frame. Weight from five pounds upwards. Any weight of gun or length of barrel furnished, with single or double trigger. In the 40 caliber guns, 70 to 90 grains of powder, and in the 45 caliber 70 to 150 grains of powder are used. It is entirely new in its design, being handsomely proportioned and cannot be excelled in these respects or in workmanship. In all cases the very best material is used, and they are adjusted and finished in the Wyoming Armory.

F. W. FREUND,
WYOMING ARMORY,
54 STEVENS AVE.,
JERSEY CITY, N. J.

Prices of F. W. Freund's Pat. Improved hand-made Rifle, $200 to $300.

one a great deal, but I never had a set of open sights that suited my fancy in every particular until I got yours.

On my recent trip, the first two shots I had was at a running deer; one struck the right fore shoulder, and the other passed through the deer's heart. I also fired at a target from 50 to 150 yards, and I was astonished and delighted at the almost perfect sighting of the rifle and the certainty with which I could strike the object aimed at up to 150 yards by taking a fine or coarse bead through the sight and without raising the graduated leaf.

I have shown the sights to a number of excellent hunters and expert shots, and all pronounce them the most perfect they have ever seen.

[From General G. Crook.]

HEADQUARTERS DEPARTMENT OF THE PLATTE, COMMANDING GENERAL'S OFFICE,
FORT OMAHA, NEB., Dec. 7, 1880.

The gun arrived yesterday, and after carefully examining it I take pleasure in saying that all the repairs have been made to my complete satisfaction, and in a manner to demonstrate not alone skillful workmanship, but a thorough knowledge of your business. Your patent sight is so valuable an improvement and yet so simple that I cannot help wondering why it was not sooner discovered. It serves equally well as a "peep" and "open" sight, and can be used without reference to the manner in which the sun may be shining. In my opinion it fills a want long felt, and beyond improvement. It only needs to be known to be adopted by all sportsmen throughout the country.

In a letter of Sept. 6th, 1880, Mr. Wm. Taylor, Quartermaster Agent of Rock Creek, W. T., says the following:

You are aware I have a beautiful rifle, but I am free to say, for all practical purposes and beauty your new sights have increased its value at least fifty per cent.

How annoying it is to have a good shooting rifle with the Old Blug sights over which it is necessary to have the sun and everything else favorable to enable you to see the game for a dead shot. I find it easy to catch the object over your sights in any direction and under the most unfavorable circumstances. No good shot, after seeing your new sights, would do without one.

[From a letter of Col. F. Van Vliet, Capt. Third Cav., Br't Lt. Col. U. S. A.]

FORT D. A. RUSSELL, W. T., December 30, 1880.

I used your improved sights in my last hunting trip and found them perfect, both for hunting and target practice. The front sight gives a perfect bead, and has the peculiarity of showing dark against light objects and light against dark ones. The rear sight is extremely clear, and is the only one I have ever seen that at the time of taking sight allows you a clear view of your game.

[The following from Major Lord, Quartermaster.]

DEPOT QUARTERMASTER'S OFFICE, CHEYENNE DEPOT, WYOMING.
November 20th, 1880.

Your patent sights which I have just put on my Winchester repeating rifle, 45 caliber, have given such perfect satisfaction, that I take pleasure in indorsing them as superior to any sights heretofore used by me.

U. P. RAILROAD HOTEL, CHEYENNE, W. T., Jan. 4, 1881.

Mr. C. F. Reed says: Am very much pleased with the new patent sights which you have recently adjusted to my rifle. For the past ten years I have been testing the different patterns of open sights, and I find that yours are the only ones that will not blur even when tried under unfavorable circumstances.

It has long been considered desirable to have a guide in the rear sight below the notch. This has been attempted by inlaying a line of white metal, but as the line was so much in the dark it was found to be practically useless.

You have at last accomplished the purpose by cutting a diamond-shaped opening immediately below the notch; the front sight forming the line surrounded by light, and the two angles of the opening together with the notch serving as perfect guides. The diamond-shaped opening exposes the front sight and also the object aimed at to full view, with a clearness and accuracy equalled only by the telescope. The front sight being made of spring tempered steel, is very strong, and by its peculiar construction shows white against a dark object and black against a light colored object.

I feel that my rifle is now as perfect a sporting gun as can be produced.

Mr. P. H. McEvoy, a gunsmith and dealer in arms at North Platte, Neb., writes as follows to a firm in this city:

NORTH PLATTE, Neb., December 15, 1880.

The rifle that I sent you to have your patent sights put on came back all right. I have tried the gun at all ranges, and I will say that they are the best I have ever seen, in fact they are the only ones that ever gave me any satisfaction, and I have been in the gun business for many years and used all kinds of sights. You could not buy mine for $25 now, if I could not get others.

[The following extract is from the Cheyenne *Daily Leader* of October 17th, 1879.]

A SUCCESSFUL HUNT.—The English gentlemen, Messrs. J. M. Michelson and J. Michelson, who left our city on the 7th of last May for a pleasure hunt through the country north of Laramie Peak, have returned to this city. These gentlemen are from Yorkshire, England, and came to this country to enjoy the sports of hunting antelope, deer, elk, bear and such other game as abound in this country. On their arrival here they were fortunate enough to secure the valuable services of Harry Yount, our well known guide and scout, who led them over the wild plains and back again to Cheyenne. They speak in the highest terms of praise of their hunting trip, and relate many incidents of the hunt both exciting and amusing.

They relate an instance of the killing of an old grizzly bear which they came upon near the foot of the mountain. He was a very large one, as his hide fully attests, and had evidently done a great deal of mischief. They state that the animal, after it was fatally shot, grasped with its mouth small trees measuring two or three inches in diameter, and severed them as quickly as though they were mere pipe stems. In its final death struggles it seized its own paw and crushed it to a jelly. When dressed they found two old rifle bullets, which had been apparently shot into him a long time ago, Each bullet, too, was from a different make and caliber of rifle, showing that he had been shot formerly by two persons. These gentlemen, however, armed with the F. W. Freund's Patent Rifle, found no difficulty in putting a quietus upon this savage animal, though in its death agonies it

exhibited its wonderful strength and the savage ferocity for which it is most to be dreaded of all the wild beasts found on the American continent.

MILES CITY, Jan. 10, 1881.

After much delay my gun was received on October 1, 1880. Owing to my great hurry to get out among the buffalo, I did not write to you as I should have done. I have killed over 700 buffalo since then with your gun, and it worked like a charm during the coldest weather. I have shown it to a great many hunters, and they are all very much pleased with it. You will soon receive more orders from this place.

Very respectfully,
D. H. BABBIT.

The rifle above referred to is Freund's 40 caliber.

The following names are selected out of a large list of our customers and given as reference, having used the rifle and sights:

PIERRE LORILLARD, New Jersey; FRED. GEBHARD, Union Club, New York; CLARENCE KING, 62 Centre St., New York; VAN BUREN, Union Club New York; PETERS & ALSON, Powder River, Wyo., late of New York; HENRY OELRICH, Powder River, Wyo., late of New York; CHARLES OELRICH, Powder River, Wyo., late of New York; L. MAYER, Powder River, Wyo., late of New York; FREWEN BROS., Powder River, Wyo., late of England; RICHARD ASHWART, Powder River, Wyo., late of England; L. SATERIS, Laramie City, Wyo., late of England; C. FINCH, Big Springs, Tex., late of England; JOHN A. BENSON, 507 Montgomery St., San Francisco, Cal.; F. B. BIDDLE, Powderville, Mont., late of New York; J. A. CHANLER, 55 W. Twenty-sixth St., New York; GEO. H. GOULD, No. 5 E. Twenty-sixth St., New York; Mr. GEO. M. GREER, 47 W. Fifty-second St., New York; E. REMINGTON, Jr., Ilion, N. Y.; General STRONG, Chicago; General GIBBENS, U. S. A., commander Fort Laramie, Wyo.; General MASON, U. S. A., commander Fort Russell, Wyo.; General PH. SHERIDAN, General GEO CROOK.

MR. FREUND, Gunmaker, Cheyenne, Wyo.:

LEIPSIC, OHIO, Aug. 30, 1883.

Dear Sir—The rifle you made for me gives perfect satisfaction. I am now glad I did not have a telescope put on the rifle. I can shoot very well with your patent sight. I am sixty-five years of age, and I believe that on a clear day I can keep within an eight inch circle, shooting all day. I have also written to Mr. Mayer, of Newport, who gave me your address, the result of such a good rifle. I remain yours, very respectfully,

JOHN KURTZ.

E. Remington, son of the great manufacturer of firearms, at Ilion, N. Y., whose name is famous throughout the world for gun-making, had a rifle made by Mr. Freund, of this city. This gun has all the latest improvements, including the sights. After a hunting trip he wrote to Mr. Freund as follows:

MR. FREUND:

BUFFALO, WYO. TER., Aug. 30th, 1883.

Dear Sir—"I consider your rifle and sights a great success." With best regards, etc.,

I am yours, respectfully,
E. REMINGTON.

[More about Freund's Patent Sights.]

Mr Freund, Cheyenne, Wyoming:

FORT D. A. RUSSELL, August 28, 1883.

DEAR SIR—I have tried your illuminated rifle sights placed by you on my Sharp's sporting and hunting rifle, by firing the rifle with a bright sun light shining in my face, on my back, across the line of sight from either side; also in firing in cloudy, dark weather; in the imperfect light of torch light; also firing at dark and light colored objects, and under all these diverse conditions they have given uniformly satisfactory results. They are correctly placed and wonderfully accurate, and seem to me to be about perfect in all respects. I would not give them up and attempt to do without them, or others like them, for three times the price charged.

Respectfully yours,
W. E. HOFMAN, Lieutenant 9th U. S. Infantry.

The Lieutenant has been for many years on the frontier, and knows the value of what is superior and practical.

C. FREUND:

September 20th, 1883.

Dear Sir—I have your improved rifle with new sights, and take pleasure in stating that it excels in every respect any and all of the rifles I have ever before used, and I have used all kinds of the latest magazine and breech-loading rifles. The sights are well adapted to the various changes of light. I cannot speak in too high praise of the skillful workmanship employed by you in preparing for me a rifle beautiful in design and most reliable in accuracy of shooting. Yours respectfully, etc.,

GEO. M. GREER, 47 W. Fifty-second St., New York City.

[The Middleton Affair.]

It will be remembered by the *Leader* that Detective Lykins, in his fight with the outlaw Middleton, snapped his rifle upon four different cartridges, and each one missed fire. Considerable surprise was manifested by all who read the account of the affair, inasmuch as it was thought strange that Mr. Lykins should provide himself with an arm or ammunition that should be the least defective. It seems not to have been his fault, however, for he had an arm which is reported to be reliable—the Sharp's rifle. Nor was it the fault of the cartridges, for he brought this ammunition to Mr. Freund, of this city, and introduced the cartridges into one of their improved Sharp's, and each one of the supposed defective cartridges was exploded. Thus it seems that it was the fault of the so-called "Old Reliable Sharp's," which in this case proved itself very unreliable, and came very near costing him his life, and in all probability was the means of the non-arrest and capture of one of the most desperate outlaws that ever traveled the plains. Mr. Lykins says he has enough of the "Old Reliable" now, and accordingly ordered one of Freund's Patent Rifles.

[Solid Shots—Perforating the target—Return rifle match between the Fourth Infantry and Wyoming rifle teams—The Wyoming wins.]

At the conclusion of the rifle match the members of the Fourth Infantry team announced that they were ready to challenge any team, provided they be allowed to use the Freund's improved rifle. They believe themselves the best marksmen, but their guns are faulty. It was conceded by everybody who witnessed the match that the rifle, with Freund's improvement, is the best arm in existence. As this is a Cheyenne production, it is a matter of pride to note the fact.—*Cheyenne, Wyoming Ter., May 16, 1879.*

14

The following testimonials are offered as those received within the last few weeks, when it became known that there was a possibility that the Freund Rifle was to be manufactured on a large scale, and thus be put upon the market at a price within the reach of all. These recommendations are but a small portion of what could have been obtained if more time could have been had in getting them, and are mostly given by customers in the immediate vicinity of Cheyenne, who have had a practical experience with the Freund Gun, and who have given them cheerfully and voluntarily. The Freund Rifle being so well recommended in the West, where it has every chance of being thoroughly and practically tested, need not fear anything from any other. Many of those who have given their recommendations to the Freund Rifle are the leading stock-growers in the world, and employ many men. All of these are anxiously waiting the time when the Freund Rifle can be manufactured at some prominent point in the East, where labor and material are cheap, and thus being able to produce a really valuable and superior rifle at a price within the reach of all.

The West has long since approved the single breech-loader as the best, and now experienced sportsmen and marksmen in the East, as well, endorse it as the best. It combines the elements of strength, simplicity and reliability, and peculiarly so in the Freund Rifle, which is emphatically the most perfect rifle in the world. The leading Powers of the world have adopted the single breech-loader for the military service. The Freund Rifle is the best and most perfect in existence, and to-day successfully challenges a comparison with any rifle made.

There are to-day in the hands of dealers and manufacturers a vast quantity of poor, inferior and antiquated guns, and the public is buying them because it can get nothing better. The reason is that the introduction of a new and better rifle, as the Freund Rifle, would cause all of the present stock, and many patented inventions, to be of little or no value. But it would certainly be a gain to purchasers to be furnished with a first-class rifle for the price they now pay for an inferior one.

It is not our purpose to assail or decry the business of anyone else, but it is contended that the Freund Rifle is the best, and facts are set forth to the world to that effect. If anybody don't agree with this statement, let them bring forth their proofs and arguments to disprove the claim.

The Freund Rifles are now on exhibition at the American Institute, in New York. Many of the original references from leading banks, merchants, stock growers, and leading men of this section of the country, can be seen at the exhibit of Richards & Co., in the Institute.

In presenting the above claims, facts have simply been stated, and with the greatest confidence in the merits of the case, it is believed that the final verdict must be unanimous in favor for the Freund Rifle. A good gun is always appreciated, and the one most used and sold.

With these few remarks we trust we have impressed the reader with some idea of our claims, and with the belief that they are true, hoping that he will add his influence in securing the results aimed at—the manufacture of this superior gun in large quantities, and at a low price.

Very respectfully yours,

F. W. FREUND.

15

CHEYENNE CITY, Oct. 10th, 1883.

The undersigned, after a personal and business acquaintance of several years with Mr. F. W. Freund, proprietor of the Wyoming Armory, and originator of many valuable and important improvements in fire-arms, has great pleasure in recommending both him and his inventions to the public.

I know it to be the opinion of prominent military men, sportsmen, hunters and ranchmen throughout the Rocky Mountain region, that the gun manufactured by him, and furnished with his improved sights, breech block, and double ejector, is without a rival, and that if manufactured on a large scale at some central point, it would soon come into very general use throughout the country. Besides being a successful inventor and skillful mechanic, Mr. Freund has an excellent standing as a competent, efficient and honorable business man, fully entitled to the respect and confidence of any and all persons who may have transactions with him.

JOHN W. HOYT,
Late Governor Wyoming Ter.

☞ The undersigned, having had personal knowledge of Mr. Freund's improvements and inventions in fire-arms, sights, etc., acquired by trial, use or inspection thereof, unite in strongly indorsing the merits and value of such inventions and improvements, and fully and freely concur in the above recommendations as will appear following:—

A. H. SWAN, } Vice President First National Bank, Cheyenne,
} President Swan Land & Cattle Company, limited.
N. R. DAVIS, Cattle and Horse Breeder, Owl Creek, Wyo.
JOHN B. THOMAS, of Thomas & Page, Stock Growers, Cheyenne.
JOHN HUNTON, Stock Ranch and Range, Bordeaux, Wyo.
JAMES DATER, New York City, Ranches on Cheyenne River, Wyo.
CHAS. F. MILLER, Stock Grower, Ex-Judge Probate, Laramie Co., Wyo.
E. W. WHITCOMB, Stock Grower and Merchant, Cheyenne, Wyo.
F. W. LAFRINTZ, Sec'y Swan Land & Cattle Company, limited.
WM. GUITERMAN, Stock Grower, Cheyenne, Wyo.
J. PALMER, Cattle and Horse Raiser, Cheyenne.
H. H. VEGNOLES, Stock Grower, Cheyenne.
EDWARD C. DAVID, Surveyor Gen'l, Wyoming Ter.
F. E. WARREN, { Fine Cattle and Horse Raiser, and
{ V.-P. First Nat'l Bank, Cheyenne, Wyo.

RICHARD FREWEN,
Manager of Frewen Cattle Company, of England.
I think Mr. F's Rifle about the best in America.

S. M. LEHMER,
Twin Mountain Ranch, Wyo.
Mr. Freund's patent rifle is the most salable rifle in the West.

RITCHIE BROS.,
4 B 4 Ranch, Little Powder River, Wyo. Ter.
Mr. Freund's rifle and sights have the best reputation in this country.

D. SHUDY,
Ranch Camp Clark, Nebraska.
I can highly recommend Mr. Freund's inventions and improved sights, as I have used them and found them all he claims for them.

J. H. PRATT,
Pres't Pratt & Ferris Cattle Company.
I have always heard Mr. Freund's rifle and sights spoken of in the highest terms.

{ Sec'y Wyoming Live Stock Association,
THOS. STURGIS, } Pres't Stock Growers' National Bank.
Freund guns are the best we ever used.

W. C. LYKINS, Cheyenne, late Stock Inspector of Wyoming Territory.
I consider Freund's rifle and sights the Boss.

E. W. MADISON,
Wyoming.
The Freund gun and sights are perfection.

R. B. ANDERSON,
Stock Grower.
Only try it.

A. W. HAYGOOD,
Ranch on Lone Tree Creek, Wyo.
Have used rifle for five years. 'Tis the best I ever had.

R. M. SEARIGHT,
Stock Grower, Cheyenne River.
Have known Mr. Freund for eight years; have full confidence in his mechanical skill and business integrity; have used his patent breech-block and sights since their inception, and have seen many of them in the hands of skillful sportsmen; know that the Freund Improved Rifle has never been equaled.

THOMAS MOORE,
Chief Packer Military Division of the Missouri, U. S. A.

J. H. GODDARD,
Foreman U. P. Ry Shops.
They (the Freund guns) are the favorite on the line of the U. P. Ry.

JNO. SPARKS,
Rancho Grande, Nevada.
The Freund Rifle is the best in Wyoming and Nevada.

W. A. WYMAN, M. D.,
Wyoming.
So say I.

THE HUNTER'S SIGHTS.

A SPECIALTY.   ·   **THE LONG FELT VACANCY.**

NO HUNTING RIFLE IS NOW COMPLETE WITHOUT THE

F. W. FREUND PATENT SIGHTS.

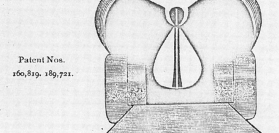

Patent Nos.
160,819. 189,721.

Patent No. 268,090.
AND
Patents applied for.

Front and Rear Sights as they appear while taking aim.

The Advanced Invention in Gun Sights of the Nineteenth Century.

The Sights for old and young, and the cheapest in the end. The best on horseback or a-foot for quick shooting. They do not blur and are clear and distinct under the most unfavorable light. The best for aged eye sight, and quick in catching to shoot at standing, running or flying objects, and will be a saving on ammunition, can be used later in the evening than the old style sights.

These Sights have lately been introduced, and prove to fill a long felt vacancy. They are now put on all the leading Rifles with the greatest success. They excel all other kinds of sights, for hunting, sporting and frontier use, beyond all comparison. Some of the best experts wonder why these sight have not been invented long ere this. Gun makers and Sportsmen have felt the necessity of different open sights, but all their shaping and constructing amounted to the same old sights. In ordering sights, without having the guns, it would be well to send the old sights of the guns as guides, giving a description of the gun, length of barrel, calibre and amount of powder the gun shoots. Any person can put these sights on themselves, providing they are for standard Rifles.

The *optical focus* of the rear sight is a *wonderful change* from the old style, and it is impossible for any one to illustrate or to understand its full value unless seen on a rifle, and tried under unfavorable circumstances.

The eye has a clear track. In the old sight the centre of the eye's rays occurred in the notch. The lower half of the rays were interrupted at the near sight below the notch, because the rays could not penetrate the solid matter there ; of course the result was, that the eye did not receive a clear image, but the notch was always blurred. With this new sight, however, the rays form a perfect focus in the eye, because of the open spaces above and below the notch. Light comes in on all sides; the rays are therefore perfect, and there can consequently be no blurring whatever. Again, the bead is under perfect control; should it chance to fall below the notch it can readily be seen through the angular shaped opening and at once adjusted in the notch, and quickly levelled to the object aimed

They are now used in preference to any other on the Plains and in the Rocky Mountains. Acknowledged to be the best in the world by military men of highest rank, long range shootists, miners, stockmen, tourists, and frontiers of the Far West.

In order to give the public at large the benefit of my Sights, and bring them within the reach of the masses, I am now furnishing three different grades, comparatively speaking, of different grades of guns, all good and first class, and of good material and same merit.

All my sights are of first class work and material, and under the same principle of my patent of rear sight, and "Patent of Muzzle Sight, and come as follows :

No. 1. First class work and material substantially as shown under my patent, front sight made of fine steel with bright projecting centre line, and dark shoulders. Price, $5.00 per set.

No. 2. Ditto nicely furnished, front sight with solid   projecting centre line. Price, $7.50.

No. 3. Ditto fine finished changeable adjustment on rear sight, front sight with projecting   silver centre line. Price, $10.00 per set.

These sights are marked according to the grade.

Any of the cheaper grades will be taken back in exchange for higher price sights at par, by paying the difference. Rear sights of standard make can be sent to me for alteration to my patent, in which case I will credit for sight what it is worth to me, or charge for the work according to the grade, thus one can be sure that the sight will make a good fit. Although I make different patterns to fit the leading rifles now in the market.

Whenever the rifle is sent to put the sight on, and the owner wishes to have it tried at the target, I will only charge for ammunition, if none was sent with the rifle. If any work or repair on the rifle should be desired, it can be done at the same time in the most skillful and experienced manner, and to those whom I am not personally acquainted, I would, in addition to reference state that I have spent a life time in fine art throughout gun making, and in the best fire arm factories in the world, such as Vienna, Paris, London, and in America. I have been located in the Western Frontier for some years, engaged in repairing fire arms from almost all parts of the world, and manufacturing fine rifles, and sights, and have now permanently located East for the reason of better facilities of labor and material to enable me to fill all orders at short notice, and at the lowest figures, and all work is done under my instruction, and goes through my hands before sending it to the customers.

Parties ordering rifles from the city or from factories, can order the guns to be sent to me for sighting and shooting at a target. In this case they should inform me directly, giving instructions.

THE INTERCHANGEABLE INTERCONVERTIBLE

NEW REAR SIGHT.

The above Sight is an *interchangeable and interconvertible Rear Sight* such as has never been invented, seen or heard of before; and in addition to its other many superior qualities combines and makes up *the best in the world*. By this new construction a Rear Sight is produced which can be instantly changed from one style to another without removing the Gun from the shoulder. It is furnished with two different Plates, by which it can be changed from a bright contrast to a dark shade Sight or reverse. Any Person ordering Sights can also send their own Design for which I charge 50 cents extra for each additional Plate. The Sights are made of different sizes according to the Rifle for which they are intented for. This new Sight is not alone very convenient but also economical if any further or additional changes should be desired.

All rear sights are attachable to the barrel, and are made to raise for longer distance shooting, and of two different styles hinge leaf, or step notch Spring sight, which latter is generally preferred by sportsmen for hunting sights, if constructed under my patents, which allows a much higher elevation on account of the double step being out of the centre, and in which the notches are so constructed that they will not move unless the sight is raised, this makes an elegant and quick adjusting rear sight, for sporting purposes, as shown in the cut above.

THE FAMOUS MUZZLE SIGHT.

The workings of this sight are somewhat as follows : The gun is, in the first place, generally sighted so as to shoot with a very fine bead a distance of one hundred yards, and when the patent sight is used with the white center line, the rear sight does not need to be rammed up so as to point out the distance of three or four hundred yards. The sight can be taken coarser even clear down to the barrel. A great saving of time is thus effected. It combines four different styles of sights in one, as follows, a pin head shape on light, a straight line on dark, light on dark, and dark on light.

When using the above Sights be sure to direct the aim so that the *top* of the *bead* will appear over the top sight notch, and *not through the angular opening.*

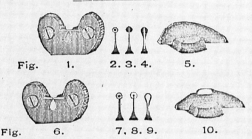

Fig.　1.　　2. 3. 4.　　5.

Fig.　6.　　7, 8, 9.　　10.

FIG. 1.—Is a rear view of rear sight with adjustment by a line of contrasting color; it can be made visible instantly, extending from the notch on either side on a horizontal plan for special use in dim light, cloudy weather or in the woods, and to show a contrasting color of light color when aiming against a dark object, or instantly changeable to dark color sight for use in bright sun light; it is entirely new and original and cannot be equalled. It is the admiration of the sportsmen of the day.

FIG. 2.—A special Target Bead of black centre and light color outside, by which it shows distinctly anywhere on a target.

FIG. 3.—A Sporting and Hunting Bead, black outside and of contrasting color when aiming against any object made medium fine.

FIG. 4.—Regular Hunting Bead of contrasting color, special for out door work.

FIG. 5.—Side view of bead and Block.

FIG. 6.—Rear Sight, showing opening and bright line.

FIGS. 7 and 8 —Rear view of Bead, with silver or ivory head of contrasting color

FIG. 9.—Hunting Bead; enlarged white centre.

FIG. 10.—Side view of Beads No. 2, 7 and 8, with Block.

Adjustable Peep Sight to Nos. 1 and 2, grade $2.50, for No. 3, grade $1.50 extra. Something new and good.

All Sights can be refinished equal to new at little expense. They are made of knife blade shape, best adapted for general use. All the above Beads can be fitted to the same Block, and changed without disturbing the Block from the Barrel. The Bead of contrasting color has never been equalled by any other device.

Price of Separate Beads, $2.50　　Block, 50 Cents.
A set of Sights consist of one Front and one Rear Sight.

These Sights are a production of years of patient and thorough study in the field and shop, by a practical Rifleman.

———————

The following names are selected out of a large list of my customers and given as references, having used the sights, and only a few of the letters are here given for want of space.

PIERRE LORILLARD, New Jersey ; FRED. GEBHARD, Union Club, New York ; CLARENCE KING, 62 Cedar St., New York ; VAN BUREN, Union Club, New York, PETERS & ALSON, Powder River, Wyo., late of New York; HENRY OELRICH, Powder River, Wyo., late of New York; CHARLES OELRICH, Powder River, Wyo., late of New York; L. MAYER, Powder River, Wyo., late of New York ; FREWEN BROS., Powder River, Wyo., late of England ; RICHARD ASHWORTH, Powder River, Wyo., late of England ; L. SATERIS, Laramie City, Wyo , late of England ; C. FINCH, Big Springs, Tex., late of England ; JOHN A. BENSON, 507 Montgomery St., San Francisco, Cal.; F B. BIDDLE, Powderville, Mont., late of New York ; J. A. CHANLER, 55 W. Twenty-sixth St., New York ; GEO. H. GOULD, No 5 E. Twenty-sixth St., New York ; Mr. GEO M. GREER, 47 W. Fifty-second St., New York ; E. REMINGTON, Jr., Ilion., N. Y.; General STRONG, Chicago ; General GIBBENS, U. S. A., Commander Fort Laramie, Wyo.; General MASON, U. S. A , commander Fort Russell, Wyo.; General PH. SHERIDAN, General GEO. CROOK, General WINGAT, 20 Nassau St, New York ; Mr. BISHOP, 15 Broad St., New York ; Mr. H. B. LECKLER, 107 Willow St., Brooklyn, N. Y.

———————

Mr. Freund.　　　　　　　　　　　　　　　　　　　　　　Fort D. A. Russell.

Dear Sir—I have tried your illuminated rifle sights placed by you on my Sharp's sporting and hunting rifle, by firing the rifle with a bright sun light shining in my face, on my back, across the line of sight from either side ; also in firing in cloudy, dark weather; in the imperfect light of twilight ; also firing at dark and light colored objects, and under all these diverse conditions they have given uniformly satisfactory results. They are correctly placed and wonderfully accurate, and seem to me to be about perfect in all respects. I would not give them up and attempt to do without them, or others like them, for three times the price charged.　　　Respectfully yours,

W. E. HOFMAN, Lieutenant 9th U. S. Infantry.

Gen. W. E. Strong, of Chicago, president of the Peshtigo Lumber Company. That gentleman writes as follows, regarding the sights :
Mr. Freund,

Dear Sir—I have just returned from a hunting expedition on the head waters of the Peshtigo river, in Northwestern Wisconsin, and during the trip I tested thoroughly your patent sights recently put on my Winchester rifle, model of 1873. In my opinion your sights are by long odds the best ever invented for hunting purposes. From my earliest boyhood I have been fond of a rifle, and have used one a great deal, but I never had a set of open sights that suited my fancy in every particular until I got yours.

On my recent trip, the first two shots I had was at a running deer; one struck the right fore shoulder, and the other passed through the deer's heart. I also fired at a target from 50 to 150 yards, and I was astonished and delighed at the almost perfect sighting of the rifle and the certainty with which I could strike the object aimed at up to 150 yards by taking a fine or coarse bead through the sight and without raising the graduated leaf.

I have shown the sights to a number of excellent hunters and expert shots, and all pronounce them the most perfect they have ever seen.

Mr. Freund,

Dear Sir—The rifle you made for me gives perfect satisfaction. I am now glad I did not have a telescope put on the rifle. I can shoot very well with your patent sight. I am sixty-five years of age, and I believe that on a clear day I can keep within an eight inch circle, at four hundred yards, shooting all day. I have also written to Mr. Mayer, of Newport, who gave me your address, the result of such a good rifle.

I remain yours, very respectfully,   JOHN KURTZ.

[From General G. Crook.]
HEADQUARTERS DEPARTMENT OF THE PLATTE, COMMANDING GENERAL'S OFFICE.

The gun arrived yesterday, and after carefully examining it I take pleasure in saying that all the repairs have been made to my complete satisfaction, and in a manner to demonstrate not alone skillful workmanship, but a thorough knowledge of your business. Your patent sight is so valuable an improvement and yet so simple that I cannot help wondering why it was not sooner discovered. It serves equally well as a "peep" and "open" sight, and can be used without reference to the manner in which the sun may be shining. In my opinion it fills a want long felt, and beyond improvement. It only needs to be known to be adopted by all sportsmen throughout the country.

Mr. Wm. Taylor, Quartermaster Agent of Rock Creek, W. T., says the following :
Mr. Freund,

Dear Sir—You are aware I have a beautiful rifle, but I am free to say, for all practical purposes and beauty your new sights have increased its value at least fifty per cent.

How annoying it is to have a good shooting rifle with the Old Blug sights over which it is necessary to have the sun and everything else favorable to enable you to see the game for a dead shot. I find it easy to catch the object over which your sights in any direction and under the most unfavorable circumstances. No good shot, after seeing your new sights, would do without one.

[From a letter of Col. F. Van Vliet, Capt. Third Cav., Br't Lt. Col. U. S. A.]
Mr. Freund.   Fort D. A. Russell, W. T.

Dear Sir—I used your improved sights in my last hunting trip and found them perfect, both for hunting and target practice. The front sight gives a perfect bead, and has the peculiarity of showing dark against light objects and light against dark ones. The rear sight is extremely clear, and is the only one I have ever seen that at the time of taking sight allows you a clear view of your game.

[The following from Major Lord, Quartermaster.]
DEPOT QUARTERMASTER'S OFFICE, CHEYENNE DEPOT, WYOMING.
Mr. Freund,

Dear Sir—Your patent sights which I have just put on my Winchester repeating rifle, 45 caliber, have given such perfect satisfaction, that I take pleasure in indorsing them as superior to any sights heretofore used by me.

Be careful of imitations; my Sights are all stamped with my name.

All orders should be accompanied with remittance to cover amount. Care should be taken to give name and address clear.

All orders from abroad will receive prompt and careful attention, and satisfaction guaranteed or money returned. Send your orders at an early date to avoid delay.

Send remittances by check, draft, P. O. order, or express, to my order, addressed as below.

It is very important in ordering to state if you are used to fine or coarse sights.

Packages from New York City can be sent by Zimmermann's Express, office, 84 Courtlandt Street.

Thanking for the past liberal patronage, and hoping for a continuance of the same, I remain, very respectfully yours,

F. W. FREUND,
JERSEY CITY, N. J.

P. S.—Since the publication of the above references, letters and testimonials of many eminent hunters and sportsmen from all parts of the country have been received, who having used these Sights, highly recommend them for their superior merits and fine workmanship, as well as the promptness and care with which each one has been suited to the eyesight and test; as well as other complimentary encomiums given in the past year in the leading sporting papers, all of which will appear in a new circular.

Colorado Armory.

Established—Denver, 1869; Durango, 1881.
COLORADO.

Geo. Freund, Prop'r.

CATALOGUE NO. 389.

Guns, Ammunition, Sportsmen's Goods,

PATENT SIGHTS, ETC.

INTERIOR.

Geo. Freund, Durango, Colo.,

Manufacturer, Importer and Dealer in Guns, Pistols, Ammunition, Gun Materials, Cutlery, Fishing Tackle and Sportsmen's Articles of all kinds.

First Street, Durango, Colo.   Price List on Inside.

Don't forget to tie a board on one side of all guns you send to us for repairs. It will save two-thirds of the express charges.

The Medal of Superiority

Awarded to

GEO. FREUND. ○ DURANGO. COLO.,

For Breech-loading Rifles, Sights and Ammunition.

MINERS, ATTENTION!

The most Complete and Successful Miner's Article ever Invented.

Can be carried in Vest Pocket. Knife with one large blade, cap-crimper, fuse-cutter and can-opener, $3.25. Without can-opener, $2.75, by registered mail on receipt of price. The entire Knife is made of Metal with Nickel Finish.

MINERS POCKET KNIFE
GEO. FREUND'S PATENT

REDUCED TO ¼ SIZE.

Address GEORGE FREUND,

Colorado Armory, Durango, Colo

Read this circular through before asking questions.

the amount of express charges both ways, which will be deducted from the bill when sent.

All bills not paid at maturity are subject to interest from time of purchase until paid, at the rate of twelve per cent. per annum.

All communications should be addressed to
GEO. FREUND, Durango, Colo., U. S. A.

P. S. Persons desiring sights will save time and trouble by carefully reading this circular as to particulars how to order. Care should be taken to give name and address clear.

All the component parts—springs, screws, hammers, stocks, etc., for the leading guns, rifles and pistols, kept on hand for repairs and sold at slight advance on factory prices. Persons wishing such will send broken parts, if possible, with correct name of maker of arm wanting same, which can be sent by registered mail at one cent per ounce, in addition to cost of registering, being the cheapest way to send small packages long distances. Postage stamps can be remitted for fractional parts of one dollar.

Read this circular through before asking questions.

No Hunting Rifle is now Complete Without the

Geo. Freund Patent Sights,

The Advanced Invention in Gun Sights of the Nineteenth Century.

Explanation and Description.

The sights for old and young, and the cheapest in the end. The best on horseback or afoot for quick shooting. They do not blur and are clear and distinct under the most unfavorable light. The best for aged eyesight, and quick in catching to shoot at standing, running or flying objects, and will be a saving on ammunition, and can be used later in the evening than the old style sights.

Read this catalogue through before asking questions.

✳ THE ✳

Old Reliable Gun House

OF THE WEST.

NOTICE.

This Catalogue contains only a partial list of the large and varied assortment I constantly keep in stock, yet you will find it contains as much or more than many other catalogues much larger in size. Anything in my line not mentioned will be furnished at bottom prices and satisfaction guaranteed.

REPAIRING

And Fine Gun Work and Sighting a Specialty.

I have a complete set of most approved tools of the latest pattern, and am prepared to accommodate my patrons needing repairs with skillful and first-class work. Having had over twenty years' experience in the work, I am thoroughly acquainted with the needs of my customers and defy competition in this line.

Headquarters for Freund's Patent "More Light" Sights.

REFERENCES—First National and Colorado State Banks, or any business house in the city, and the thousands of persons in the West with whom I have done business, from Omaha, Neb., to the Pacific coast.

Complying with numerous requests, I have permitted a few of the leading business houses in the city to insert their cards in this catalogue. I cheerfully testify to their merits and bespeak for them a share of the patronage from a generous public.

Wishing all a prosperous season,

I am respectfully,

GEO. FREUND.

TERMS CASH.

Any one desiring goods sent C. O. D. can be accommodated, providing a remittance accompanies their order to

These sights have lately been introduced, and prove to fill a long felt want. They are now put on all the leading rifles, with the greatest success. They excel all other kinds of sights for hunting, sporting and frontier use, beyond all comparison. Some of the best experts wonder why these sights have not been invented long ere this. Gun makers and sportsmen have felt the necessity of different open sights but all their shaping and constructing amounted to the same old sights. In ordering sights, without having the guns, it would be well to send the old sights of the guns as guides, giving a description of the gun, length of barrel, calibre and amount of powder the gun shoots. Any person can put these sights on themselves, providing they are for standard rifles.

The OPTICAL FOCUS of the rear sight is a WONDERFUL CHANGE from the old style, and it is impossible for any one to illustrate or to understand its full value unless seen on a rifle and tried under unfavorable circumstances.

The eye has a clear track. In the old sight the centre of the eye's rays occurred in the notch. The lower half of the rays were intercepted at the near sight below the notch, because the rays could not penetrate the the solid matter there. Of course, the result was, that the eye did not receive a clear image, but the notch was always blurred. With this new sight, however, the rays form a perfect focus in the eye, because of the open spaces above and below the notch. Light comes in on all sides; the rays are therefore perfect, and there can consequently be no blurring whatever. Again, the bead is under perfect control; should it chance to fall below the notch, it can quickly be seen through the angular shaped opening and at once adjusted in the notch and quickly leveled at the object aimed.

They are now used in preference to any other on the Plains and in the Rocky Mountains. Acknowledged to be the best in the world by military men of highest rank, long range shootists, miners, stockmen, tourists and frontiers of the Far West.

In order to give the public at large the benefit of my sights, and bring them within the reach of the masses, I am now furnishing three different grades, comparatively speaking, for different grades of guns, all good and first class, and of good material and same merit.

All my sights are of first class work and material, and under the same principle of my patent Rear and Muzzle Sight.

(8)

These sights are marked according to the grade.

Rear sights made in all styles to order, including Rocky Mountain Buckhorn.

Any of the cheaper grades will be taken back in exchange for higher price sights at par, by paying the difference. Rear sights of standard make can be sent to me for alteration to my patent, in which case I will credit for sight what it is worth to me, or charge for the work according to grade. Thus one can be sure that the sight will make a good fit, although I make the different patterns to fit the leading rifles now in the market.

Whenever a rifle is sent to put the sight on, and the owner wishes to have it tried at the target, I will only charge for ammunition, if none be sent with the rifle. If any work or repair on the rifle should be desired, it can be done at the same time in the most skillful and experienced manner; and to those whom I am not personally acquainted, I would, in addition to reference, state that I have spent a lifetime in fine art gun making, and in the best fire-arms factories in the world, such as PARIS, LONDON and AMERICA. I have now located on the Western Frontier for some years, engaged in repairing fire-arms from almost all parts of the globe and manufacturing fine rifles and sights, and have now permanently located at Durango, Colo.

The new and only original "MORE LIGHT" patent Gun Sights, with their improved styles and many additional superior qualities combined, makes it the best in use.

All rear sights are attachable to the barrel, and are made to raise for longer distance shooting, and of two different styles, Hinge Leaf or Step Notch Spring Sight, which latter is generally preferred by sportsmen for hunting sights.

A shooting range is connected with the establishment, where all guns purchased will be thoroughly tested before leaving the Armory. A great advantage to the purchaser of guns at this Armory is having the new sights put on and regulated.

Parties who purchase or order rifles from abroad and have patent sights adjusted at the same time, a discount will be made on sights from the regular price.

Any infringement upon these sights will be prosecuted to the full extent of the law.

(9)

— THE —
Famous Muzzle Sight.

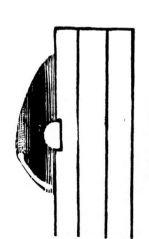

SIDE VIEW. **END VIEW.**

This sight is made of fine steel, and is spring tempered, so as to give it strength and elasticity. It may bend, but it will not break with ordinary usage, nor will it stay bent.

It has a white centre line, and dark shoulders, so that under all circumstances the sight will show plain in taking aim, and is now being brought into general use throughout the West.

The workings of the sight are somewhat as follows: The gun is, in the first place, generally sighted so as to shoot with a very fine bead a distance of one hundred yards, and when the patent sight is used with the white centre line, the rear sight does not need to be rammed up so as to point out the distance of three or four hundred yards. The sight can be taken coarser, even clear down to the barrel. A great saving of time is thus affected. It combines four different styles of sights in one, as follows: A pin head shape on light, a straight line on dark, light on dark, and dark on light.

WHEN USING THE ABOVE SIGHTS BE SURE TO DIRECT THE AIM SO THAT THE TOP OF THE BEAD WILL APPEAR OVER THE TOP SIGHT NOTCH, AND NOT THROUGH THE CIRCULAR OPENING.

The following names are selected out of a large list of my customers and given as references, having used the sights. Only a few of the letters are here given, for want of space:

Pierre Lorillard, New Jersey; **Fred Gebhard,** Union Club, New York; **Clarence King,** 62 Cedar st., New York; **Van Buren,** Union Club, New York; **Peters**

(10)

& Alson, Powder River, Wyo., late of New York; **Henry Oelrich,** Powder River, Wyo., late of New York; **L. Meyer,** Powder River, Wyo., late of New York; **Erewen Bros.,** Powder River, Wyo., late of England; **Richard Ashworth,** Powder River, Wyo., late of England; **L. Materis,** Larimie City, Wyo., late of England; **C. Finch,** Big Springs, Tex., late of England; **John A. Benson,** 507 Montgomery st., San Francisco, Cal.; **F. B. Biddle,** Powderville, Mont., late of New York; **J. A. Chanler,** 55 W. Twenty-sixth st., New York; **Geo. H. Gould,** No.5 E Twenty-sixth st., New York; **Geo. M. Greer,** 47 W. Fifty second st., New York; **F. Remington Jr.,** Illion, N. Y.; **General Strong,** Chicago; **General Gibbens,** U. S. A. Commander Fort Larimie; Wyo.; **General Mason,** U. S. A. Commander Fort Russell, Wyo.; **General Ph. Sheridan,** General Geo. Crook, **General Wingat,** 20 Nassua st., New York; **Mr. Bishop,** 15 Broad st., New York; Mr. **H. B. Leckler,** 107 Willow st., Brooklyn, N. Y.

FORT D. A. RUSSELL.

MR. FREUND:—

Dear Sir:—I have tried your Illuminated Rifle Sights placed by you on my Sharp's sporting and hunting rifle, by firing the rifle with a bright sunlight shining in my face, on my back, across the line of sight from either side; also in firing in cloudy, dark weather; in the imperfect light of twilight; also, firing at dark and light colored objects; and under all these diverse conditions they have given uniformly satisfactory results. They are correctly placed and wonderfully accurate, and seem to me to be about perfect in all respects. I would not give them up and attempt to do without them, or others like them, for three times the price charged. Respectfully yours,

W. E. HOFMAN,
Lieutenant 9th U. S. Infantry.

MR. FREUND—

Dear Sir:—The rifle you made for me gives perfect satisfaction. I am now glad I did not have a telescope put on my rifle. I can shoot very well with your patent sight. I am sixty-five years of age, and I believe that on a clear day I can keep within an eight inch circle at four hundred yards, shooting all day. I have also written to Mr. Major, of Newport, who gave me your address, the result of such a good rifle. I remain yours, very respectfully,

JOHN KURTZ.

FRANK KING, Chief Engineer D. & R. G. Ry., Denver, Colo.; **SAN JUAN HARDWARE CO.,** Durango and Silverton, Colo.; **LAWSON & BEILL,** New York City; **W. DAUGHERTY,** Capt. 22d Infantry, Fort Lewis, Colo.; **J. H. ERNEST WATERS,** Supt. Sheridan Mine, Telluride, Colo.; **HERR & HERR,** Coal Dealers, Durango, Colo.;

(11)

perfect satisfaction that I take pleasure in indorsing them as superior to any sights heretofore used by me.

Be Careful of Imitations. My Sights are all Stamped With my Name.

All orders should be accompanied with remittance to cover amount. Care should be taken to give name and address clear.

All orders from abroad will receive prompt attention, and satisfaction guaranteed. Send your orders at an early day to avoid delay.

Send remittances by draft, P. O order. or express, to my order, addressed as below.

It is very important in ordering to state if you are used to fine or coarse sights.

Thanking for the past liberal patronage, and hoping for a continuance of the same, I remain very respectfully yours,

GEO. FREUND,
Durango, Colo.

Our facilities for doing all kinds of Gun or Pistol Repairing are by far the best in the country.

H. B. ADSIT, County Clerk San Juan County, Colo.; F. J. DUDDLESON, Alamosa, Colo.; L. H. VAN NOSTRAND, Manager Grand View Mining and Smelting Co., Rico, Colo.. J. H. DOWELL, San Miguel, Colo.; H. G. HEFFRON, Silverton, Colo.; C. F STOLLSTEIMER, U. S. Indian Agent, Ignacio, Colo.

[From General G. Crook.]

HEADQUARTERS DEPARTMENT OF THE PLATTE, COMMANDING GENERAL'S OFFICE.

The gun arrived yesterday, and after carefully examining it I take pleasure in saying that all of the repairs have been made to my complete satisfaction, and in a manner to demonstrate not alone skillful workmanship, but a thorough knowledge of your business. Your patent sight is so valuable an improvement, and yet so simple, that I cannot help wondering why it was not sooner discovered. It serves equally well as a "peep" or an "open" sight, and can be used without reference to the manner in which the sun may be shining. In my opinion it fills a want long felt, and beyond improvement. It only needs to be known to be adopted by all sportsmen throughout the country.

Mr. Wm. Taylor, Quartermaster Agent of Rock Creek, W. T., says the following:

MR. FREUND—
Dear Sir:—You are aware I have a beautiful rifle, but I am free to say, for all practical purposes and beauty. your new sights have increased its value at least fifty per cent. How annoying it is to have a good shooting rifle with the old plug sights, over which it is necessary to have the sun and everything else favorable to enable you to see the game for a dead shot. I find it easy to catch the object over your sights in any direction and under the most unfavorable circumstances. No good shot, after seeing your new sights, would do without them.

[From a letter of Col. F. Van Vliet, Capt. Third Cavalry, Br't Lt. U. S. A.]

MR. FREUND—
Dear Sir:—I used your improved sights on my last hunting trip and found them perfect, both for hunting and target practice. The front sight gives a perfect bead. and has the peculiarity of showing dark against light objects and light against dark ones. The rear sight is extremely clear, and is the only one I have ever seen that at the time of taking sight allows you a clear view of your game.

[The following from Major Lord, Quartermaster.]
DEPOT QUARTERMASTER'S OFFICE, CHEYENNE DEPOT, WYOMING.

MR. FREUND—
Dear Sir:—Your patent sights which I have just put on my Winchester repeating rifle, 45 calibre, have given such

ABRAM RAPP,

Wholesale and Retail

FINE CLOTHING,

FURNISHING GOODS,

BOOTS, SHOES, HATS, TRUNKS, BLANKETS, ETC., ETC., ETC.

Mail Orders Solicited.

Durango, Colorado.

DURANGO

Foundry and Machine

WORKS.

Manufacturers and Repairers of Mining Machinery, Stamp Mills, Concentrators, Saw Mills, Boilers, Engines, Mining Pumps, etc.

Airy, Pocock & Airy

DURANGO, COLORADO.

Parsons & Thorp,

Druggists.

Stationery, Fancy Goods,

Paints, Oils and Brushes,

Wines, Liquors, Cigars and Smokers' Articles.

Cor. First and G, : Durango, Colo.

J. C. Sullivan,

NOTARY PUBLIC.

Office in STRATER HOTEL BUILDING,

With Burgess & Blake,

Durango, Colo.

WINDSOR

Short Order House,

First Street, Between H. and I,

DURANGO, * COLORADO.

Joseph Baumeister, Prop'r.

OPEN DAY AND NIGHT.

SMALLEY & HOAGLAND,

—— Dealers in ——

Second-Hand Goods.

OPPOSITE GRAND CENTRAL HOTEL,

Durango, Colorado.

Naeglin Bros.,

Pioneer Blacksmiths,

—— DEALERS IN ——

Farming Implements

Blacksmiths and Wagon Makers.

First Street, Near Railroad Crossing,

DURANGO, * COLORADO.

St. Clair Saloon,

—— DEALERS IN ——

Wines, Liquors and Cigars.

WHEELER & REINHEIMER,
PROPRIETORS.

Pawnbroker's Office

in connection with the above saloon; also

A GENERAL INTELLIGENCE OFFICE

run in connection.

Trimble Hot Springs.

The Hermosa House

The Best Accommodations Given to Patients of Any Hot Springs in Colorado.

R. M. GALLAGHER, Prop'r,

Herr & Herr,

Miners and Dealers in

PORTER COAL.

Office, One Door North of Colorado Armory,

FIRST STREET,

DURANGO, ✳ COLORADO.

CHAS. J. BOHRER, Pres.   E. F. BOHRER, Sec

⊱ THE ⊰

Durango Brewing Co.

PURE LAGER BEER.

Prompt Attention Given to All Orders for Keg and Bottled Beer.

GEM NOVELTY Theatre.

A. B. COLBURN, ○ PROPRIETOR.

↠ FIRST STREET, ↠⋯

DURANGO, - COLORADO.

Rodgers & Carr,

RANCH SALOON

Durango, Colo.

⊷ T. P. ARCHDEKIN, ⊶

Wholesale and Retail
Dealer in

All Kinds of Fresh, Salt and Dried Meats,

BUTTER, EGGS, FRESH FRUIT,

Fresh Vegetables, Fish, Oysters, Poultry, Etc., Etc.

First Street,

DURANGO, COLORADO.

The San Juan

Smelting and Mining

COMPANY,

DURANGO, ✳ COLORADO.

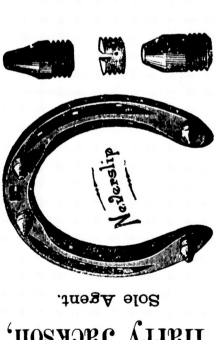

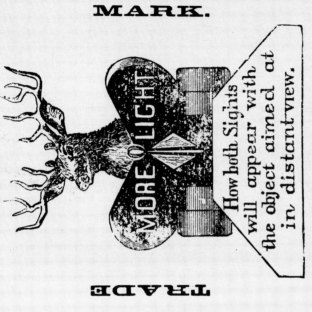

CHAPTER TWELVE

SURVIVING SPECIMENS
OF THE FREUND BROTHERS'
ALTERATION AND GUNMAKING

A number of collectors and writers have attempted to categorize the variations of Freund-altered and -manufactured rifles. Some have assigned dates to each variation, while others have tried various other types of attribution.

However, those methods of classification proved neither consistent or reliable. So, for want of a better system, this author has devised another, based on where the particular work was completed. In the case of Cheyenne there exists a time division: the earlier period when both brothers worked there, and the later period after George had departed for southwestern Colorado and Frank alone remained. Primarily rifles are classified here although the Freunds advertised that they also made shotguns, and at least one example of a long gun having both rifle and shotgun barrels made by Freund is known. In addition, Frank held a patent for an alteration to the Colt Single-Action Army revolver and several specimens survive, although they are recognized primarily for their Freund engraving.

This list is understandably incomplete. No doubt future research will reveal additional specimens of the Freund brothers' superb gunsmithing work. Descriptions of these classifications follow:

DCB   The period of the several Denver, Colorado stores. *Circa* 1870–1875.

CFB   The Cheyenne, Wyoming Territory period, when both brothers worked there. Guns generally are marked "Freund & Bro." or "Freund Bros." *Circa* 1875-1880.

CFA   The period after the brothers' partnership was dissolved, and Frank alone remained in Cheyenne. Guns generally are marked "Wyoming Armory." *Circa* 1880-1885.

NJF   The period after Frank and his family moved East, to New Jersey. Rifles generally are so marked. *Circa* 1885-1910.

DCG   The period during which George Freund lived and worked in Durango, Colorado. Rifles generally are so marked, sometimes with his "skull-and-crossbones" trademark. *Circa* 1881-1911.

With the exception of the rifles marked from Durango, Colorado, no classification is assigned to those pieces stamped only "Freund & Bro." (or with another Freund mark) on their barrel or elsewhere, as they simply passed through the Freunds' stores and were not altered by them. As such, those rifles are not relevant to this study.

Numbers are added to the letter combinations, based on their apparent order of alteration or manufacture, as well as to their serial numbers (where present).

DCB.01   *Freund-Made Plains Rifle.*

Caliber .50 percussion, 33-inch octagonal barrel marked "Cast Steel Warranted, Freund & Bro., Denver, Colorado." Ramrod held under barrel by two ferrules. One-piece American walnut stock and forearm held to barrel by one escutcheon. Double set-triggers, fancy triggerguard, crescent buttplate. Front and rear sights set in dovetails. Probably made by either Frank or George Freund, and used by the brothers in Scheutzen matches around the Denver area.

(Frank Sellers collection)

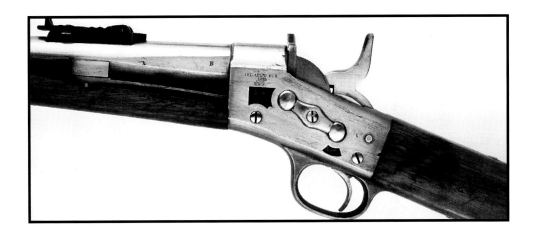

DCB.02   *Model for US. Patent Number 162224, Remington Rolling-Block Rifle.*

This Remington Rolling-Block action was altered to prevent the full cocking of the hammer while the breechblock is partially or fully open. (Courtesy Smithsonian Institution)

DCB.03   *F.W. Freund Custom Alteration of a Remington Rolling-Block Rifle.*

Caliber .40-70, 28-inch octagonal barrel marked on top flat "Freund & Bro., Denver, Colorado, USA" in script. Bottom of barrel marked ".40-70 D586." Has carved fluted nose on forearm, carved fluted areas extending rearward from right and left side panels, and fine checkering at wrist and forearm. Double set-triggers, fancy triggerguard with metal pistol grip. Front sight set into dovetail, elevated type rear sight is unmarked, 4-inch tang sight also unmarked. (Author's collection)

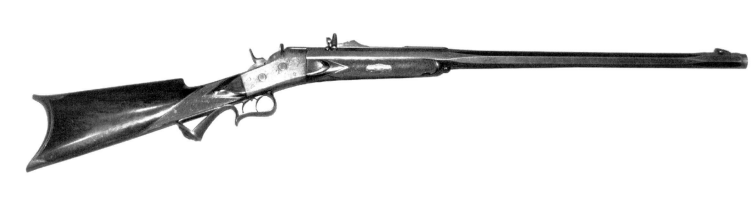

DCB.04   *F.W. Freund Custom Alteration of a Remington Rolling-Block Rifle.*
Caliber .44-70, 28-inch barrel octagonal for 11 inches forward of breech, then positioned so that edge between flats faces upward, until 3 inches from the muzzle when top flat again is horizontal; barrel has ¼-inch round turning at the muzzle. Left side of action engraved in script, "Improved by F.W. Freund, Denver, Col., Pat. July 28, 1874 & Pat. Apl'd. For." Action fitted with very fine double Scheutzen style set-triggers, fancy steel triggerguard ends in broad, finely-checkered pistol grip. Hammer nose reshaped. Custom blade front sight; open rear sight a masterpiece of custom gunsmithing, marked "Pat. Apl'd. For." One-inch high bracket slotted into barrel; 7½-inch spring-steel flat ending in a peep sight rests within this bracket; thin steel plate with raised sections slides beneath "sight spring" to adjust for elevation. Front of sighting mechanism engraved in 100-yard graduations to 500; as elevation plate slides back, middle zero fills in yard markings. Script engraved ahead of triggerguard, "Pat. No. 153432." Dark wood stock has very fine checkering. *(See* color section)
(Glenn Marsh collection)

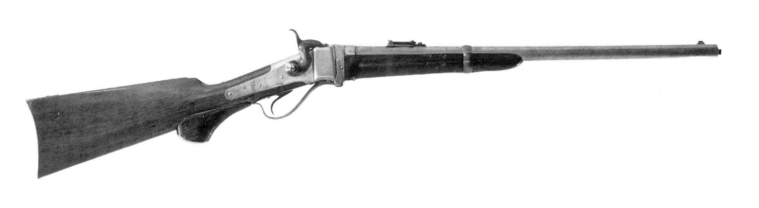

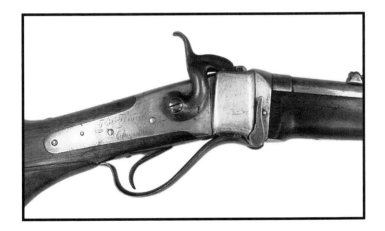

CFB.01 *Freund Alteration of a Sharps Percussion Carbine.*

Caliber .52, 22-inch round barrel. Breechblock altered to Freund's patent camming action, single extractor. No serial number. Regular front and rear sights. Marked on Freund patent pistol grip, "F.W. Freund, Pat. Oct. 19th, 1875, No. 168843 (20)"; on lockplate, "F.W. Freund, Cheyenne, W.T."; and "F.W.F." ahead of breech atop receiver. It is believed that this is the carbine that Frank Freund received from the Sharps Rifle Company after a trip to the East, and which he altered with his camming action. This modification generated voluminous correspondence between the Sharps Rifle Company and Freund. In one of Frank Freund's letters to Sharps, he stated that a second extractor could be added. (Author's collection)

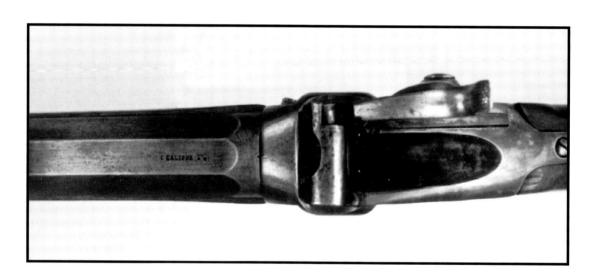

CFB.02   *Freund Alteration of a Meacham-Style Conversion of a Percussion Sharps Rifle.*

Caliber .40, 30-inch half-round, half-octagonal barrel. Serial number C45298 on tang; "45298" and "156608" marked on barrel. Freund patent camming action breechblock, double extractors. Freund-style hammer with added extension to fit into breechblock slot. Freund patent dates stamped on both sides of receiver; "156608" stamped in wood of standard Sharps forearm. Standard Sharps buttstock. Freund-marked front sight, unmarked thick buckhorn rear sight.

(Author's collection)

 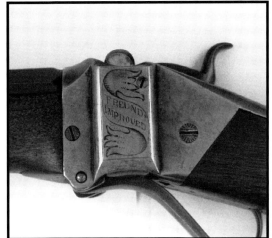

CFB.03   *Freund Alteration of a Model 1874 Sharps Rifle.*

Caliber .45-90, 24-inch half-round, half-octagonal barrel marked with Freund patents, "Fine Cast Steel" between rear sight and breech. Serial number C53982. "Famous American Frontier" engraved on right side of receiver, "Freund's Improved" on left side. Freund-altered with double extractors. Freund-style hammer with extension. Screws engraved with sunburst design. Standard Sharps forearm with schnabel; buttstock has checkered pistol grip with horn cap; Freund-style trapdoor in buttplate; wiping rods contained inside in canvas pouch. Standard Sharps sights. This rifle is pictured in *Sharps Firearms,* and described as having belonged to Boone May, a driver and messenger for the Cheyenne-Deadwood Stage Line.

(Author's collection)

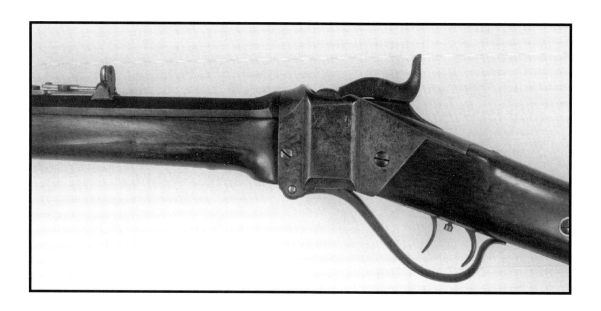

CFB.04 *Freund Alteration of a Model 1874 Sharps Rifle.*

Caliber .44-90, 32-inch octagonal barrel. Serial number 155045. Freund patent dates stamped on both sides of receiver. Freund-altered action, double extractors, Freund-altered hammer with extension, double set-triggers. Standard Sharps forearm with pewter cap, walnut buttstock. Freund front sight set into dovetail, Freund "More Light" buckhorn-style rear sight. Originally shipped to John P. Lower, Denver, Colorado, on February 15, 1876. (Author's collection)

CFB.05   *Freund Alteration of a Model 1869 Sharps Three-Band Military Musket.*

Caliber .50, 30-inch Sharps Model 1869 round barrel marked "Freund & Bro." Serial number 155418. Freund-altered camming action breechblock, double extractors. Freund patent markings on both sides of receiver. Buttstock has patchbox. Front sight dovetail has been changed; rear sight changed to buckhorn style. This gun originally was owned by buffalo hunter Henry Gerdel of Big Horn, Wyoming, and donated by his family along with a cartridge belt and cartridges to the Buffalo Bill Historical Center. Shipped to John P. Lower in Denver on February 22, 1876 as a military rifle chambered for the .50 caliber long case, part of a large shipment of guns.
(Courtesy Buffalo Bill Historical Center)

CFB.06   *Freund Alteration of a Model 1874 Sharps Rifle.*

Caliber .40-70, 30½-inch round barrel. Serial number 155801. Freund patented camming action breechblock, double extractors, double set-triggers. Lock engraved "Freund Bros. Manufact. Cheyenne, Wyo., U.S. of A", "Famous American Frontier" engraved on right side of receiver, "Freund's Improved" on left side, "Boss Gun" across rear of breech, cornucopia on hammer. Stock with detachable pistol grip and forend are replacements. Peep and open sights. A rather plain rifle that has seen much use. (Not pictured)
(from John Barsotti, *Gun Digest,* 1958, page 64.)

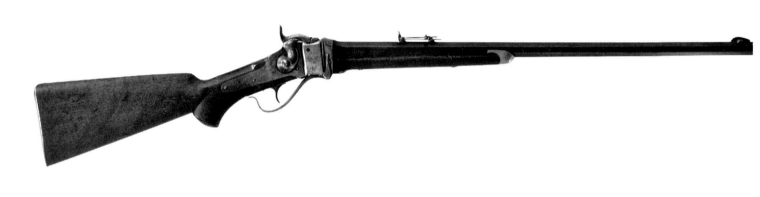

CFB.07 *Freund Alteration of a Model 1874 Sharps Rifle.*

Caliber .45-70, 26-inch octagonal barrel. Serial number 156428. Freund-altered camming action breechblock, double extractors, Freund hammer with extension. Freund patent dates stamped on both sides of receiver; "5" stamped on bottom of barrel and inside forearm. Checkered pistol grip stock, standard forearm with pewter nosecap. Freund front sight, Freund "More Light" rear sight.

(Author's collection)

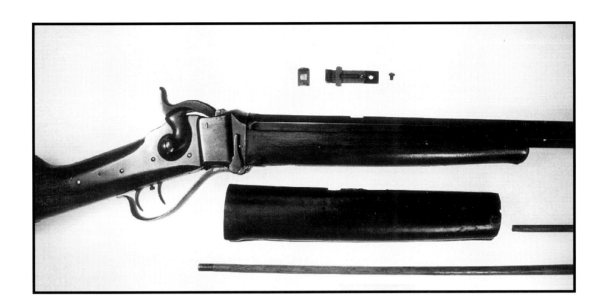

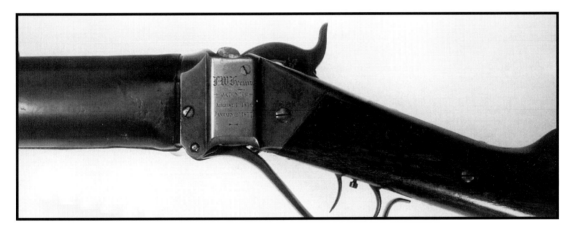

CFB.08 *Freund Alteration of a Model 1874 Sharps Rifle.*

Caliber .45-2⅞, 28-inch octagonal barrel marked "8X" on bottom near spring base. Serial number 156478 on tang, 157404 on barrel. Freund-altered camming action breechblock, double extractors, Freund hammer with extension. Left side of receiver engraved, "F.W. Freund, Patented, August 1st, 1876, January 2nd, 1877." Stock is a replacement, forearm adapted to take wiping rod, crescent buttplate. Freund-marked front sight set into dovetail; rear appears to be a modified Sharps sight. An outstanding feature is this rifle's leather forearm cover. To remove it, one must (1) slide out the rod, (2) remove the rear sight, (3) remove the wooden key at front of forearm, and (4) slip off the cover. This rifle originally belonged to Montgomery C. Meigs (1848-1930), Surveyor General of the United States. It went to his daughter, a Mrs. Orr, then Dede Jasiman, then the present owner.

(Robert Holter collection)

CFB.09   *Freund Alteration of a Model 1874 Sharps Rifle.*
        Caliber .40-90, 26-inch barrel. Serial number 156868. Deeply hand-cut into the right side of receiver is "American Frontier", "Freund Improved" on left side, "Boss Gun" engraved on tang behind action. Sharps-type rear sight a possible replacement. This rifle was a pick-up from central Montana, burned and rusted when found. An attempt was made at restoration in the 1950s, when it was fitted with a Bannerman stock and replacement lockplate and hammer. (Gary Roedl collection)

CFB.10   *Freund Alteration of a Model 1874 Sharps Rifle.*
        Caliber .45, 26-inch octagonal barrel stamped "Freund & Bro." on top flat. Serial number 157433 on tang. Freund-altered camming action breech-block, double extractors, double set-triggers, hammer with extension. Freund patent dates stamped on right side of receiver. Standard Sharps stock. Freund front sight set into dovetail, rear sight a thick, unmarked buckhorn type. Originally shipped to Ben Kittredge in Cincinnati, Ohio, on August 2, 1876. (Not pictured)
(Dave Carter collection)

CFB.11 *Freund Alteration of a Model 1874 Sharps Rifle.*

Caliber .40. Serial number 158120. Freund-altered camming action breechblock, double extractors, Freund-altered hammer with extension. Freund patent dates stamped on both sides of receiver. Freund front sight, Freund "More Light" rear sight malformed by heat. This rifle's original stock was burned in a fire; it is said that before the fire it had a small cheekpiece on the left buttstock, a pewter nosecap, a checkered forearm and pistol grip, and a silver shield engraved with the initials "H.D.L.", for H.D. Lovell. It was given to Lovell's son, who gave it to the father of the present owner. (Not pictured)
(Lloyd Tillett collection)

CFB.12 *Freund Alteration of a Model 1874 Sharps Rifle.*

Caliber .45, 2.4-inch case length. 28-inch half-round, half-octagonal barrel marked "5 6/16" on bottom; barrel has rings for swivels set into left side. Serial number 158513. Freund-altered flat-top breechblock, double extractors, Freund-altered hammer with extension. Freund patent dates stamped on left side of receiver. Pistol grip stock has a patched cut-out at heel for a tang sight. Checkered pistol grip and checkered forearm with schnabel; sling mounts set into buttstock and barrel ahead of forearm. Freund-marked front sight, Freund-marked high buckhorn rear sight, Marbles-type tang sight.
(Author's collection)

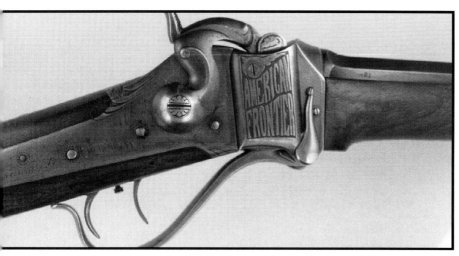
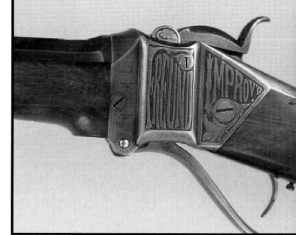

CFB.13 *Freund Alteration of a Model 1874 Sharps Rifle.*
Caliber .40-70, 30-inch half-round, half-octagonal barrel stamped "J.P. Lower, Denver, Colo." (partially obliterated). "Caliber 40" stamped atop barrel, "70" on right side. Serial number 159255. "American Frontier" deeply cut into metal on right side of receiver, "Freund Improved" on left side, "Freund & Bro., Cheyenne, W.T." engraved on lockplate. Freund-altered camming action breechblock, double extractors, Freund-style hammer with extension. Standard Sharps stock with schnabel forend, crescent buttplate. Freund-marked front sight set into a dovetail, Freund "More Light" rear sight.
(Charles Grimes collection)

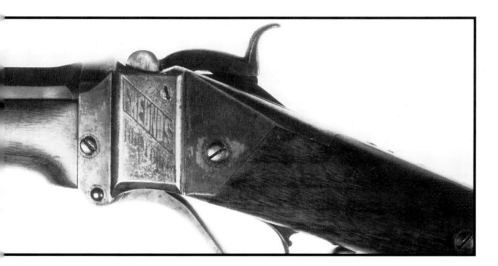

CFB.14   *Freund Alteration of a Model 1874 Sharps Rifle.*
        Caliber .40-90, 26-inch octagonal barrel. Serial number 159313. Right side of receiver engraved "Famous American Frontier" with figure of Indian holding raised tomahawk, "Freund's Improved" on left side, "Boss Gun" at front of tang near breech. Freund-altered camming action breechblock, double extractors, Freund-style hammer with extension has cornucopia engraved on side. Checkered pistol grip stock, checkered schnabel forend, buttplate with special Freund trapdoor containing double cavities for wiping rod and cartridges. Freund-marked front sight set into dovetail, Freund "More Light" rear sight. Originally shipped to H.C. Squires, Sharps agent in New York City.
(Lloyd Dietrich collection)

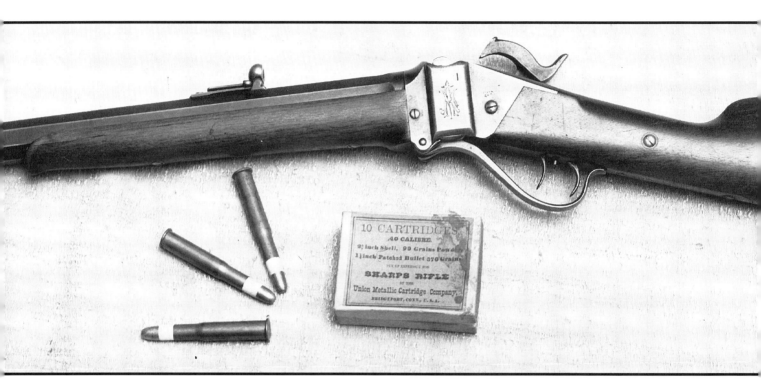

CFB.15 *Freund Alteration of a Model 1874 Sharps Rifle.*

Caliber .40-90, 28-inch octagonal barrel has Freund patent dates stamped on top, "MA" on bottom in front of forend. Serial number 160025. Right side of receiver stamped with Freund patent dates, entwined initials "MA" (for Morris Appel) engraved on left side. Freund-altered camming action breechblock, double extractors, Freund-style hammer with extension. Standard Sharps stock. Freund-marked front sight, unmarked Rocky Mountain-style rear sight has a very thick buckhorn.

Morris Appel (of Ames & Appel) hauled freight between Cheyenne and Deadwood during 1876-77, when outlaw and Indian depredations were the worst. When regular shipments were halted due to the Sioux raiding and burning the stations at Rawhide Buttes, Hat Creek, and Cheyenne River, Appel and his men (experienced bullwhackers, buffalo hunters, and Indian fighters, all) hauled 80,000 pounds of freight in one train to relieve the towns of Custer and Deadwood.

Appel died in Rapid City in the early 1930s. Mortician Don Hobart received this rifle as part of Appel's funeral expenses; later he sold it to the present owner. Originally shipped to J.P. Lower, Denver, on February 10, 1877. (Dick Hammer collection)

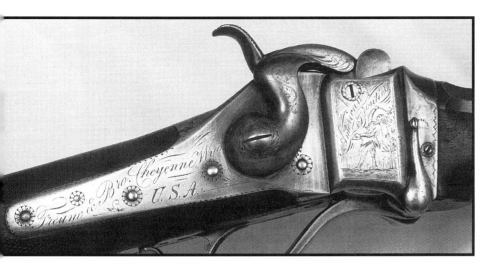

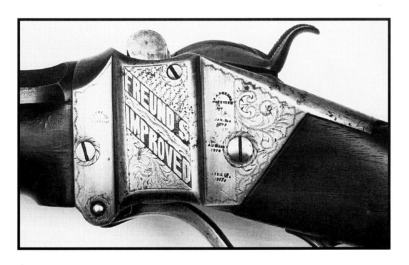

CFB.16   *Freund Alteration of a Model 1874 Sharps Rifle.*

Caliber .40-70, 28-inch octagonal barrel with serial number 155795 struck out on underside. Serial number 160291 on barrel and stock. Right side of receiver engraved "American Frontier" with figure of Indian holding a raised tomahawk, "Freund's Improved" and stamped Freund patent dates on left side, "Boss Gun" above serial number on tang near breech. Engraved screw holes; lockplate stamped "Freund & Bro., Cheyenne, Wyo. U.S.A." Balance of gun lightly engraved. Freund-altered camming action breechblock, double extractors, Freund-style hammer with extension. Checkered forend and pistol grip with cap having unidentified crest, buttstock has Freund trapdoor (with small button release) engraved "Freund's Inventions" containing two cavities for wiping rod and cartridges. Freund-marked front sight set into dovetail, Freund-marked "More Light" rear sight graduated to 1000 yards.

(Private collection)

CFB.17 *Freund Alteration of a Model 1874 Sharps Rifle.*
Caliber .40-70 Winchester, 26-inch octagonal barrel marked "40 Caliber" and serial number 161424. Right side of receiver stamped with Freund patent dates. Freund-altered camming action breechblock, double extractors, Freund-style hammer with extension. Standard Sharps stock with checkered wrist and forend. Standard Sharps front and rear sights, sporting-style tang sight marked "New Model 1869." (Not pictured)
(Burke Johnson collection)

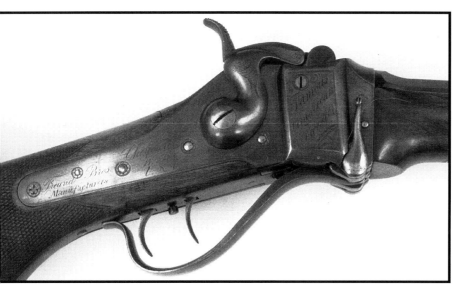 

CFB.18 *Freund Alteration of a Model 1874 Sharps Rifle.*
Caliber .40-70, 28-inch octagonal barrel. Serial number 161424 on tang, 156852 on barrel. Right side of receiver engraved "Famous American Frontier", "Freunds Improved" on left side, "Boss Gun" on tang above serial number. Engraved screw holes; lockplate engraved "Freund Bros., Manufacturers, Cheyenne, W.T. U.S.A." Freund-altered camming action breechblock, double extractors, Freund-style hammer with extension has engraved cornucopia; double set-triggers. Sharps walnut stock with checkered pistol grip and forend; buttstock has Freund trapdoor, rods lacking. Freund-marked front sight set into dovetail, Freund-marked "More Light" rear sight graduated to 1000 yards. Metal refinished.
(Author's collection)

CFB.19 *Freund Alteration of a Model 1874 Sharps Rifle.*
Caliber .40, 30-inch octagonal barrel. No serial number. Right side of receiver engraved "American Frontier", "Freund Improved" on left side, "Freund Bros., Cheyenne, Wyo. U.S.A." on lockplate. Balance of receiver lightly scroll and border engraved. Checkered wrist and forend. Target front sight adjustable for windage, no rear sight, vernier peep tang sight. Sling swivels on barrel and buttstock. Metal refinished. (Pictured in color section)
(William Ruger collection)

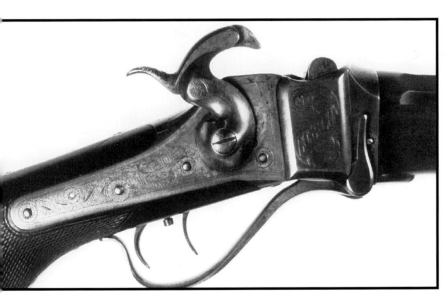 

CFB.20 *Freund Alteration of a Model 1874 Sharps Rifle.*
Caliber .40-90, 28-inch octagonal barrel. Serial number 162453 stamped on barrel and tang. Right side of receiver engraved in riband "Boss Gun", "Freund's Improved" in riband on left side, "Freund Bro., Cheyenne Wyoming Armory" in riband on lockplate. Balance of lockplate, hammer, and triggerguard medium engraved. Freund-altered camming action breechblock, double extractors, Freund-style hammer with extension. Standard Sharps stock with checkered pistol grip and schnabel forend. Standard Sharps front sight, standard Sharps rear sight with thick buckhorn.
(Dave Carter collection)

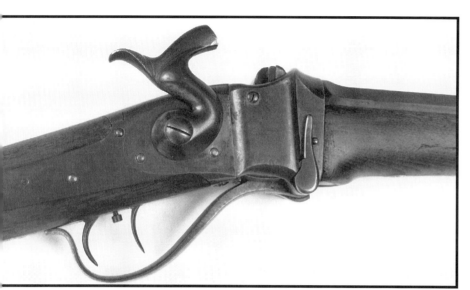

CFB.21 *Freund Alteration of a Model 1874 Sharps Rifle.*

Caliber .45-70, 26-inch octagonal barrel. No serial number. Left side of receiver engraved "F.W. Freund, Pat'd., August 1st, 76, No. 180586, Jan. 2nd, 77, No. 185911." ("180586" is in error; it should read "180567," Freund's patent for a breechloading arm. Patent number 180586 was issued on the same date [August 1, 1876], but was for a cigar box. The error probably was made by the engraver, or by whoever ordered the engraving.) Freund-altered camming action breechblock, single extractor. Hammer lacks Freund extension and appears unaltered, double set-triggers. Standard Sharps stock. Freund-marked front sight set into dovetail, Sharps-R.S. Lawrence rear sight.

The rifle's original owner, Bartlett Richards, died in prison in 1911, following the Nebraska Land Conspiracy trials of the early 1900s. Richards had assembled the Spade Ranch, ranging cattle over about 500,000 acres of deeded and public lands. The government alleged that Richards encouraged homesteaders who didn't intend to occupy the land to file claims, so that he could buy them out.

"He should never have gone to jail," said the present owner of the Spade, Lawrence Bixby. "He was open about his fences, and didn't do anything different from other ranchers. He just had this fight with Teddy Roosevelt, and Roosevelt was bigger…he was president. They made an example of Richards because he was one of the big ranchers. He was charged with land fraud because he brought Civil War widows out here to homestead. Hell, the government gave those women seven dollars and a piece of paper saying they could homestead on 160 acres. They'd have starved to death. Richards loaned them milk cows, horses, let them charge at his store, and bought their land when they left. It didn't cost the government anything…they'd taken it away from the Indians."

Bixby, the Spade's second owner, found this rifle when he moved onto the ranch. He gave it to his granddaughter and her husband.
(Bruce Graham collection)

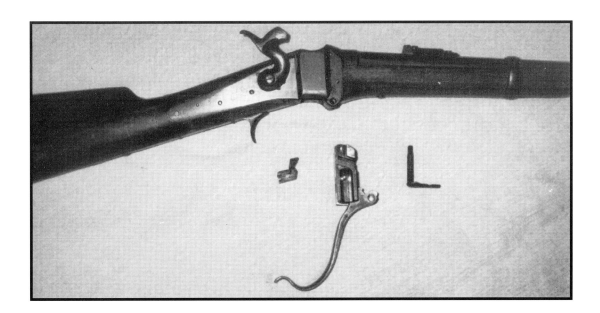

CFB.22 *Freund Alteration of a Percussion Sharps Carbine.*

This is a model gun made from a percussion Sharps carbine, believed to be the gun that Frank Freund supplied to the U.S. Patent Office. Nelson King of the Sharps Rifle Company also had applied for a similar patent around the same time, which resulted in charges of patent infringement on both sides *(see* chapter ten). Freund's application provided several improvements to the Sharps action:

To provide a gas check
To half-cock the hammer
To move the firing pin in a straight line
To prevent accidental firing when the hammer bears on the firing pin.

Apparently Freund and Sharps settled their differences, as Frank Freund was granted his patent on January 3, 1877.
(Courtesy Chris Schneider)

CFB.23 *Freund Alteration of a Model 1874 Sharps Rifle, for General George Crook.*

A description of General Crook's rifle is not available, but the following is correspondence between Crook and Frank Freund regarding this gun:

Headquarters Department of the Platte,
Commanding General's Office
The gun arrived yesterday, and after carefully examining it, I take pleasure in saying that all the repairs have been made to my complete satisfaction, and in a manner to demonstrate not alone skillful workmanship, but a thorough knowledge of your business. Your patent sight is so valuable an improvement and yet so simple that I cannot help wondering why it was not sooner discovered. It serves equally well as a "peep" and "open" sight, and can be used without reference to the manner in which the sun may be shining. In my opinion, it fills a want long felt, and beyond improvement. It only needs to be known to be adopted by all sportsmen throughout the country.

Headquarters Dep't. Platte;
Com'dg. General's Office
Omaha, Neb., April 30, 1878
I have received the Sharp's *[sic]* rifle left in your hands for repairs and alteration. The double ejector spring and rounding off of the inner edge of the breech-block are improvements which only need to be seen by practical men to commend themselves for general use. The beauty and perfection of workmanship in this, as in other repairs I have had made at your establishment, are entitled to the highest praise.

Cheyenne, Wyo., Feb. 27th, 1878
Sharps Rifle Company
Bridgeport, Conn.
Gents
Enclosed, we send you the action of Gen. Crook's rifle to put a new Pistol grip stock on. The Rifle stock was broken by Gen. Crook on the last campaign onto the Big Horn and sent to us to be restocked. He will be here in Cheyenne shortly and will want his rifle to use. Please make it without delay. And if you have any charges, send the bill to us. We will not make any charges to him.
Respt. yours
of
Freund & Bros.

On March 19, 1878, the Freund brothers received a bill from the Sharps Rifle Company in the amount of $16.00 for the above restocking of General Crook's rifle.

Following his retirement from the Army, General Crook continued to make hunting trips into Wyoming for a number of years. On those jaunts he was accompanied by President Rutherford B. Hayes' son, which likely explains why the General's rifle is exhibited at the Hayes Presidential Center today.
(Courtesy Gil Gonzales, Photography Resources, Rutherford B. Hayes Presidential Center)

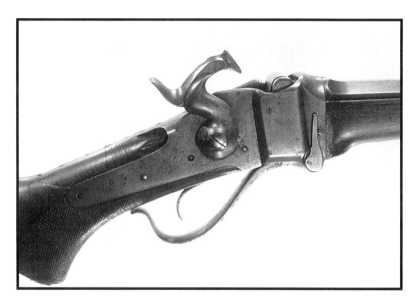

CFB.24   *Freund Alteration of a Model 1874 Sharps Rifle.*
Caliber .40-90, 26-inch octagonal barrel has stamped Freund patent dates. No serial number. Freund altered camming action breechblock, double extractors. Standard Sharps stock has checkered pistol grip with entwined silver initials "J.W.D." (for original owner J.W. Driscoll), uncheckered forend with pewter nosecap, German-style buttplate of crescent type. Freund-marked front sight set into dovetail, unmarked ladder-type rear sight graduated to 1300 yards.
J.W. Driscoll trailed cattle from Texas to Wyoming during the 1870s and '80s. He gave this rifle to a man who worked with him, J.A. "Tennessee" Vaughn. Vaughn's son sold it to the present owner. *(See* color section)
(Dennis Brooks collection)

CFB.25   *Freund Alteration of a Model 1874 Sharps Rifle.*
Caliber .40-70, tapered octagonal barrel has "Improved Old Reliable" stamped on top flat. No serial number. Freund altered camming action breechblock, double extractors, double set-triggers. Right side of receiver engraved "Model 1879", "Freund Imp'd." on left, "F&B, Cheyenne, W.T." in riband on lockplate. Balance of receiver border-engraved in floral design. Stock has finely-checkered Sharps pistol grip engraved "to E.L. Chester by W.E. Shaw", checkered forend held with escutcheon is cut for ebony wiping rod held in two ferrules under barrel, German-style crescent buttplate. Freund-marked front sight set into dovetail, standard Sharps rear sight, stock likely had tang sight at one time. (Pictured in color section)
(Private collection)

CFB.26   *Freund Alteration of a Model 1874 Sharps Rifle.*

Caliber .40-70, 30-inch round barrel. No serial number. Right side of receiver engraved "Famous American Frontier", "Freund's Improved" on left side, "Freund Bro., Manufact., Cheyenne, Wyo. U.S. of A." on lockplate, "Boss Gun" on tang behind breech. Engraved hammer; double set-triggers. Straight Sharps stock has Freund patent detachable pistol grip; checkered schnabel forend, unusual carving near breech. Checkered steel buttplate with sliding trapdoor. Beach-type globe front sight has ivory base, Freund improved Sharps rear sight.

Illustrated in *The American Gun Quarterly,* first issue, and in *The Fireside Book of Guns,* first issue.
(Offered by Alan Kelley in *The Gun Report,* June 1974)

CFB.27   *Freund Alteration of a Model 1874 Sharps Rifle.*

Caliber .45-70, 28-inch octagonal barrel one inch wide at muzzle. No serial number. Freund patent dates stamped on both sides of receiver, right side engraved "Freund Gun", "Freund Improved" on left side, "Freund Bros., Cheyenne, Wyoming Armory" on lockplate. Replacement pistol grip stock and forend. Bought by E.M. Farris for $3.50 from an Ironton, Ohio junkyard following World War II. During wartime many old guns were donated to scrap drives and this one, though in good condition, had been stripped of its wood. Farris sold the gun to railroader C.R. "Bull" Ramsay, who later sold it to the present owner.
(Not pictured)
(John Barsotti collection)

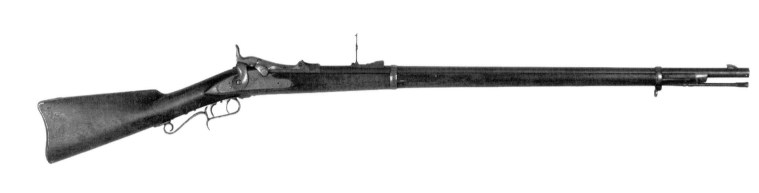

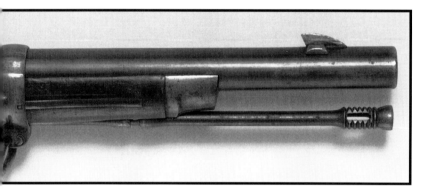 

CFB.28   *Freund Alteration of a U.S. Model 1873 Springfield Trap-Door Rifle.*
       Caliber .45-70, 32-inch round barrel secured by two bands (one a re-
placement showing traces of nickel plating). Serial number 13894, an early example. Original full walnut stock, and iron ramrod. Freund altered for sporting
use by thinning the wrist of the stock to change the contour of the comb and the
areas behind the lock and lock-screws, the addition of a copper blade to the
front sight, and the installation of new double set-triggers and "Freund" marked
iron triggerguard with decorative spur and loop. This was accomplished by removing the center section of the old trigger-plate along with the old mechanism, and replacing it with a new Freund-made mechanism and plate. The new
triggerguard is keyed to the new trigger-plate, and screwed to the stock. The
remaining piece of the original trigger-plate is stamped "Freund."
(Courtesy Minnesota Historical Society)

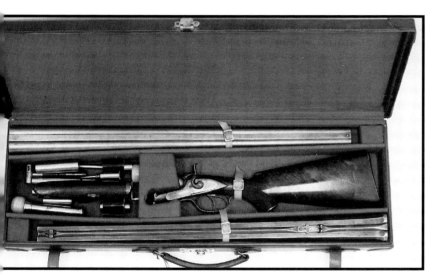 

CFB.29 *Freund Alteration of a German Shotgun to a Double-Barrel Rifle and Double-Barrel Shotgun.*

Caliber .45-90 rifle barrels marked "Fine Cast Steel", 10 gauge shotgun barrels marked "F.W. Freund, Maker, Cheyenne, Wyo."; both barrel sets are 30 inches long. No serial number. Both sides of finely engraved German action marked "Freund & Bro." in gold; engraved hammers. Lever below triggerguard operates to install or remove barrels. Fancy walnut stock has finely-checkered pistol grip, matching checkered forend is keyed shotgun type. Freund-marked rifle front sight set into dovetail, Freund-marked "More Light" rifle rear sight. Restored carrying case holds both sets of barrels, action, forearm, and tools. (Author's collection)

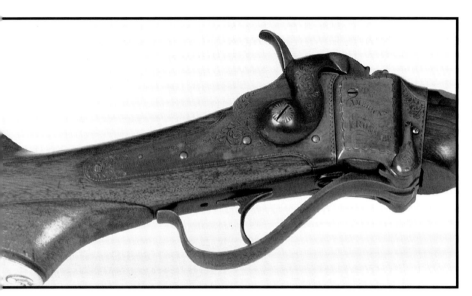
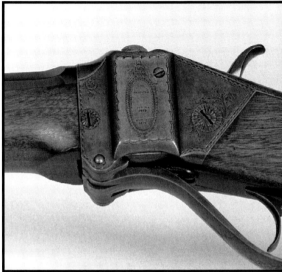
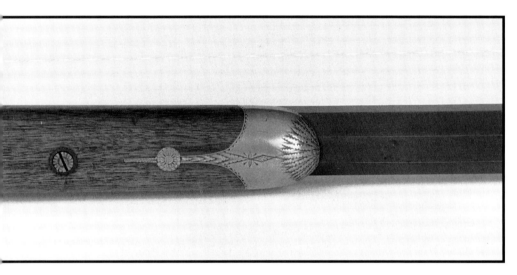

CFA.01 *Freund Alteration of a Model 1874 Sharps Rifle.*

Caliber .45, 28-inch octagonal barrel stamped with Freund patent dates. Serial number 154756. Right side of receiver engraved "The American Frontier", Freund patent dates inside ornate panel on left side, "Freund's Pat., Wyoming Armory, Cheyenne, Wyoming Ter." on lockplate. Lock, frame, hammer, triggerguard, and forend cap profusely scroll and floral engraved. Silver cap on Freund patent detachable pistol grip engraved with large old-English "C" over "1883." Provenance from William F. "Buffalo Bill" Cody.
(Courtesy Heritage Plantation of Sandwich, Massachusetts)

CFA.02 *Freund Alteration of a Model 1874 Sharps Rifle.*

Caliber .45-70, 26-inch half-round, half-octagonal barrel. Serial number 156269 on tang. Freund altered camming action breechblock, double extractors, double set-triggers. "Freund Patent, Wyoming Armory, Cheyenne, Wyo. T." engraved on lockplate. Sunbursts engraved around screw heads. Freund altered buttplate with trapdoor containing wiping rods. Freund "More Light" rear sight. (Not pictured)
(Formerly in the Bill Peace collection)

CFA.03 *Freund Alteration of a Model 1874 Sharps Rifle.*

Caliber .40, 26-inch half-round, half-octagonal barrel has stamped serial number 156629, concave muzzle. Freund altered camming action breechblock, double extractors, engraved Freund-style hammer with extension. Right side of receiver engraved with Indian shooting bow and arrow, "American Frontier" on left side, "Freund's Patent, Wyoming Armory, Cheyenne, Wyo., T., U.S.A." on lockplate. Action has scroll border engraving. Walnut stock has checkered Freund patent detachable pistol grip, checkered target-type forearm with schnabel.
(Private collection)

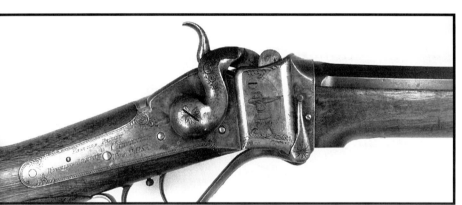

CFA.04   *Freund Alteration of a Model 1874 Sharps Rifle.*
Caliber .45-90, 28-inch octagonal barrel stamped with serial number 157405 and turned round at muzzle. Freund altered camming action breechblock, double extractors, engraved Freund-style hammer with extension. Right side of receiver engraved with Indian shooting bow and arrow, "American Frontier" on left side, "Freund's Patent, Wyoming Armory, Cheyenne, Wyo. T. U.S.A." on lockplate. Action has scroll border engraving. Walnut stock has checkered Freund patent detachable pistol grip, plain forearm with schnabel. Unmarked front sight set into dovetail, unmarked Walter Cooper style rear sight. "Ed F. Stahl" (a United States deputy mineral surveyor) stamped on barrel under forearm (*see* Appendix I).
(Author's collection)

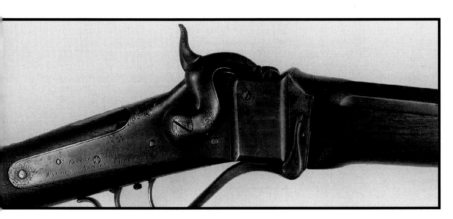

CFA.05   *Freund Alteration of a Model 1874 Sharps Rifle.*
    Caliber .40-90, 30-inch half-round, half-octagonal barrel. No serial number. Freund altered camming action breechblock, double extractors, engraved Freund-style hammer with extension. Right side of receiver engraved "American Frontier" in small letters, "Freund's Patent" in small letters on left side, "Freund Patent, Wyoming Armory, Cheyenne, Wyo." on lockplate. Action has light scroll border engraving. Straight Sharps stock with Germanic-style buttplate, plain Sharps forearm has small schnabel. Freund-marked front sight set into dovetail, Freund ladder-type rear sight. "W. Delaney" (packmaster of Bradley's command in June 1878) engraved on tang behind breech (*see* Appendix I).
(Author's collection)

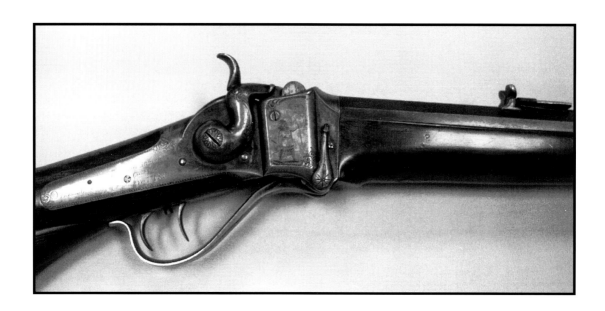

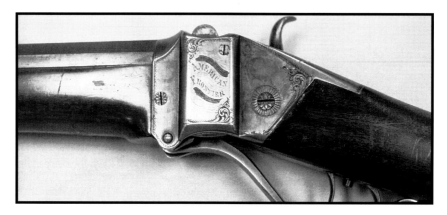

CFA.06   *Freund Alteration of a Model 1874 Sharps Rifle.*
     Caliber unknown, 28-inch octagonal barrel marked on top "Remington Cast Steel." Serial number 15914 (last digit illegible). Freund altered camming action breechblock, double extractors, engraved Freund-style hammer with extension, double set-triggers. Right side of receiver engraved with Indian holding bow, "American Frontier" on left side, "Freund's Pat., Wyoming Armory, Cheyenne, Wyo. T. U.S.A." on lockplate. Action has scroll border engraving, engraved screws. Straight Sharps stock with Freund patent detachable pistol grip, plain Sharps forearm, checkered shotgun-style buttplate with Freund-style trapdoor containing wiping rods. Freund-marked front sight set into dovetail, unmarked thick buckhorn ladder rear sight with replacement slide.
(Frank Sellers collection)

CFA.07   *Freund Alteration of a Model 1874 Sharps Rifle.*

Caliber .40, 28-inch octagonal barrel. Serial number 162271. Freund altered camming action breechblock, double extractors, engraved Freund-style hammer with extension. Right side of receiver engraved with Indian holding bow, "American Frontier" on left side, "Freund Patent, Wyoming Armory, Cheyenne, Wyo. T." on lockplate. Action has scroll border engraving. Checkered, semi-pistol grip walnut stock. Freund-marked front sight, unmarked ladder-style rear sight. "Pierre Lorillard" engraved on upper tang. (Pictured in color section)
(William Ruger collection)

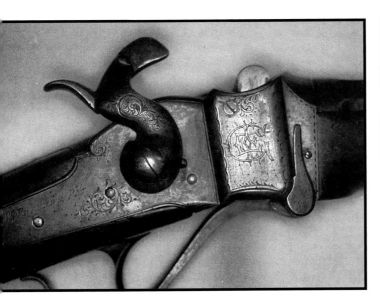 

CFA.08   *Freund Alteration of a Model 1874 Sharps Rifle.*

Caliber .40-90BN, 30-inch octagonal barrel. Serial number 162365 on tang, 162398 on barrel. Freund altered camming action breechblock, double extractors, engraved Freund-style hammer with extension. Right side of receiver engraved with entwined initials that appear to be "E.E.G.", "Famous American Frontier" in three scrolls on left side. Action has scroll border engraving. Stock is replacement from Civil War-era rifle, has crudely-checkered pistol grip, patch box filled in. Standard Sharps front and rear sights. Originally shipped to N.R. Davis, Cheyenne, Wyoming Territory on March 17, 1879 in a shipment of sixty rifles in six cases, along with 6000 cartridges. Davis was a prosperous businessman who evidently received rifles for the Freunds and Dammann much as Dow did for Walter Cooper.
(Lance Peterson collection)

CFA.09 *Freund Alteration of a Model 1874 Sharps Rifle.*
Caliber .40-90, 25-inch octagonal barrel. Serial number 162398 on tang, 162365 on barrel. Single trigger. Right side of receiver engraved with entwined initials "G.W.S." for Wyoming cattleman G.W. Stanley (*see* Appendix I), "Famous American Frontier" in scroll riband on left side, "F.W. Freund Pat." on lockplate. Receiver, hammer, and lever are scroll panel and border engraved. Walnut stock with Freund patent detachable pistol grip, steel buttplate has Freund-style trapdoor containing wiping rod. Freund front sight, standard Sharps rear sight. Originally shipped to N.R. Davis, Cheyenne, Wyoming Territory on March 17, 1879 in a shipment of sixty rifles. (Refer to previous entry; note exchanged barrels and tangs, and same serial number.) (*See* color section) (Courtesy Autry Museum of Western Heritage)

CFA.10 *Freund Alteration of a Model 1874/77 Transition Sharps Rifle.*
Caliber .45, unmarked 29-inch octagonal barrel. No serial number. Flat-top Freund altered camming action breechblock, double extractors, hammer with extension. "F.W.F." engraved atop breech. Fancy grain walnut stock with finely-checkered pistol grip and forend, checkered hard composition buttplate. Freund-marked front sight set into dovetail, Freund "More Light" rear sight is bridged thin buckhorn type. Style of stock and forend suggest the work of C.E. Overbaugh, perhaps one of twenty rifles completed by him in March of 1879 and sold direct by Frank Hyde of the Sharps Rifle Company to New York dealer Homer Fisher. (Pictured in color section)
(Private collection)

CFA.11 *Freund Alteration of a Model 1874 Sharps Rifle.*

Caliber .40-90BN, 32-inch octagonal barrel marked "F.W. Freund, Pat." No serial number. Freund altered camming action breechblock, double extractors, engraved Freund-style hammer with extension. Right side of receiver engraved with Indian holding bow, "American Frontier" on left side, "Freund Patent, Cheyenne Armory, Cheyenne, Wyo. T. U.S.A." on lockplate. Lockplate, receiver, and screws are engraved. Standard Sharps stock with Freund patent detachable pistol grip, Sharps forend with ebony cap, checkered steel buttplate with Freund-style sliding trapdoor containing original five-piece wiping rod. Original Sharps front and rear sights. Owned at one time by the Miller Brothers, of Wild West Show fame.

(Dave Carter collection)

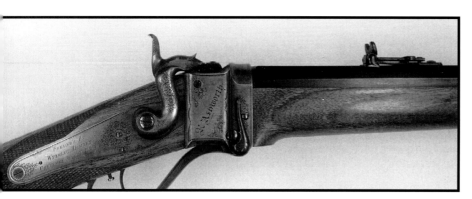
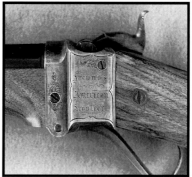

CFA.12   *Freund Alteration of a Model 1877 Sharps Rifle.*
Caliber .40-90, 26-inch half-round, half-octagonal barrel marked "Fine Cast Steel." Freund altered camming action breechblock, double extractors, engraved Freund-style hammer with extension. Right side of receiver engraved "R. Ashworth", "Freund American Frontier" on left side, "Freund's Pt., Wyoming Armory, Cheyenne, Wyo. T. U.S.A." on sideplate. Receiver, hammer, lever, and screws scroll engraved. Freund-marked front sight, Freund "More Light" rear sight marked "Freund's Patent 1880."
Richard Ashworth was an English rancher in Wyoming (*see* Appendix I). This rifle was burned in a fire at the home of arms author John Amber; subsequently restored by Ed Webber including pistol grip stock and forend, and with engraving freshened by Lynton McKenzie. (*See* color section)
(Author's collection)

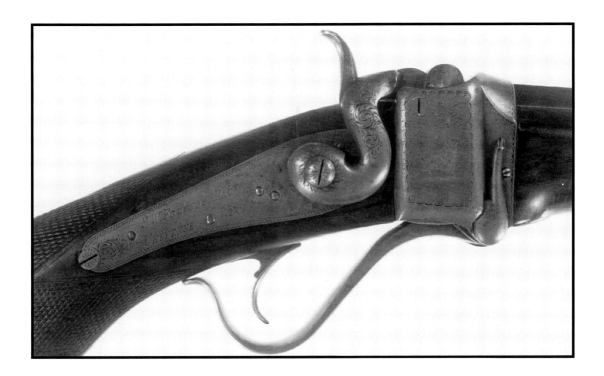

CFA.13   *Freund Alteration of a Model 1877 Sharps Rifle.*

Caliber .40-90, 28-inch half-round, half-octagonal barrel. No serial number. Freund altered camming action breechblock, double extractors, engraved Freund-style hammer with extension. Right side of receiver engraved with bull buffalo tossing Indian with bow into air, script initials "E.B.B." for Edgar Beecher Bronson on left side, "The Famous American Frontier Rifle" on upper tang, "F.W. Freund Patt., Cheyenne, Wyo." on lockplate. Border engraving on receiver and lever, all screw- and bolt heads engraved. Circassian walnut stock with finely-checkered pistol grip, matching forend has black horn cap, checkered Freund-style steel buttplate with sliding trapdoor for wiping rod. Freund-marked front sight, unmarked heavy buckhorn type rear sight set into dovetail. Originally shipped to Edgar Beecher Bronson at his Texas ranch in 1883.

Bronson was a cattle rancher in Wyoming and Texas who made hunting trips to Mexico, South America, and Africa. He was born on September 20, 1856, maternally related to famed pre-Civil War abolitionists the Reverend Henry Ward Beecher, and Beecher's sister Harriet Beecher Stowe, author of *Uncle Tom's Cabin*. Bronson's nephew was 1st Lt. Frederick H. Beecher, who was mortally wounded in Colorado on a small Arickaree River island that later became known as Beecher's Island. (*See* color section)
(Dennis Brooks collection)

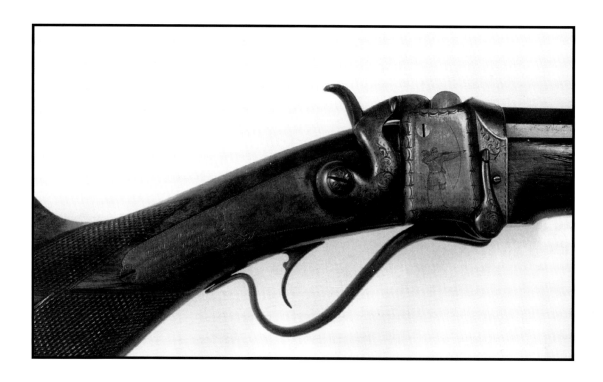

CFA.14   *Freund Alteration of a Model 1877 Sharps Rifle.*
Caliber .45-100, 28-inch half-round, half-octagonal barrel. Serial number 106 engraved on lower tang. Freund altered camming action breechblock, double extractors. Right sidc of receiver has engraved Indian holding drawn bow, bear on left side, "Freund Patent, Wyoming Armory, Cheyenne, Wyo. Ter. U.S.A." and deer grazing on foliage in panel on lockplate, entwined initials "PWS" on upper tang presumed for Phillip W. Schuyler (*see* Appendix I). Receiver, hammer, lever, and screws elaborately engraved. Checkered pistol grip walnut stock, checkered forend with ebony cap, buttplate with sliding trapdoor. Freund-marked front sight set into dovetail, Freund "More Light" rear sight. (*See* color section) (Private collection)

CFA.15   *Freund Alteration of a Model 1877 Sharps Rifle.*
Caliber .45-2.4, 28-inch half-round, half-octagonal barrel. Single trigger. Serial number 152. Right side of receiver has engraved Indian, "American Frontier" on left side, "Freund's Patent, Cheyenne, Wyo. Ter." on lockplate. (Not pictured)
(Formerly in the M.C. Clark collection)

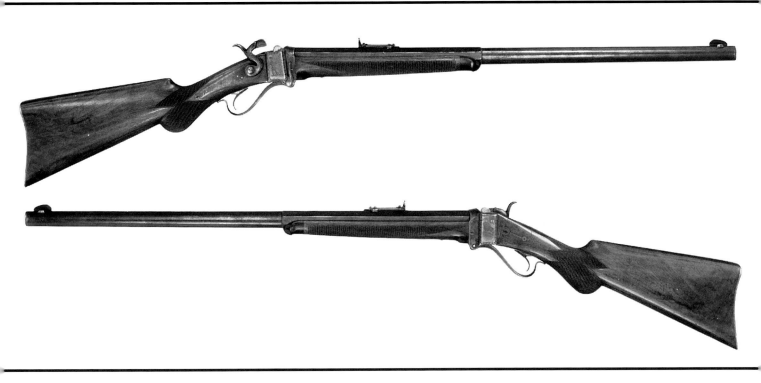

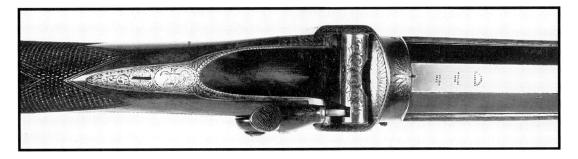

CFA.16 *Freund Alteration of a Model 1877 Sharps Rifle.*

Caliber .45-2⅞, 28-inch half-round, half-octagonal barrel stamped with Freund patent dates near breech. Serial number 202. Freund altered camming action breechblock, double extractors. Right side of receiver has engraved Indian holding drawn bow, bull buffalo on left side, "Freund Patent, Wyoming Armory, Cheyenne, Wyo. T." on lockplate, entwined initials "W.D." on tang near breech for W.T. Dawson, editor of *American Field* magazine. Elaborately engraved receiver, hammer, trigger bar, tang, and screws. Checkered European walnut pistol grip stock, checkered forend with ebony cap. Trigger bar extends over pistol grip. Buttplate with trapdoor containing wiping rods. Freund-marked front sight, Freund-marked converted Sharps rear sight. (*See* color section) (Dr. R.L. Moore collection)

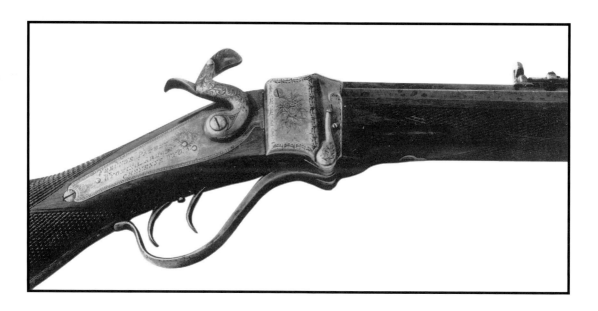

CFA.17 *Freund Alteration of a Model 1877 Sharps Rifle.*

Caliber .40-90 necked, 27-inch half-round, half-octagonal barrel stamped with Freund patent dates. Serial number 203 engraved on lower tang. Freund altered camming action breechblock, double extractors. Right and left sides of receiver engraved in floral designs, "Freund's Patent, Wyoming Armory, Cheyenne, Wyo. T." on lockplate, entwined initials "L.F.S." on tang near breech. Engraved receiver, hammer, trigger bar, tang, and screws. Checkered pistol grip stock, checkered forend, buttplate with trapdoor containing wiping rods. Inlaid silver bead front sight, Freund "More Light" rear sight.
(John Barsotti collection)

CFA.18 *Freund-Designed and Hand-Built Wyoming Saddle Gun.*

Caliber .40-90, 18-inch octagonal barrel has "Freund Patent" stamped near breech. No serial number. Sharps-type action, modified with Freund altered camming action breechblock, double extractors. Right side of receiver engraved "Freund Patent", entwined gold initials "C.K." (for Clarence King) on left side, "Freund's Pat., Wyoming Armory, Cheyenne, Wyo." on lockplate, scene of fox stalking game on breech lever. Elaborate floral and scroll designs engraved on receiver, lock, hammer, tang, and screws. Checkered pistol grip stock, checkered forend with ebony cap, buttplate with sliding trapdoor containing jointed wiping rod. Freund-marked front sight, engraved rear sight marked "Freund's Patent."

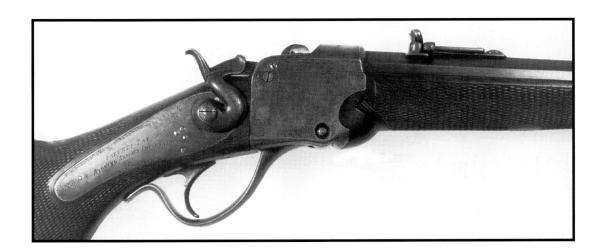

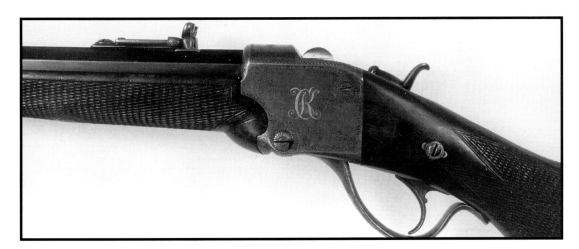

Clarence King was the first head of the U.S. Government Geological Survey, and was in charge of the survey of the 40th Parallel. Illustrated in color in the February 1970 issue of *Guns* magazine. Research indicates that this is the only surviving specimen of Freund's Wyoming Saddle Gun. (*See* color section)
(Private collection)

CFA.19   *Freund Custom-Embellished Colt Single-Action Army Revolver.*
        Caliber .45, 7½-inch barrel. Serial number 75282. Profuse scroll engraving on barrel, frame, cylinder, backstrap, and triggerguard, fern motif on loading gate. Engraved panel on right side of frame has cowboy with quirt on his wrist mounted on horse galloping after two steers. Front sight is half of an 1882 dime. Mother-of-pearl grips, left side has "T.F. Bryson" inscribed in Old English letters. Although not so marked, the panel scene and other engraving strongly suggest the source of the work was Freund's Wyoming Armory in Cheyenne; a letter from Colt authority R.L. Wilson acknowledges the same. (Private collection)

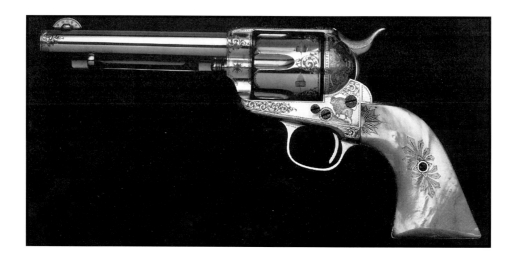

CFA.20   *Freund Custom-Embellished Colt Single-Action Army Revolver.*
        Caliber .45, 5½-inch barrel. Serial number 83639. "Freund's Armory, Cheyenne, Wyo." engraved on topstrap, entwined initials "J.S." (or "S.J.") on backstrap behind hammer. Profuse scroll engraving in the style of Sharps rifles

done at Freund's Wyoming Armory appears on barrel, frame, cylinder, hammer, and backstrap. Hammer faceted and gold plated, gold plated trigger and ejector rod; balance of revolver silver plated. Engraved panel on right side of frame has cowboy mounted on galloping horse firing revolver at four bull buffalo, one animal going down; panel on left side of frame has standing bull buffalo. Front sight is half of an old dime set into dovetail. Mother-of-pearl grips engraved around grip screw and where grips meet frame. (*See* color section)
(Ray Bentley collection)

NJF.01   *F.W. Freund-Made False-Hammer Rifle.*

Caliber .45, 28-inch octagonal barrel turned round at muzzle with dished crown, engraved on top flat "F.W. Freund, Patentee and Maker, Jersey City, N.J." and "Cast Steel" within riband surrounded by tight scroll engraving repeated on angled flats at breech. Beautifully-made action somewhat resembles experimental Sharps Model 1875; tight scroll border engraved including loading groove. Both sides of receiver engraved "Freund's Patent" within banner surrounded by tight scrolls. Action cocks automatically when breech opened; small external hammer or cocking indicator also may be cocked manually. Cocking mechanism and trigger may be removed from action by pulling down grooved section at bottom of pistol grip. Well-grained walnut stock has raised cheekpiece and pistol grip with engraved steel cap, matching checkered forend with black horn cap is secured by knurled blued-steel rod (also serves as fourth section of wiping rod) that locks into notched fitting on barrel at forend front. Black horn buttplate has trapdoor containing three-section nickel-plated wiping rod. Freund-marked front and large folding buckhorn rear sights. Made as a take-down rifle that can be disassembled in the field without tools.
(Author's collection)   ♦351♦

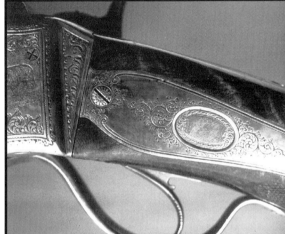

NJF.02   *Freund Alteration of a Model 1877 Sharps Rifle.*

Caliber .45 Sharps, 30-inch octagonal barrel. Serial number 1. Single trigger. Sling rings mounted on barrel and stock. Right side of receiver engraved with bull moose, bear on left side, "F.W. Freund, Jersey City, N.J." on lockplate, "F.W.F. Patd." on both sides of triggerguard. Added left sideplate has uninscribed inlaid gold nameplate. Elaborately engraved on receiver, sideplate, hammer, trigger bar, tang, and screws. Stock of Circassian walnut has cheekpiece, matching forend held with two silver escutcheons on each side has ebony cap and flute-carved ebony block at juncture with receiver. Freund front and rear sights. Presented by Frank W. Freund to Theodore Roosevelt. (*See* color section) (Courtesy Theodore Roosevelt National Park)

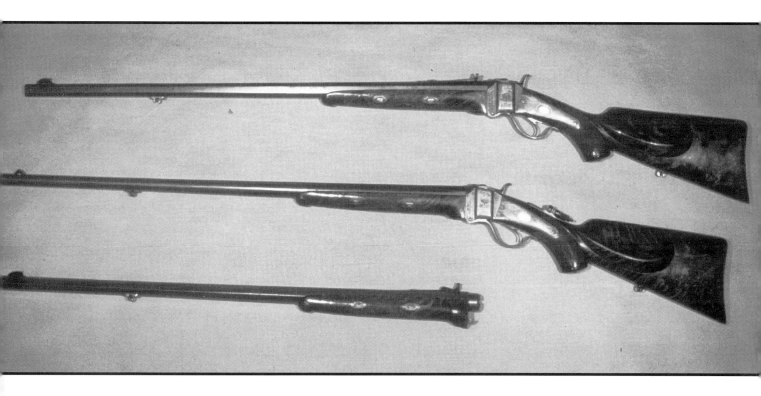

NJF.03 *Freund Alteration of a Model 1877 Sharps Rifle.*

Caliber .45 2⅞, 30-inch round barrel engraved on top flat portion "F.W. Freund, Patentee & Mftr., Jersey City, N.J.", "Nickel Steel, Cal. 30 Smokeless" stamped on bottom. Serial number 2. Single trigger. Sling rings mounted on barrel and stock. Right side of receiver engraved with bear, bull moose on left side, "F.W. Freund, Jersey City, N.J." on lockplate. Added left sideplate has uninscribed inlaid gold nameplate. Elaborately engraved on receiver, sideplate, hammer, trigger bar, tang, and screws. Stock of Circassian walnut has cheekpiece and finely-checkered pistol grip, matching checkered forend held with two silver escutcheons on each side has ebony cap and flute-carved ebony block at juncture with receiver. Freund-marked front sight, Marbles-type peep tang sight. Accompanying this and the previous rifle are a spare 30-inch octagonal barrel with Freund-marked front and rear sights, and ebony-capped forend with flute-carved ebony block at juncture with receiver. Apparently all were presented by Frank W. Freund to Theodore Roosevelt, though it is doubtful T.R. used them on his hunting trips to the West, to Africa, or the Amazon. Both rifles along with the spare barrel and forend were presented to White House staff by T.R. when he left the presidency. (*See* color section)
(Courtesy Theodore Roosevelt National Park)

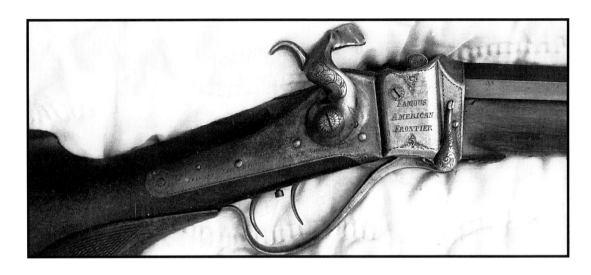

DCG.01   *Freund Alteration of a Model 1874 Sharps Rifle.*

Caliber .45, 24-inch octagonal barrel has Freund patent dates stamped behind rear sight, followed in larger letters by "Freund Patent, June 29, 1880 [date of 'More Light' sight patent] Durango, Colo." Serial number 160856. Freund altered camming action breechblock, double extractors, double set-triggers, Freund-style hammer with extension. Right side of receiver engraved "Famous American Frontier", "Geo. Freund, Manufacturer Dealer of Fire Armes [sic], Durango, Colo." on left side, "Freund Pat., Wyoming Armory, Cheyenne, W.T." on lockplate. Scroll engraved on receiver, lockplate, hammer, lever, trigger bar, and screws. Standard Sharps stock has Freund patent detachable pistol grip with horn cap, matching forend with horn cap, steel buttplate has sliding trap-door containing jointed wood wiping rod. Freund-marked front sight, Freund "More Light" rear sight. Probably originally worked on in Cheyenne, then taken to Durango, Colorado where George Freund finished the engraving. From Carlisle Ranch in the Four Corners area west of Durango; in original family for 103 years until purchased by present owner in 1985.
(David Tawney collection)

DCG.02   *George Freund Repaired Model 1874 Sharps Rifle.*
Caliber .44-77, 30-inch half-round, half-octagonal barrel marked in several places with George Freund's "skull-and-crossbones" registered trademark. Serial number C54381. Double set-triggers. Circular metal plaque set into right side of stock, wrist has tack-bordered, fancy-edged brass repair done in professional manner. Freund-marked front sight, J.P. Lower-style Rocky Mountain buckhorn rear sight.
(Author's collection)

DCG.03   *Freund-Marked Factory Conversion of a Sharps Rifle to Cartridge.*
Caliber .45-90, 28-inch round barrel marked "Business 45", "Old Reliable", "Freund's Patent, June 29,1880", serial number 161667, and with George Freund's "skull-and-crossbones" trademark in several places. Standard Sharps stock and forend, wrist has old rawhide-wrapped repair. Standard Sharps front sight set into dovetail, standard Sharps rear sight.
(Author's collection)

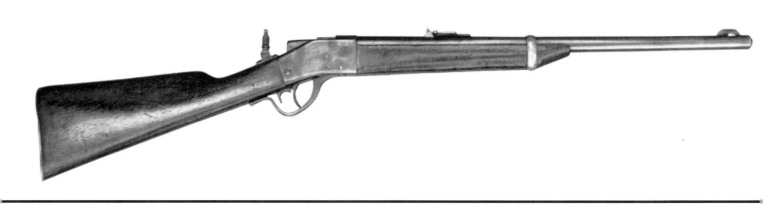

DCG.04 *Freund Alteration of a Sharps 1878 Musket to a Sporting Carbine.*
Caliber .45-70, barrel cut to 22 inches and stamped "Durango, Colo." on top near breech. "J.P. Lower, Denver, Col." originally stamped on top of receiver, overstamped with "Freund's Patent, June 29,1880, Durango, Colo." Serial number 15782. Standard Sharps stock, shortened forend held by single barrel band. Freund-marked front sight set into dovetail, Freund-marked rear sight, small Marbles-type tang sight.
(Author's collection)

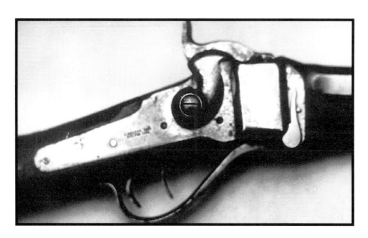

DCG.05 *Factory Conversion of a Sharps Rifle from Percussion to Cartridge.*
Caliber .40-70 straight, 30-inch octagonal barrel stamped with George Freund's "skull-and-crossbones" trademark two inches from muzzle, and again between receiver and rear sight. Serial number C40863 on tang, 40863 on barrel. Right and left sides of barrel have stamped initials "JJA." Government walnut stock. Freund-marked front sight, Freund-marked "More Light" rear sight.
(Dennis Brooks collection)

APPENDIX I:
ENDORSEMENTS

The following list was begun as an alphabetical collection of period endorsements of the Freund Brothers' guns and sights, including excerpted letters from the Freunds' advertisements.

Initially, the driving force was to try to identify some of the original owners of existing Freund guns, as well as to answer some of the questions oftheard by collectors, such as, "If only that old gun could talk."

Of course, inanimate objects don't speak. But it was thought that a name or number might open an avenue to further research, and once the original owner was identified, further investigation might then be undertaken to learn more about a particular gun's ownership, history, and use.

Antique weapons are among the most important and meaningful historical artifacts in existence today. One definition of an historic weapon is that it is an arm for which there is a valid basis for specific association with a person, organized unit, event, or unique historical circumstance.

As only temporary custodians of the antique weapons we possess, gun collectors have both an inherent right and a responsibility to preserve, develop, and promote the history of firearms in America. Additionally, a properly identified historic weapon can be worth many times the value of an identical but unattributed weapon.

A certain few of the identified historic weapons are among the most important and meaningful arms in existence. This became clear when this same list of names was used in the restoration of a rifle that had been severly damaged by fire.

Several years ago the author was asked if he was interested in an engraved Freund-marked Model 1877 Sharps rifle. At that time I owned several Freund pieces, but none was engraved. I learned that the gun in question had once been owned by writer John Amber, and that it had been in a fire that destroyed a number of his rifles. Fortunately, Amber had had the presence of mind to put the guns' metal parts into kerosene as quickly as possible afterward.

The engraved rifle's sideplate and buttplate were missing, as well as its forestock and buttstock. But everything else was there, including Freund sights. On acquiring it, one of the first tasks was to go through existing books and other publications to see if pictures of this rifle's appearance before the fire might be found. We hit the bullseye right off the bat. In *Cartridges of the World* this particular gun was illustrated in the section on .40-90 cartridges. It was only natural, as John Amber had been the editor of that publication in its early years. Further research turned up the gun again, in part two of John Barsotti's article on the Freund Brothers, as published in the 1958 issue of *Gun Digest*.

A description of the gun in that publication stated the following: "A name engraved on the right side is now obliterated." The name apparently had been crossed out with chisel cuts. No clear letters were evident in the mass of chisel marks, and it seemed a hopeless task to determine the name of the man to whom the gun had belonged.

I took the Freund rifle to gunsmith Ed Webber, of Big Timber, Montana, for restoration (Ed had built 1877 Sharps rifles in the past). Because the name had been engraved into the receiver there was no stamped impression left in the metal under the chisel marks. But Ed said he would try planing away the defaced metal and taking smoke tracings at regular intervals. Unbelievably, enough letters and their placement appeared that by using our list we were able to determine the name to be "R. Ashworth." The next step was to obtain information leading to the identification of this Mr. R. Ashworth.

From Appendix I the following was learned:

ASHWORTH, Richard, Powder River Wyoming, late of England.

The next contact was Roger Joyce, historian with the Wyoming Division of Parks and Cultural Resources, who provided us with the following from the *Annals of Wyoming*, Volume 19, number 2, July 1947:

"In 1881, a young Englishman came to the Big Horn Basin to seek his fortune in the cow business. The cow business was being advertised extensively in England. It attracted millions of capital, from the titled gentry to the stable boy, who spent their savings on stocks or shares in the new 'Free Grass Country.' Dick Ashworth, as he was glad to be called, was a good mixer with this raw land and was well liked.

He brought British money and spent well at the only three spots in which to spend, one being Arlend, a new town that was getting started that year and now is a ghost town. Then there was the new town of Meeteetse, a few miles closer to his ranch on the Greybull. He adopted the Double Mill Iron brand, which was a good one, as were most early brands. Men knew how to brand and knew that an intricate brand would blotch and some were tough to work over, while some were easy. You will note that the list of brands in the article were all sensible, fine brands.

Richard Ashworth purchased his cattle in Oregon and a second herd from Sparks and Tinnan in Nevada.

Ashworth later took on an English partner named Johnson and they purchased the Wise brand, which was what was called 'pig pen' and, of course, would be illegal nowadays.

These two, now prosperous cowmen, started a ranch on the head of Sage Creek, known today as the Hoodoo Ranch and owned by U.S. Senator E.V. Robertson of Wyoming. The Englishmen returned to England in the early '90s."

In another section of the *Annals of Wyoming*, the following was found:

"Milo Burke made two trips to Oregon to buy cattle and one trip was for Dick Ashworth of the old Double Mill Iron."

Apparently Ashworth susequently returned to the United States, as Bob and Terry Edgar stated that he owned the Pitchfork Ranch near Meeteetse, Wyoming. I later confirmed this with Jack Turnell of the above ranch.

It seemed only fitting that the Freund rifle belonging to Richard Ashworth should be restored. But why had his name been chiseled out? We can only speculate. Possibly Richard Ashworth was a so-called "remittance man," perhaps a black sheep who had been sent to America by his family and paid to stay here. A number of remittance men were known to have lived in the nineteenth century American West.

The next opportunity to use this data in the identification of a Freund rifle occurred on a visit to Dr. R.L. Moore, Jr., of Philadelphia, Mississippi. Dr. Moore, owner of the original Sharps Rifle Company records, provides information on factory specifications and shipments to the owners of Sharps rifles today. He also owns an extremely fine Model 1877 Sharps rifle, converted by the Freund brothers, for which he had no information about its original owner. There are three entwined initials on the tang of this rifle, which, after considerable scrutiny, appeared to be a "W," a "T," and a "D." Only one combination of these letters could be found on our list:

"W.T. Dodson, Editor, American Field."

The fact that Dodson was editor of *American Field* magazine probably explains why Dr. Moore's Sharps rifle is such a fine specimen and has so many extra features.

The third opportunity to use this list came after I had traded for a very fine Freund rifle, and traced its history back to Pleasant Valley, Arizona. The gun had belonged to the previous owner's great-grandfather, and there was a chance it could have been used in the Pleasant Valley War, also known as "Arizona's Dark and Bloody Ground." The only other information was that his great grandfather said he had won the gun in a poker game. They must have had some real games back in those days, as I'm sure that anyone who has collected for any length of time has owned a gun with the same provenance— "won it in a poker game."

When the gun was brought home and disassembled, I found the name "Ed F. Stahl" engraved under the barrel. On consulting the appendix, I found:

"Ed F. Stahle, surveyor and engineer, United States Deputy Mineral Surveyor, Surveys of Mines, Ranches and Ditches."

Stahl was a specialist from Cheyenne, Wyoming, who on October 27, 1883 had written a letter of recommendation for the Freund patent rifle (*see* the following listings).

A puzzling difference between the name on the rifle and as it appears in the appendix is the "e" at the end of Stahl's last name. However, I have been told by historians that this was a common practice at the time. Another collector has a rifle on which its maker spelled his name once with an "e" and once without— both ways on the same gun.

With the help of the information contained in this still-growing appendix, in the future we should be able to identify more original owners of Freund rifles not only for ourselves but for the many generations of historians and collectors yet to come.

ENDORSEMENTS OF FREUND GUNS AND SIGHTS

The following names of individuals appear in the Freund brothers' catalogs and/or the various advertising broadsides used by both Frank W. and George Freund. This alphabetical list was compiled to aid in identifying the original owners of Freund guns, and it as well contains names of those who wrote to endorse the Freunds' sights.

Adamson, A.D., Stock Range, Horse Creek, late of Montrose, Scotland

Adsit, H.B. County Clerk, San Juan County, Colorado

Altman & Weber, Post Traders, Cheyenne, Wyoming Territory

Anderson, R.B., Stock Grower

Appel, Morris, Freighter

Ashworth, Richard, Powder River, Wyoming, late of England

Babbitt, A.T., General Manager, Standard Cattle Co.

Babbitt, D.H., Miles City, Montana Territory

Baggs, George, Stock Range, Snake River, Wyoming

Bailey, G.E., Territorial Geologist, Cheyenne, Wyoming

Baird Jr., Jesse, Delaware

Barton, Hugh, Range Old Hunting Ground near Ft. Collins, Colorado

Bathorll, John R., Central Association of Wyoming, Cheyenne

Beach, A., Cattle Grower, Laramie River, Wyoming Territory

Benson, John, 507 Montgomery St., San Francisco, California

Bergman, Isacc, Judge of Probate, Cheyenne

Biddle, F.B., Powderville, Montana, late of New York

Bishop, M., 15 Broad St., New York

Boch, S., Veterinary Surgeon, U.S. Army

Booth, Geo. M., Salt Lake City, Utah

Bothwell, John R., President, Central Association of Wyoming

Bradley, P.C., Editor, *American Field* magazine

Brainard, Wm. E., Sacramento, California

Breed, Geo. F. & Hogarth, 32 Broadway, New York

Brown, W.J., Humboldt Wells, Nevada

Bronson, Edgar Beecher, Cattle Grower, Wyoming Territory

Bull, Jos., V.P. Railroad Land Agent

Bubb, Capt I.W., United States Army

Buren, Van, Union Club, New York

Butler, Frank E., Buffalo Bill's Wild West Show

Caribean, Edw., Stockton, California

Chandler, J.A., 55 W. Twenty-Sixth St., New York

Charlton, G.C., The Bonton Livery, Sale and Feed Stables, Cheyenne

Chase, John, Proprietor, Inter-Ocean Hotel, Cheyenne

Clark, Wm., Tourist from England

Cody, Wm., "Buffalo Bill", Nebraska, Hunter, Showman

Collins & Petty, Arms and Ammunition, Omaha, Nebraska

Cotton & Thebbetts, Cattle Growers, Powder River, Wyoming Territory

Crook, Brig. Gen. George, United States Army

Dammann, F.A., Wyoming Armory, Cheyenne

Dater, James, New York City, Ranches on Cheyenne River, Wyoming

Daugherty, W.W., Capt. 22nd Infantry, Ft. Lewis, Colorado

David, Edward C., Surveyor General, Wyoming Territory

Davis, N.R., Cattle Grower, Cheyenne, Wyoming Territory

Deadwood *Times*, Black Hills, South Dakota

Delaney, Wm., Packmaster, U.S. Army

Denver *Tribune*, Denver, Colorado

Dodson, W.T., Editor, *American Field* magazine

Doenhoff, Count, Secretary of the German Legation

Dole, T.R.B., Dole & Bro. Stock Growers, Laramie, Wyoming

Douglass-Willan, L.H., Douglass-Willan, Sartoris and Co., Laramie City, Wyoming

Dowell, J.H., San Miguel, Colorado

Draper, George R., Wholesale and Retail Groceries, Cheyenne

Ducat, Geo., Home Insurance Co., New York

Duddleson, F.J., Alamosa, Colorado

Durbin Bros., Stock Growers, Wyoming

Emmons, G.F., Cattle Grower, Cheyenne, Wyoming Territory

Endlich, Doctor, Smithsonian Institution, New York

Ferris, Major S.P., United States Army

Foster, J.T., Tourist from London, England

Finch, C., Big Spring, Texas, late of England

Frewen Bros., Cattle Growers, Powder River, W.T., late of England

Gardener, James T., Director of the New York State Survey

Gebhard, Fred, Union Club, New York

Gibbons, General, U.S. Army Commander, Ft. Sheridan, Wyoming

Goddard, J.H., Foreman, Union Pacific Railway Shop

Goodell, of Sturgis, Goodell, and Hane

Goodwin, O.P., Stock Raiser, Ranch & Range, Horse Creek, Wyoming

Greer, Geo. N., 47 W. Fifty-Second St., New York

Griswold, C.P., Medicine Bow, Wyoming Territory

Gould, Geo. H., No. 5 E. Twenty-Sixth St., New York

Gutterman & Whittcomb, Cattle Growers, Cheyenne, Wyoming Territory

Gwin, R.S., Cattle Grower and Miner, Wyoming Territory

Haas, Herman, Manufacturer & Dealer in Wagons, Cheyenne, Wyoming

Hammond, J.W., Cattle Grower, Cheyenne

Harrison, S., Cattle Grower, Fort Fetterman

Hartley & Graham, Wholesale Guns, New York City

Hasbrouck, R.E., Ranch on Muddy Creek, Wyoming

Hay, Henry G., Cashier, Stock Growers Bank, Cheyenne, Wyoming

Hayes, Webb Jr., Ohio

Haygood, A.W., Ranch on Lone Tree Creek, Wyoming

Heffron, H.G., Silverton, Colorado

Herr & Herr, Coal Dealers, Durango, Colorado

Hepaley, W.N., Vice President New Jersey Gun Club

Hofman, W.E., Lieutenant 9th U.S. Infantry

Holman, Paul, Indiana

Hoyt, George W., Commissioner, Laramie County, Wyoming

Hoyt, John W., Governor of Wyoming Territory

Hunton, John, Cattle Grower, Brodeaux, Wyoming Territory

Hutchinson, John W., 81 Chambers St., New York

Irwin, Richard, Cattle Grower, Rosita, Colorado

Irwin, Wm. C., Cattle Grower, Cheyenne, Wyoming Territory

Jones, G.M., Proprietor U.P.R.R. House, Cheyenne, Wyoming Territory

Kent, T.A., Banker, Cheyenne, Wyoming

Kent, T.A., Cattle Grower, Cheyenne, Wyoming Territory

King, Clarence, U.S. Geologist, 62 Cedar St., New York

King, Frank, Chief Engineer D.& R.G.R.R., Denver, Colorado

Kingman, Judge I.W., Sheep Grower, Cheyenne, Wyoming Territory

Kurtz, John

LaFreutz, F.W., Secretary Swan Land & Cattle Company, Ltd.

Lane, John, Sherman, Wyoming Territory

Lawson & Bell, New York City

Lawyes, S.E., Oberne, Hoslick & Co., LaSalle St., Chicago

Leach, M.F., Detective U.P.R.R.

Leckler, H.B., 107 Willow St., Brooklyn, New York

Lee & Co., Post Traders & Merchants in Kansas, New Mexico, & Texas

Lehmer, S.M., Twin Mountain Ranch, Wyoming

Longhurst, F.E., Pacific Express Co., Cheyenne, Wyoming

Lord, Major, Quartermaster Cheyenne Depot, Wyoming

Lorillard, Pierre, New Jersey

Lower, John P., Arms and Ammunition, Denver, Colorado

Lykins, W.C., Stock Inspector of Wyoming Territory

Lykins, Wm. C., Detective, Wyoming Territory

McCook, Gen. Ed. M., Denver, Colorado

McFarland, 1st Lieut., 11th Infantry

Madison, E.W., Wyoming

Mason, General, U.S. Army Commander, Fort Russell, Wyoming

Mason, Col. F.M.S., Fort D.A. Russell, Wyoming

May, Boone, Messenger Black Hills Stage Line

May, James D., Miner, Leadville, Colorado

Mayer, L., Powder River, Wyoming, late of New York

Mayer, Max & Co., Wholesaler in Guns, Omaha, Nebraska

Mayer & Oelrichs, Rawhide Butte, Wyoming Territory

Maynard, H.J., M.D., County Physician, Laramie County, Wyoming

Meline, J.A., Buffalo Hunter, Kansas and New Mexico

Merritt, Brig. Gen., United States Army

Michelson, J.M. & J., English Gentlemen

Miller, C.C., Agent U.P.R.R., Tipton, Wyoming Territory

Miller, Charles F., Stock Grower, Ex-Judge Probate, Laramie, Wyoming

Mitchell Bros., Tourists from Pickering, England

Moore, Thomas, Chief Packer, Military Division of the Missouri, U.S. Army

Moores & Sons, John P., Guns, Fishing Tackle, Pistols, New York

Morgan, George F., Importer of Hereford Cattle

Murrin, Col. Luke, Cheyenne, Wyoming Territory.

Oelrichs, Chas., Cattle Dealer, Cheyenne and Powder River, Wyoming, late of New York

Oelrich, Henry, Powder River, Wyoming, late of New York

O'Neil, J.P., Deadwood, Dakota

Organ, C.P., Hardware, Cheyenne

Page, E.A., Thomas & Page, Stock Growers, Wyoming

Palmer, J., Cattle and Horse Raiser, Cheyenne

Peasley, I.C., Burlington, Iowa

Peters, F., Cattle Grower, Powder River, Wyoming Territory

Peters & Alson, Powder River, Wyoming, late of New York

Phillips, Portugee, Cattle Grower, Wyoming Territory

Plessner, Henry, Sidney, Nebraska

Post, E.M., Banker, Cheyenne, Wyoming Territory

Postt, M.E., U.S. House of Representatives, Washington, D.C.

Pratt, J.H., President Pratt & Ferris Cattle Company

Pratt, Joseph G., Union Pacific Railway Land Department

Price & Jenks, Cattle Growers, Cheyenne, Wyoming Territory

Reed, C.F., U.P.R.R. House, Cheyenne, Wyoming Territory

Reel, A.W., Stock Grower

Reel, Att., Stock Grower, Harris Fork, Wyoming

Remington, E., Jr., Ilion, New York

Reynolds, D.H., Big Horn Cattle Company

Riley, George, Hunter & Trader, Wyoming Territory

Ritchie Bros., 4R4 Ranch, Little Powder River, Wyoming Territory

Roosevelt, Theodore, President of the United States, Hunter, Outdoorsman

Russell, H.W., Late Capt. Company C, 99th Regiment Pennsylvania Veteran Volunteers

San Juan Hardware Co., Durango and Silverton, Colorado

Sateris, L., Laramie City, Wyoming, late of England

Schoemaker, A., Association, Evanston, Wyoming Territory

Schuyler, Philip, 18 Washington Street, New York

Schurz, Carl, United States Secretary of the Interior

Schweickert, F., Hardware Dealer, Cheyenne, Wyoming

Seabury & Gardner, Ranch & Range, Bear Creek, Wyoming

Searight, Robert, Cattle Grower, Cheyenne, Wyoming Territory

Sharpless, Oscar, Conductor U.P.R.R.

Sheedy, D., Ranch Camp, Clark, Nebraska

Sheridan, Lieut. Gen. P.H., United States Army

Shilling, W.N., Ross Fork, Idaho

Shiner, M.C., Cattle Grower, San Antonio, Texas

Shudy, D., Ranch Camp, Clark, Nebraska

Smith, Major J.H., United States Army

Spalding, A.G. & Bros., Guns and Sporting Goods, Chicago

Sparks, John, Cattle Grower, Cheyenne, Wyoming Territory

Sparks, Jno., Rancho Grande, Nevada

Stahle, Ed. F., Surveyor and Engineer, Cheyenne, Wyoming Territory

Stanley, J & G.W., Cattle Growers, Cheyenne, Wyoming Territory

Stanton, Fred J., Surveyor, U.S. Land Attorney, Cheyenne

Stollsteimer, C.F., U.S. Indian Agent, Ignacio, Colorado

Strain, J.R., Pioche, Nevada.

Strong, General, Chicago

Sturgis, Thos., Secretary Wyoming Live Stock Growers Association

Sturgis & Goodell, Cattle Growers, Cheyenne, Wyoming Territory

Sturgiss & Lane, Cattle Growers, Cheyenne, Wyoming Territory

Stuyvesant, Rutherford, 246 East 15th Street, New York

Swan, A.H., Vice President First National Bank, Cheyenne

Swan Bros., Cattle Growers, Cheyenne, Wyoming Territory

Talbott, Major John, Cheyenne, Wyoming Territory

Taylor, Wm. H., Leadville, Colorado

Taylor, Wm., Quartermaster Agent of Rock Creek, Wyoming Territory

Terry, John, Hunter, Wind River Valley, Wyoming Territory

Thayer, G.D., General Merchandise, Rock Creek, Wyoming Territory

Thayer, Gen., Sheep Grower, Cheyenne, Wyoming Territory

Thomas, John B., Thomas and Page Stock Growers, Cheyenne

Teschemacher & DeBillier, Stock Growers, Range North Platte, Wyoming

Van Nostrand, L.H., Manager Grandview Mining & Smelting Co., Rico, Colorado

Van Vliet, Col. F., Third Cavalry, Fort D.A. Russell, Wyoming

Vegnoles, H.H., Stock Grower, Cheyenne

Voorhees, Luke, Superintendent Cheyenne & Black Hills Stage Line

Wakefield, John, Tourist from England

Wallace, D.H., Peoples Bank, New Castle, Pennsylvania

Warren, F.E., V.P. First National Bank, Cheyenne, Wyoming

Waters, J.H. Ernest, Superintendent Sheridan Mine, Telluride, Colorado

Whightman & Co., Fort McKinney, Wyoming Territory

Whitcomb, E.W., Stock Grower and Merchant, Cheyenne, Wyoming

Wild, John E., Cashier, 1st National Bank, Cheyenne, Wyoming Territory

Wilson, Prof. A.D., Geological Surveyor

Wilson, Posey S., Fort Collins, Colorado

Wingate, General George W., President National Rifle Association 1886-1906

Winn, A.J., Big Horn Cattle Company

Witsuk, Charles, Steward County Hospital, Laramie County, Wyoming

Wolcott, Major F., Cattle Grower, Fort Fetterman, Wyoming Territory

Wright, Charles W., Attorney General State of Colorado

Wyman, O.D., M.D., Wyoming

Wyoming Meat Company

Yeath, John, Camp Stambaugh, Wyoming Territory

Young, Captain Robert H., United States Army

Yount, Harry, Scout and Hunter

Zimmerman, F.C., Gun Maker & Dealer in General Merchandise, Dodge City

The following testimonials were excerpted from the Freund Brothers' catalogs, broadsides, and other period advertising:

I have much pleasure in recommending Mr. Freund's patented sights, it being the best rifle sight for huntsmen or sportsmen I have ever seen.
—A.D. Adamson
Stock Range, Horse Creek, late of Montrose, Scotland

* * *

H.&M.D. Altman
Wholesale Dealers in Fine Whiskeys, Wines, Cigars, Tobacco, Pipes and Fancy Goods.
Cheyenne, Wyo. Terr.
Oct. 20, 1883
Wyoming Armory Cheyenne Wyoming
Dear Sir:

We have had in constant use at our store here and at our sutler store at Ft. Fetterman your improvement on Rifle for some time. It gives perfect satisfaction as it will stand and make a reputation on its own merits. It does not require much attention and so simple that a ten year old boy can take care of it. All stockmen that have seen your pattern at our northern store only regret that the Rifle is out of their reach as the present prices are too high. We seen a Indian at our Ft. Fetterman store which had one of your improved Rifle & sight. We asked him what he thinks of it, he said I will continue to use it as long as I stay on this globe and when I leave for the next, I will take one along, as I expect to locate where they do have plenty of game. He also said all good Indians who anticipate emigrating to that glorious country should have F.W. Freund's Improvement on Rifles. The only Indians who have no use for this Rifle are those who are preparing to visit the country not called the happy hunting ground.
Yours truly,
Altman & Webel Post Traders
H. Altman & Co.

* * *

The Freund gun and sights are perfection.
—R.B. Anderson
Stock Grower

* * *

Fort Laramie, Wy. Ter., June 1st, 1878
Messrs. Freund Bros., Cheyenne, Wy.:

Gentlemen, —*I have been for many years engaged in the freighting business, and while on the road, am at all times handling arms. The first time that I really felt lost, with a good gun in my hands, was in 1876. I had a good Sharps rifle, but had used up all the factory cartridges, and after reloading them I attempted to load my rifle, and found it impossible, as I could not force them in. I am not a mechanic, but still I know that there was a defect in that gun. Since having my rifle changed to your system, it gives me perfect satisfaction in every particular, and I shall be pleased to make my ideas of the gun known to all freighters and hunters of my acquaintance, and recommend it for their use.*

Very respectfully yours,

M. Appel

* * *

OFFICE OF TERRITORIAL GEOLOGIST AND MINING ENGINEER OF WYOMING
Cheyenne, Wyo Oct. 12, 1883

To Whom It May Concern:
This is to certify that I am well acquainted with the rifle manufactured by F.W. Freund of Cheyenne, Wyoming, having used it both at the target and in the field. After an experience of fifteen years on the frontier I am free to say that in the patent sight and breech action, as well as in other points, it is superior to any other rifle in this country. In my journeys I constantly meet frontier men who have these rifles, and they, without exception, give unstinted praise as to their uniform character and quality.

—G.E. Bailey, Territorial Geologist

* * *

I have known Mr. Freund for many years, and know him to be a good workman. He has given me every satisfaction, and his rifle has no equal.

—Hugh Barton
Range Old Happy Hunting Ground near Ft. Collins, Colorado

* * *

Freund's guns and sights are good enough for me, or I would not pay the price.
—George Boggs
Stock Range, Snake River, Wyoming

* * *

Messrs. Geo. Breed & Hogarth, Exporters
New York, Oct. 17, 1879
The rifle arrived safely yesterday. We must say the improvement you have added to it increases its original value more than twofold, and as our correspondent in Africa is an old hunter, he will at once appreciate your skillful and practical alteration; and we wish you the success which you so justly deserve.

* * *

Miles City, Jan. 10, 1881
After much delay my gun was received on October 1, 1880. Owing to my great hurry to get out among the buffalo, I did not write to you as I should have done. I have killed over 700 buffalo since then with your gun, and it worked like a charm during the coldest weather. I have shown it to a great many hunters, and they are all very much pleased with it. You will soon receive more orders from this place.

Very respectfully,

D.H. Babbitt
[The rifle above referred to was Freund's .40 caliber.]

* * *

Headquarters 5th U.S. Cavalry
Fort D.A. Russell, Wyoming, May 16
Messrs. Freund Bros., Cheyenne, Wy.
Gentlemen, —*The Sharps rifle, model of 1878, altered by you, is a perfect success. Your new breech-block works to perfection. I fired sixty-five shots in ten minutes without cleaning, the cartridges being more or less dirty, nor did any shell stick in loading or being expelled, the old trouble of forcing the shell in with the thumb being done away with. Trouble was always had on account of the cavity in the breech being so narrow that it made it difficult to force all of the cartridge into the barrel, especially if the shells were not perfectly clean.*

Yours respectfully,

S. Bock
5th U.S. Cavalry
[See also the letter from P.C. Bradley, dated April 13, 1886]

* * *

From General G. Crook
Headquarters Department of the Platte,
Commanding General's Office.

The gun arrived yesterday, and after carefully examining it, I take pleasure in saying that all the repairs have been made to my complete satisfaction, and in a manner to demonstrate not alone skillful workmanship, but a thorough knowledge of your business. Your patent sight is so valuable an improve-

ment and yet so simple that I cannot help wondering why it was not sooner discovered. It serves equally well as a "peep" and "open" sight, and can be used without reference to the manner in which the sun may be shining. In my opinion, it fills a want long felt, and beyond improvement. It only needs to be known to be adopted by all sportsmen throughout the country.

* * *

(WHAT GEN. GEO. CROOK SAYS)
Headquarters Dep't. Platte; Com'dg. General's Office
Omaha, Neb., April 30, 1878

I have received the Sharp's [sic] rifle left in your hands for repairs and alteration. The double ejector spring and the rounding off of the inner edge of the breech block are improvements which only need to be seen by practical men to commend themselves for general use. The beauty and perfection of workmanship in this, as in other repairs I have had made at your establishment, are entitled to the highest praise.

* * *

No one in the West can dispute the superiority of Freund's Patent Rifle.

—T.R.B. Dole
Dole & Bro., Stock Growers, Laramie, Wyoming

* * *

Sundance Hill, June 12th, 1878
Gen'l. Bradley's Camp Expedition
Deadwood, Dakota

Messrs. Freund Bros., Cheyenne, Wyoming:

Gentlemen, —I have been for a long time engaged on the frontier, and for many years I used a Sharps Rifle, but always found considerable difficulty, while out, in forcing the cartridge into the barrel of the rifle, after firing it a few times, and thereby permitting the chamber to become foul, and cartridges that I reloaded without a swedge, I could not use at all.

I have had my rifle changed to your new improved breech movement, and had it with me and in constant use all through the Indian trouble in the Big Horn country, and find that I have entire confidence in my rifle now. I, like all other persons that are engaged in out-door life, depend on my rifle a great deal of my time for subsistence, and often for life. I hope it will not be long 'ere the Sharps Rifle, with the Freund's improvement, will be the only gun called for and in use on the frontier.

Yours respectfully,
William Delaney
Packmaster of Bradley's Command

* * *

Mr. Freund's rifle is the best single barrel rifle I have ever seen.

—L.H. Douglass-Willan, Douglass-Willan, Sartoris & Co., Laramie City, Wyoming

* * *

TO ALL WHOM IT MAY CONCERN
(From the Cheyenne *Daily Sun*, June 20, 1878)

To illustrate the perfection of the Freund Bros. mechanism, when applied to the "Sharps rifle," the following trial was made yesterday, June 19th:

To find out whether there would be any possible chance for a cap to burst, so as to throw the gas rearward, (which has been such a serious defect in the Sharps rifle), and also to show the resisting strength and perfect operation of the improved rifle, a 40-calibre Sharps rifle was loaded with 150 grains powder and three bullets, Sharps make, swedged and patched, each weighing 370 grains, making 1,110 grains of lead, and fired. No defect in the mechanism was shown, the cap did not break, and not a particle of gas escaped rearward. After this trial, without cleaning the gun barrel or breech, ten shots were fired in less than a minute, and conveniently taking aim from the shoulder with 90 grain cartridges, and the result was that the last as well as the first cartridge was inserted and extracted without any difficulty, and I am convinced that if a thousand shots should be fired out of the rifle without cleaning the same, there would be no more difficulty encountered. Therefore, I would say to those that wish their rifles changed, that every rifle will undergo the same test, and in their presence if desired.

The Freund Brothers have explained to me the merits of their improved firing-pin, which was patented January 2nd, 1877, and have convinced me that it is an impossibility for a cap to break, and it also has an additional cover on the breech-block, closing the same gas tight rearward. A rifle with this improved breech mechanism is in my opinion superior to any other gun, and I therefore would recommend it to all who desire a perfect and reliable arm.

—F.A. Dammann
Wyoming Armory
Cheyenne, W.T., June 19, 1878

* * *

In a recent issue of the Denver *Tribune* we find the following notice of an improved Sharps rifle, the work of Mr. F.W. Freund, of this city:

On calling upon Mr. J.P. Lower at his gun store yesterday, a Tribune *reporter was shown the latest improved Sharps rifle, which all who have seen it have pronounced the most perfect*

arm that it has ever been their pleasure to examine. The improvements are the inventions of Mr. F.W. Freund, formerly in the gun business here, but now at Cheyenne. To say that the improvements are very valuable would be simply to reiterate the conclusions of the many who were present and examined the arm with the reporter. It certainly deserves at our hands more than a passing notice.

The rifle as changed under these improvements, does away with all the objectionable features of the old Sharps breech system, leaving only the good, the perfect resisting qualities and movements undisturbed. This is done, too, without the addition or use of a single extra piece. The improvement on the breech action so modifies the old model Sharps as to let the breech block swing back when out of the direct line of resistance, affording ample space to admit the use of double extractors, permitting the cartridge to project rearward some distance, which on the closing of the breech is gradually and with ease forced into the chamber. This is accomplished by the same movements of the lever as under the old system, and the cartridge, too, being pressed over the breech-block is held in position by an eccentric, so that the gun, held in any position, will not cause the cartridge to drop out of the gun before the breech is closed.

Another improvement consists of a very simple change in the construction of the firing pin, so that by either opening or closing the breech the hammer is carried to half-cock or into the safety notch, thus preventing the frequent premature discharge of the arm occasioned by the closing of the breech with hammer resting on the firing pin.

Lastly, we come to two more patents, the detachable pistol-grip and the combination front sight. The pistol-grip is made so that it can be put on any kind of an arm without the least difficulty and gives to the gun a more complete and symmetrical appearance, and to the operator better facilities for taking an accurate aim. The combination front sight is made so as to form nearly a perfect pinhead with a white center and dark shoulders, which shows light on dark objects and dark on light objects. To go into the minute details and point out the many defects of the Sharps rifle would be both tedious to the reader and a waste of time and space. For every one who has ever used a Sharps rifle, knows wherein its chief defects lie, and will consequently appreciate above all others, the improvement which affords the requisite facilities for forcing the cartridge into the chamber without even marring the strength of the old Sharps breech system for which it has a world-wide reputation.

* * *

The following is from the Deadwood *Times*, the leading paper of the Black Hills, and our readers will see from this what is thought of the Freund's inventions, etc., in "the Land of Gold":

Mr. Geo. Freund, of the firm of Freund & Bro., Cheyenne, has arrived in the city with a full and complete sample of arms, ammunition, etc. Mr. Freund is at Stebbins, Wood & Post's bank, where he may be found ready to take orders for anything in the gun line. The past reputation of Freund & Bro. throughout this section of the country is a sufficient guarantee that everything coming from their hands will be first class in every respect. Mr. Geo. Freund has with him the latest improved Sharps rifle, which has commanded the admiration and excited the curiosity of all throughout the West. Military men and hunters, alike, pronounce it the best gun in the world for army and frontier service. The Daily Tribune *of Denver and the Cheyenne* Sun *concur in the same opinion. The improved Sharps does away with the defects of the old Sharps breech, while every important feature of the old model is preserved. The gun as constructed under Freund's improvements, offered all the requisite facilities for forcing the cartridge in the chamber when projecting rearward some distance, and the hammer is carried to half-cock by either opening or closing the breech. In addition to this, Freund has two other improvements which are attached to this rifle: The combination front sight, showing dark on light objects and light on dark, and the detachable pistol-grip.*

After thoroughly testing the matter of using a plain or patchable bullet, Freund & Bro. have adopted with great success the plain and unpatched explosive bullets in the use of their new gun. This does away with the patch, which hunters and others have always found difficult to preserve intact. To prevent the patching from getting crumpled or worn by carrying, has always been a source of trouble and care to the hunters and others. In using the unpatched ball, no trouble of this kind is experienced, while the quality of shooting is just as good. It is impossible to give a minute and detailed description of the arm in question, owing to our limited space, but let it suffice to say that the gun, in our opinion, is far superior to any other gun in use for hard service, and all will find it worth their while to call on Mr. Geo. Freund, who will be pleased to show the arm and give the necessary explanations.

[*See* also the letter from W.T. Dodson, dated April 24, 1886]

* * *

I think Mr. F's rifle about the best in America.

—Richard Frewen
Manager of Frewen Cattle Company of England

* * *

(Solid Shots—Perforating the target—Return rifle match between the Fourth Infantry and Wyoming rifle teams—The Wyoming wins.)

At the conclusion of the rifle match, the members of the Fourth Infantry team announced that they were ready to challenge any team, provided they be allowed to use the Freund's improved rifle. They believe themselves the best marksmen, but their guns are faulty. It was conceded by everybody who witnessed the match that the rifle, with Freund's improvement, is the best arm in existence. As this is a Cheyenne production, it is a matter of pride to note the fact.

—Cheyenne, Wyoming Ter., May 16, 1879.

* * *

Fast Shooting

Yesterday Mr. F.W. Freund tested, at Fort D.A. Russell, a Sharps hammerless rifle of the model of 1878, which he changed to Freund Bros.' new improved breech movement. He fired fifty-four shots in five minutes, taking the gun from his shoulder, loading and sighting it at every shot. He used Government cartridges, which were taken without care or selection, and the exhibition was satisfactory in every respect, being admired by all present. This gun is made with safety trigger, which has to be moved at every shot in order to release the proper trigger. The breech movement is changed on the same principle as in the old Sharps to force the cartridge into the barrel, which it is almost impossible to do by the pressure of the thumb after the gun has been fired a few times, but goes easily into place by Freunds' improvement.

* * *

Mr. Freund makes the best rifle in the world.

—G.B. Goodell, Sturgis, Goodell & Lane
—E.A. Page, Thomas & Page, Stock Growers, Wyoming

* * *

Nothing better is wanted in the West, where firearms are most used.

—O.P. Goodwin
Stock Raiser, Ranch and Range, Horse Creek, Wyoming

* * *

C. Freund

Dear Sir —I have your improved rifle with new sights, and take pleasure in stating that it excels in every respect any and all of the rifles I have ever before used, and I have used all kinds of the latest magazine and breech-loading rifles. The sights are well adapted to the various changes of light. I cannot speak in too high praise of the skillful workmanship em-

ployed by you in preparing for me a rifle beautiful in design and most reliable in accuracy of shooting. Yours respectfully, etc.,

—Geo. M. Greer,
47 W. Fifty-second St., New York City

* * *

They [the Freund guns] *are the favorite on the line of the U.P. R'y.*

—J.H. Goddard
Foreman U.P. R'y. Shops

* * *

Office of H. Haas, Manufacturer and Dealer in Wagons, Carriages, Buggies, Hard Wood, Iron, Steel, Chains, Ox-Yokes, and Freighters' Supplies. Blacksmithing and Repairing of all kinds done to order.
Cheyenne, Wyo. Oct. 26, 1883

Mr. F.W. Freund, Cheyenne, Wyo.

Dear Sir:

With an experience of over twenty-one years on this western frontier, part of which was spent in the army on the plains fighting Indians, and part in civil life, and having used and seen almost all kinds and styles of rifles, I have come to the conclusion that your rifle is equalled by none and that it is the best in the world for strength, reliability, simplicity, and durability. The mechanism seems to be indestructible and the rifle can always be depended upon to do its work, and good and effective work, too.

I consider your Patent Sights the best upon principle and the best in practice. I will relate to you a recent experience which I had with them. We were having some fun at the ranch shooting at glass bottles. It was a very bright day with a strong eastern twilight. Several of the boys had rifles and others had shot guns. A number of shots had been fired but no bottles had come down. But when I commenced firing I brought down a bottle every time. I do not claim that I am a better shot than many who were there, but I did have a set of sights which did not deceive me. And I say that, not only from its reputation, but from my own experience that your gun and sights are perfect in construction. The material you use is of the very best and the workmanship is not excelled by any gunmaker in the world.

Yours truly,

Herman Haas

* * *

STOCK GROWERS NATIONAL BANK
Thos. Sturgis, Pres., W.C. Lane, Vice Pres.,
Henry G. Hay, Cashier, E.A. Abry, Asst. Cashier
Cheyenne, Wyo. Oct. 11, 1883
To Whom it may Concern:

This is to certify that I have been acquainted with Mr. F.W. Freund for nearly fourteen years. His inventive ability and industry have become well known traits in his character throughout this community. I have examined his various improvements in firearms from time to time with great interest, and I believe his improved sights, and improved breech mechanism are very valuable ones. My own favorable judgement has often been confirmed by the very flattering reports I have heard from hunters, Army officers and ranchmen among my acquaintances. There is no doubt in my mind, as to the value of his inventions, and as to the profit of manufacturing his guns on a larger scale.

Very Respectfully,

Henry G. Hay

* * *

Fort D.A. Russell
Mr. Freund,

Dear Sir—I have tried your illuminated rifle sights placed by you on my Sharps sporting and hunting rifle, by firing the rifle with a bright sun light shining in my face, on my back, across the line of sight from either side; also in firing in cloudy, dark weather; in the imperfect light of twilight; also firing at dark and light colored objects, and under all these diverse conditions they have given uniformly satisfactory results. They are correctly placed and wonderfully accurate, and seem to me to be about perfect in all respects. I would not give them up and attempt to do without them, or others like them for three times the price charged.

Respectfully yours,

W.E. Hofman, Lieutenant 9th U.S. Infantry

* * *

I call it the best gun in the land.
—R.E. Hasbrouck
Ranch on Muddy Creek, Wyoming

* * *

Only try it.
—A.W. Haygood
Ranch on Lone Tree Creek, Wyoming

* * *

Cheyenne City, Oct. 10th, 1883
The undersigned, after a personal and business acquaintance of several years with Mr. F.W. Freund, proprietor of the Wyoming Armory, and originator of many valuable and important improvements in firearms, has great pleasure in recommending both him and his inventions to the public. I know it to be the opinion of prominent military men, sportsmen, hunters, and ranchmen throughout the Rocky Mountain region, that the gun manufactured by him, and furnished with his improved sights, breech block, and double ejector, is without a rival, and that if manufactured on a large scale at some central point, it would soon come into very general use throughout the country. Besides being a successful inventor and skillful mechanic, Mr. Freund has an excellent standing as a competent, efficient, and honorable business man, fully entitled to the respect and confidence of any and all persons who may have transactions with him.

—John W. Hoyt
Late Governor Wyoming Territory

* * *

Ogallalla, Neb.,
January 21st, 1878
Messrs. Freund Bros., Cheyenne, Wy. Ter.:

Dear Sirs—For the last five years I have used the Sharps rifle, shooting the 40 cal. bottleneck shell, and must say that it is the strongest shooting gun, and therefore the best gun for the plains.

The objections I had against this style of gun, you have entirely done away with, by applying your breech mechanism thereto.

The last gun with your improvement which I bought of you, is a gun of which it can truly be said, that it is the best gun known up to this date, and I feel secure with this rifle in hand against any danger, even in the roughest handling of the same, and therefore highly recommend the same to those that pass through countries where a reliable gun is their best friend.

Yours very respectfully

W. C. Irvine

* * *

T.A. Kent, Banker, Cheyenne, Wyo.
Oct. 12, 1883

To Whom It May Concern:

Dear Sir:

I take pleasure in recommending to the public Mr. F.W. Freund of this city as a man in my opinion unexcelled in inventive and mechanical ability, and that I consider his patent sights and breechloading appliances the best of any in use. I have seen them put to a practical use and believe that they will eventually supersede all other inventions of the kind. I have used several kinds of breechloading guns and find his "Freund's Improved" superior to all.

Respectfully,

T.A. Kent

* * *

Fort D.A. Russell, Wyo.
Oct. 15th, 1883
F.W. Freund Esq.
Wyoming Armory, Cheyenne, Wyo.

Dear Sir:

Your patent sights which you put on my Springfield rifle are the best I have ever used. I take pleasure in stating that in my judgment they are by far the best sights in use.

I have also given your rifle a careful examination and am satisfied that if it could be manufactured at a reasonable price by machinery, that no sportsman would be without one and that it would very shortly supplant most of the breechloading rifles now on the market. It is certainly the most complete rifle I have ever seen.

Yours truly,

Your Ob't. Ser.
F.M.S. Mason
Col. 9th Inf'ty. U.S.A.

* * *

Mr. Freund,

Dear Sir—*The rifle you made for me gives perfect satisfaction. I am now glad I did not have a telescope put on the rifle. I can shoot very well with your patent sight. I am sixty-five years of age, and I believe that on a clear day I can keep within an eight inch circle, at four hundred yards, shooting all day. I have also written to Mr. Mayer, of Newport, who gave me your address, the result of such a good rifle.*

I remain yours, very respectfully,

John Kurtz

* * *

The Pacific Express Company, Cheyenne, Wyo.,
F.E. Longhurst, Agent
Oct. 30, 1883

Mr. F.W. Freund, Esq.
Mfg. Freund Sights & Rifle.

Dear Sir:

It is useless for me to say much regarding your work in general as I and all can see by the amount of business transacted by you through our company, both forwarding and receiving, that it all gives good satisfaction.

Yours truly,

F.E. Longhurst

* * *

Major Lord, Quartermaster
Depot Quartermaster's Office, Cheyenne Depot,
Mr. Freund.

Dear Sir—Your patent sights which I have just put on my Winchester repeating rifle, 45 caliber, have given such perfect satisfaction, that I take pleasure in endorsing them as superior to any sights heretofore used by me.

* * *

(From John P. Lower, gun dealer, and member of the Denver Gun Club)

March 5th, 1880

It was surprising what accuracy was obtained in shooting at a target off-hand at 500 yards. W.J. Foy, the president of our club, who is second to none at long range shooting, tried it, and was well pleased with its performance. It being an eight-pound rifle, .40 cal, and 70 grains of powder the results were excellent.

* * *

(What Mr. John P. Lower says in his letter of March 3rd, 1880)

I consider the model of 1874—Sharps rifle with your improved breech action—as being the most perfect and reliable arm to be found anywhere.

* * *

The Middleton Affair

It will be remembered by the Leader *that Detective Lykins, in his fight with the outlaw, Middleton, snapped his rifle upon four different cartridges, and each one missed fire. Considerable surprise was manifested by all who read the account of the affair, inasmuch as it was thought strange that Mr. Lykins should provide himself with an arm or ammunition that should be the least defective. It seems not to have been his fault, however, for he had an arm which is reported to be reliable—the Sharps rifle. Nor was it the fault of the cartridges, for he brought this ammunition to Mr. Freund, of this city, and introduced the cartridges into one of their improved Sharps, and each one of the supposed defective cartridges was exploded. Thus it seems that it was the fault of the so-called "Old Reliable Sharps", which in this case, proved itself very unreliable, and came very near costing him his life, and in all probability, was the means of the non-arrest and capture of one of the most desperate outlaws that ever traveled the plains. Mr. Lykins says he has enough of the "Old Reliable" now, and accordingly ordered one of Freund's Patent Rifles.*

* * *

Mr. Freund's patent rifle is the most saleable rifle in the West.

—M. Lehmer
Twin Mountain Ranch, Wyoming

* * *

Max Meyer & Co.
Est. in 1866.
215 to 223 South 11th St., 1020 to 1024 Farnam St., Omaha, Neb. Wholesale Dealers in Guns, Ammunition, Cigars, Tobaccos and Pipes and Fancy Goods, Cutlery and Sporting Goods.

Omaha, Neb., Nov. 20, 1883

Mr. F.W. Freund, Cheyenne

Dear Sir:

Your Patent Rifle with Improved sights we duly received; and at your request we had it thoroughly tested by some of our finest marksmen and hunters, all of whom have pronounced it as superior to anything they have heretofore tried; for long range and accuracy it is without a rival. You have a "Bonanza" in it, as it is the coming gun.

Yours Resp.

Max Meyer & Co.

* * *

(The following is from the Cheyenne *Daily Leader* of October 17th, 1879)

A successful hunt— *The English gentlemen, Messrs. J.M. Michelson and J. Michelson, who left our city on the 7th of last May for a pleasure hunt through the country north of Laramie Peak, have returned to this city. These gentlemen are from Yorkshire, England, and came to this country to enjoy the sports of hunting antelope, deer, elk, bear and such other game as abound in this country. On their arrival here, they were fortunate enough to secure the valuable services of Harry Yount, our well-known guide and scout, who led them over the wild plains and back again to Cheyenne. They speak in the highest terms of praise of their hunting trip, and relate many incidents of the hunt both exciting and amusing.*

They relate an instance of the killing of an old grizzly bear which they came upon near the foot of the mountain. He was a very large one, as his hide fully attests, and had evidently done a great deal of mischief. They state that the animal, after it was fatally shot, grasped with its mouth small trees measuring two or three inches in diameter, and severed them as quickly as though they were mere pipe stems. In its final death struggles, it seized its own paw and crushed it to a jelly. When dressed they found two old rifle bullets, which had been apparently shot into him a long time ago. Each bullet, too, was from a different make and caliber of rifle, showing that he had been shot formerly by two persons. These gentlemen, however, armed with the F.W. Freund's Patent Rifle, found no difficulty in putting a quietus upon this savage animal, though in its death agonies, it exhibited its wonderful strength and savage ferocity for which it is most to be dreaded of all the wild beasts found on the American continent.

* * *

Mr. P.H. McEvoy, a gunsmith and dealer in arms at North Platte, Neb., writes as follows to a firm in this city:

North Platte, Neb., December 15, 1880.

The rifle that I sent you to have your patent sights put on came back all right. I have tried the gun at all ranges, and I will say that they are the best I have ever seen; in fact they are the only ones that ever gave me any satisfaction, and I have been in the gun business for many years and used all kinds of sights. You could not buy mine for $25 now, if I could not get others.

* * *

It was left for the Far West to turn out the best gun with the best sights in the world.

—H.J. Maynard, M.D.
County Physician
Laramie County, Wyoming

* * *

To Whom It May Concern:

J.A. Meline, of said county and Territory, being by me duly sworn according to law, deposes and says that he is a hunter by profession, and has been engaged in such business for twenty-two years; that he is well acquainted with the rifle known as Sharps Breech-Loading Rifle, and has used it exclusively for five years; that he has had them made to order for him at the factory, and always has considered it the best rifle in use, finding but one objection to it—that there is found a difficulty in forcing in the cartridge after the rifle has been fired a few times, and by reason of this trouble, or difficulty, he has known many accidents to occur, such difficulty being more observable where the rifle has a small bore, and a long cartridge is used.

This deponent further says that he has personal knowledge of the following accidents as having occurred from the said defect in the rifle:

While attacked by the Cheyenne Indians in the Fall of 1874, on the north fork of Smoky Hill River, one Charles Brown, a hunter, lost his life by his inability to force a cartridge into his rifle—which was one of Sharps breech-loading rifles—soon enough. The Indians succeeded in taking his life before he could put his rifle into a condition to fire, the rifle being found after his death with a portion of the cartridge exhibited out of the gun, showing the fact of his difficulty in placing it in.

Also, that one McLaughlin, a hunter of experience, was severely injured in the hand and lost one finger by reason of the explosion of a cartridge while he was in the act of forcing it in, requiring the aid of a stick.

That one Henry Campbell and another James Campbell, both hunters, were injured in the hand and eye, while in a similar act as above mentioned, and that he has knowledge of many others injured thereby, and that he has experienced considerable trouble with the rifle in that respect.

This deponent further says that he has lately been shown an improvement in the said Sharps Breech-Loading Rifle, made by F.W. Freund, of Cheyenne, Wyoming, it having an improvement of the breech block which avoids the danger of all such accidents, and that he admired the said improvement very much, for which said improvement this deponent is informed that the said inventor, F.W. Freund, has applied for a patent; and this deponent says that it is his opinion that no man using rifles will want any other kind after once seeing this improvement; that great credit is due the inventor, which all hunters will appreciate.

—J.A. Meline
Subscribed and worn to before me, this
11th day of April, A.D. 1876
[seal] E.P. Johnson
Notary Public

* * *

Miller, C.C.
Agent U.P.R.R.
Table Rock, Wyoming Territory

Table Rock
July 15, 1878

Freund Brothers Cheyenne, Wyo. Terr.

Dear Sirs:

The Sharps rifle with your new improved breech mechanism is as perfect a firearm as could be produced. I take considerable pride in owning a good rifle but I had begun to get discouraged in trying to find one. I have owned and tested all of the modern breech-loaders now in use. They all shoot well, but in the breech mechanism they are far from being perfect. I discarded all of them in preference to the factory-made Sharps sporting rifle—weight 12 pounds, .40 caliber shooting 90 grains of powder. The one I now own makes the third one of the same kind. I sold the former, and was going to get rid of this one on account of the caps in the shell bursting nearly every time the rifle was fired, and letting the gas come back through the hole of the firing bolt into my eyes nearly blinded me. Another objection to them that used to bother me badly was when I went hunting, and got up within fifty yards of four or five hundred antelope, all in a bunch, maybe the first shot fired would be one of the old shells that had been reloaded three or four times, and in trying to extract the empty shell after firing (there being an extractor on only one side), I would pull the head of the shell off, leaving the remainder sticking in the barrel; that would end the shooting right there; would have to come home and spend about half a day with a ramrod and hooks, trying to get the rest of the shell out. This would happen, too, nearly every time I went hunting; but with yours, if I had only one extractor, it could not happen, as the extractor falls back twice the distance of the Sharps, and takes a better hold of the shell; further you have not cut in the breech block for extractor, while in the Sharps there is an opening for dust to accumulate, and make the extractor in such a case useless. Another thing that would occur very often was, when trying to insert an old shell in the rifle it would get stuck just before getting it in far enough to clear the breech-block, and it would be impossible for me to get it either in any further, or out. This also ends the shooting, unless you have a ramrod along to punch the cartridge back out again, as it is very dangerous to try and force it in any further by pounding, etc. It is impossible for any of these things to happen with your improved breech-block.

Another fault with my Sharps was that after firing them and extracting the empty shell, the hammer had to be drawn back to half-cock for safety before inserting another cartridge and raising the lever. With your improvement, the simple and easy way in which the hammer is drawn back to half-cock, either in throwing the hammer down or up, making the same perfectly safe against premature discharges in loading or carrying the arm; the improved firing bolt and making the firing hole so much smaller, thereby making it impossible for a cap to ever burst, or for the firing hole to get clogged up with brass off the caps, is perfect.

The rifle I have now was open to all objection I have stated, as you well know, until I heard of your recent improvement, when I determined to try it, and I am very glad I did so, as you have made it the kind of a rifle I have long been trying to obtain, and I would not use any other. I have fired the rifle about two hundred times since I got it back from you, six weeks ago.

The most of the shells used have been reloaded as many as twenty-five times each, but the gun works exactly as well with them as new shells that have never been used.

For sportsmen, hunters, and men in an Indian country where it is an imperative necessity that a rifle should never fail them and where, if it does, they are liable to lose their lives, your improvement is invaluable to them, as it can never fail to do all that is claimed for it.

I remain,
Yours very respectfully,

C.C. Miller
Agent U.P.R.R., Table Rock, Wyoming Territory

* * *

I consider Freund's rifle and sights the Boss.

—E.W. Madison, Wyoming

* * *

Have known Mr. Freund for eight years; have full confidence in his mechanical skill and business integrity; have used his patent breech-block and sights since their inception, and have seen many of them in the hands of skillful sportsmen; know that the Freund Improved Rifle has never been equaled.

—Thomas Moore
Chief Packer Military Division of the Missouri, U.S. Army

* * *

Chas. M. Oelrichs & Co.
Cattle Dealers, Range North Platte River and Rawhide Creek.

Cheyenne, Wyo., Oct. 21st, 1883.

To F.W. Freund, Esq.
Cheyenne, Wyo. Terr.

Dear Sir:

It gives me great pleasure in saying that I have used your gun (as reissued) and it has always proved satisfactory in every respect. I think it a great improvement on the old Sharps rifle and have no doubt that it will eventually be largely used and take the place of less practical guns.

Yours truly,

Chas. M. Oelrichs.

* * *

U.S. House of Representatives,
Washington D.C.
May 23, 1884

Dear Mr. Freund:

I saw Captain Clark yesterday and had a long talk with him in relation to your matters. He tells me that he has written you fully on the subject. I am convinced that the only thing to be done under the circumstances, is that you make a carbine in accordance with his suggestions, and if it stands the test, as I have no doubt it will, the War Department will recommend an appropriation for funds to make say five hundred for a field test. With such recommendation there will be no trouble in securing the necessary appropriation.

Captain Clark tells me that some other arm will have to be substituted for the carbine now in use. I think it is a prize worth struggling for, and would suggest that you make a big effort to secure same. I shall be glad to do all I can to assist you.

Very truly yours,

M.E. Post

* * *

Cheyenne, Wyo.

Beyond a doubt Mr. Freund has invented the best rifle for use in the West. I have used many kinds of rifles, both magazines and single breech. His is superior to any and all.

—Joseph G. Pratt
U.P. R'y Land Dept.

* * *

I have always heard Mr. Freund's rifle and sights spoken of in the highest terms.

—J.H. Pratt
President Pratt & Ferris Cattle Company

* * *

U.P. Railroad Hotel, Cheyenne, W. T.,
Jan. 4, 1881

Mr. C.F. Reed says: *Am very much pleased with the new patent sights which you have recently adjusted to my rifle. For the past ten years I have been testing the different patterns of open sights, and I find that yours are the only ones that will not blur even when tried under unfavorable circumstances.*

It has long been considered desirable to have a guide in the rear sight below the notch. This has been attempted by inlaying a line of white metal, but as the line was so much in the dark, it was found to be practically useless.

You have at last accomplished the purpose by cutting a diamond-shaped opening immediately below the notch; the front sight forming the line surrounded by light, and the two angles of the opening together with the notch serving as perfect guides. The diamond-shaped opening exposes the front sight and also the object aimed at to full view, with a clearness and accuracy equalled only by the telescope. The front sight being made of spring tempered steel, is very strong, and by its peculiar construction shows white against a dark object and black against a light colored object.

My rifle is now as perfect a sporting gun as can be produced.

* * *

E. Remington, son of the manufacturer of firearms, at Ilion, N.Y., whose name is famous throughout the world for gun-making, had a rifle made by Mr. Freund, of this city. This gun has all the latest improvements, including the sights. After a hunting trip he wrote Mr. Freund as follows:

Buffalo, Wyo. Ter., August 30, 1883
Mr. Freund:
Dear Sir:— *I consider your rifle and sights a great success. With best regards, etc.*
I am yours, respectfully,
E. Remington

* * *

Mr. Freund's rifle and sights have the best reputation in this country.
—Richie Bros.
4R4 Ranch, Little Powder River, Wyoming Territory

* * *

[The following letter was written by Theodore Roosevelt to Frank Freund on June 2, 1891, when Roosevelt was with the U.S. Civil Service Commission.]

June 2, 1891
F.W. Freund, Esq.,
70 Montgomery St., Jersey City, N.J.
My Dear Sir:
Many thanks for sending me the plate. It seems to me to be just what I want, but I should like to ask if one thing could be done. Of course, understand that all this extra work I will pay extra for. What I want to know is if this line down the rear sight could not be made of vermilion paint. I have an idea that a line of vermilion would serve all the purposes of the white line and yet would not be so apt to glare or dazzle the eye and would not seem to fade into the front sight in certain light. If possible I want my rifles at 689 Madison Avenue before June 12th, as I will then want to take them out to the country to try them. Hereafter I shall always deal with you directly and not through an intermediary. Do you ever load cartridges, etc.?
Sincerely Yours,
Theodore Roosevelt

* * *

Have used [a Freund] *rifle for five years. 'Tis the best I ever had.*
—R.M. Searight
Stock Grower, Cheyenne River

* * *

The Freund Rifle is the best in Wyoming and Nevada.
—Jno. Sparks
Rancho Grande, Nevada

* * *

Gen. W.E. Strong, of Chicago, is president of the Peshtigo Lumber Company. That gentleman writes as follows, regarding the sights:

Mr. Freund,
Dear Sir— *I have just returned from a hunting expedition on the head waters of the Peshtigo river, in Northwestern Wisconsin, and during the trip I tested thoroughly your patent sights recently put on my Winchester rifle, model of 1873. In my opinion, your sights are by long odds, the best ever invented for hunting purposes. From my earliest boyhood I have been fond of a rifle, and have used one a great deal, but I never had a set of open sights that suited my fancy in every particular until I got yours.*
On my recent trip, the first two shots I had were at a running deer; one struck the right foreshoulder, and the other passed through the deer's heart. I also fired at a target from 50 to 150 yards, and I was astonished and delighted at the almost perfect sighting of the rifle and the certainty with which I could strike the object aimed at up to 150 yards by taking a fine or coarse bead through the sight and without raising the graduated leaf.
I have shown the sights to a number of excellent hunters and expert shots, and all pronounce them the most perfect they have ever seen.

* * *

We have a few of Freund's patent rifles in our outfit, the best we ever used.
—Seabury & Gardner
Ranch and Range, Bear Creek, Wyoming

* * *

(What Gen. Philip Sheridan says:)
Headquarters Military Division of the Missouri,
Chicago,
January 23, 1878.
The Sharps rifle, in which, at my request, you inserted your patent breech-block, pleases me very much, and the more I see

of it the more strongly am I impressed with the practical ideas you have developed in your invention.

* * *

I can highly recommend Mr. Freund's inventions and improved sights, as I have used them and found them all he claims for them.

—D. Shudy
Ranch Camp, Clark, Nebraska

* * *

F. Schweickert
Dealers in Hardware, Stoves and Tinware.

Cheyenne, Wyo., October 13th, 1883

To Whom It May Concern:

I wish to state that in my experience of over eighteen years in the hardware business, I have hand held many different kinds and brands of guns and ammunition and have come in contact with all classes of people. From my own experience and the information gathered from hunters, miners, sportsmen and stockmen, I have come to the conclusion that the very best gun for all such to handle is Mr. Freund's patent rifle. It does not need the repairing and adjusting that other guns do from the hands of the manufacturer, and if made in large quantities, it will undoubtedly outsell almost any gun in this section of country. It doesn't require one skilled nor an expert in the gun business to sell the gun. It sells itself. It will find a ready sale if given to the trade, on account of its many valuable qualities and will hold its own and more with any gun now upon the market. No agents nor drummers are required to sell the Freund Rifle; it has already a high and wide reputation. It will receive a continuous and steady demand and pay the manufacturer a good profit. I hope that I will see the day when this superior rifle will be sold to the trade, and I believe that when once upon the market, no dealer can afford to be without the Freund Rifle in stock. Another great merit of Mr. Freund's rifle is in its simplicity of construction; any person can take it apart and put it together, all without any previous instruction, and even if he had never seen a rifle before. This will be appreciated when one is in a hurry. In the store or on the range the breech mechanism of the rifle, in a moment's time, can be taken apart and thoroughly and easily cleaned and as quickly and easily put together.

I wish to say in regard to Mr. Freund that I have known him for nearly twenty years. He is a thorough gunmaker *having had great experience in some of the best gun factories in the world—Vienna, Paris and London—and also with the Remington's at Ilion, N.Y., all before coming West, and is well acquainted with all systems of gun making, both by hand and*

machine. I have also found his judgement the best in regard to good ammunition. He is thoroughly reliable, which can be and is attested by all his neighbors and all citizens of Cheyenne. He has a wide acquaintance all through this Western country and this fact alone would insure a large sale for his rifle—even if not pushed in any other way by agents, etc.

I am, very respectfully,

F. Schweickert

* * *

A.G. Spalding & Bros., Guns & Sporting Goods,
108 Madison St., Chicago,
Jan. 23, 1884

F.W. Freund, Esq.
Cheyenne, Wyo.

Dear Sir:

We shipped you by express, today, the two rifles sent us for inspection. We consider them the best single shot rifles, by all odds, that we have ever seen. Since we have had them, we have placed them for examination in the hands of the leading rifle men of Chicago, and the verdict has been the same in every case: "Superb and safe."

Your sights have attracted a great deal of attention and have been unanimously pronounced the best hunting sight ever placed on a rifle.

We sincerely trust you may engage capital in the manufacture of these rifles by machinery, as we feel certain it would result in their extensive sale, and at remunerative prices. Placed on the market at prices at which the masses could reach them, they could not fail to be a success, as they would virtually be without competition.

Congratulating you on your success in improving these weapons to their present high standard, we remain,

Very truly yours,

A.G. Spalding & Bros.

* * *

Office of Ed. F. Stahle, Surveyor and Engineer,
United States Deputy Mineral Surveyor,
Surveys of Mines, Ranches, and Ditches a Specialty.
P.O. Box 358
Cheyenne, Wyo.
Oct. 27th, 1883.

To All Whom It May Concern:

This is to certify that having used the "Freund Patent Rifle" for the past two years I am thoroughly acquainted with the same.

For durability, simplicity of action, beauty of workmanship, and strength in all its parts, it far excels any gun ever made for the hunters or sportsmen in the West.

In the Spring of 1882 I paid $120- for one of these Rifles, and after one year's use I sold it for the same price.

The one I am now using I paid $160- for and would not care to sell it at any price.

The double extractor and breech block certainly supply a long felt want in rifles and it cannot get out of order.

His sights are certainly perfection, for in shooting at larger game, the opening affords a person sight of the whole object.

Having come in contact with a great many hunters, I know positively that could this rifle be bought for a lower figure it would be preferred to all guns now in use, as in no respect does it lack the requirements of western hunters and Sportsmen. In conclusion I will say that it is one of the most perfect pieces of mechanism ever attained in the way of a rifle.

Respectfully,

Ed. F. Stahle

* * *

Freund guns are the best we ever used.
—Thos. Sturgis,
(Sec'y. Wyoming Live Stock Association,
Pres't. Stock Growers' National Bank)
—W.C. Lykins,
(Cheyenne, late Stock Inspector of Wyoming Territory)

* * *

General P. H. Sheridan's Gun
With Freund's Improvements

We stepped into Freund Bros.' gun store to inspect this much talked of gun, and found it to be a splendid piece of mechanism, not only in the material, but in the beauty and simplicity of the improvements, which are the following: first, for forcing the cartridge into the chamber of the gun by the closing of the breech; second, it "half-cocks" the hammer by opening or closing the breech; third, it has two extractors working simultaneously; fourth, the construction of the firing pin; fifth, the construction of all these movements with the guard lever. The object of having the breech with the improved movement is to prevent any accident by the premature discharge of the cartridge by forcing it into the chamber with the breech-block or clubbing it in with a stock or hammer when the gun becomes foul from recent discharges. The improved movement will force the cartridge into the barrel by the simple pressure on the lever, even if the cartridge does extend out quite a distance, so that the gun can be fired an unlimited number of times without cleaning.

The half-cock improvement is also done with the movement-lever, which, when it opens and closes, sets the hammer on a safety notch, and in no case can the hammer set on the firing pin, as in the old style of rifle which is liable to a premature

discharge by a sudden jar on the ground or by any other causes that may give a sudden jar or blow on the hammer; but prevents all accidents to which other guns are subject between the act of loading and firing, and obviates the necessity of the operator half-cocking the gun, where in a case of emergency, such as an attack by Indians, quick work is wanted.

For military purposes this movement cannot be surpassed, as in a battle men will become excited; some will put their guns at "half-cock", some at "full-cock", and most men will not cock them at all, and in either of the two last-mentioned cases, the arm in the hands of an excited man becomes dangerous either to himself or those about him, as it is impossible for an officer to watch the guns of his men continually to see that they put them on the "half-cock" before loading. We will further illustrate: An officer gives the order to a company of soldiers, either of infantry or cavalry, to "load", so as to be prepared to use their guns at any time after the order is given. It would require a great deal of valuable time of each officer to inspect the arms of the company, to see that each man has his gun on half-cock, to prevent accident; but by the above improvement on the Sharps rifle, the operation of loading puts the gun on half-cock, and so it remains until it is again cocked for the purpose of firing. It also has two strong and simple extractors, which work simultaneously. They take hold of the cartridge on each side, and by the guard lever power (which is the greatest possible power) extract the cartridge. These extractors are placed on each side of the chamber and fit in close, and no openings are left, as in the old gun, so that no dirt or dust can enter the barrel of the gun by the breech.

The firing pin has a gas cup that prevents the force of the gas escaping from the cap in case of a breakage of the cap in the cartridge.

All the above improvements are combined with the guard lever of the Sharps rifle, which is constructed with a link giving the whole movement a power that cannot be surpassed on a scale so small as a gun breech. The resisting power of the Sharps breech, against a large charge of powder, is greater than any other gun now made, and the above entire mechanism attached to a Sharps .40-calibre rifle will surpass any military gun in the world. All the work on this gun displays a perfect knowledge of the art of gun making, and cannot be excelled for the grand essentials of strength, effectiveness, simplicity and beauty.

The Freund Brothers inform us that they will have this magnificent rifle on exhibition at their armory for several days. Of course, they will be glad to welcome sportsmen and others interested, and exemplify the advantages of the arm that has already received the highest possible compliment—that of being adopted by such eminent authority as the Lieutenant-General of the United States Army.

* * *

Evanston, Wy. Ter.,
July 28th, 1878

Messrs. Freund Bros., Cheyenne, Wy. Ter.:

Dear Sirs:— *The 40 cal. long-range rifle I received of you is the strongest shooting gun I ever saw. We gave it a splendid trial on July 4th, when several of us shootists went up to Echo, and were surprised over the accurate shooting of the gun, and the truly wonderful working of your breech mechanism. We tried the gun over 1,600 yards, and found that it carried up accurately, and all present came to the conclusion that the rifle with your improvement was a great achievement.*

The wedge motion of the breech-block gives your rifle at least 100 per cent advantage over the Sharps perpendicular motion in sending the cartridge home. Your extractors fall back further, thereby clearing the shell entirely from the chamber, and are entirely covered, so as to exclude all sand and dirt.

We made a splendid score at a bull's-eye, 109 inches square, 300 yards off-hand with your gun, three of us, shooting five shots each, making nine bull's-eyes out of fifteen shots, at 1,600 yards, with peep sight. We made some excellent shots; the most of them would have killed a deer, as they struck in a space the size of one.

The Editors Evanston Age, *also the* Argus *and Ogden* Freeman, *of Utah were there. These gentlemen, and all others present, expressed themselves in the most favorable terms, and agreed that your gun is the most perfect of all they ever saw used on trial shooting.*

Yours, &c.,

A.S. Schoemaker

* * *

Mr. Wm. Taylor, Quartermaster Agent of Rock Creek, W.T., says the following:

Mr. Freund,

Dear Sir— *You are aware I have a beautiful rifle, but I am free to say, for all practical purposes and beauty, your new sights have increased its value at least fifty per cent.*

How annoying it is to have a good shooting rifle with the old blue sights over which it is necessary to have the sun and everything else favorable to enable you to see the game for a dead shot. I find it easy to catch the object over with your sights in any direction and under the most unfavorable circumstances. No good shot, after seeing your new sights, would do without one.

* * *

No better gun or sights can be found than those of Mr. Freund. It includes all strength, reliability and simplicity which can be confined in the smallest space, as that of any gun; nor have the sights any equal. I have used one for five years. It is my pet.

—Maj. John Talbot

(Major Talbot has served in the Regular Army for a number of years, and is well known in the Army, and is considered as one of the bravest Indian fighters known. He is a very fine rifle shot, and can be equaled by only a few in pistol shooting.)

* * *

Freund guns are the best we ever used.

—Tischemacher & DeBillier
Stock Growers, Range North Platte, Wyoming

* * *

(From a letter of Col. F. VanVliet, Capt. Third Cav., Br't. Lt. Col. U.S.A.)

Fort D.A. Russell

Mr. Freund.

Dear Sir— *I used your improved sights on my last hunting trip and found them perfect, both for hunting and target practice. The front sight gives a perfect bead, and has the peculiarity of showing dark against light objects and light against darkness. The rear sight is extremely clear, and is the only one I have ever seen that at the time of taking sight allows you a clear view of your game.*

* * *

Cheyenne & Black Hills Stage & Express Co.
Cheyenne, Wy.
Oct. 15, 1883

To Whom It May Concern:

Having had great use and good advantages for testing F.W. Freund's improved firearms, as we used them or had our messenger *use them* against the Indians, Road Agents *and Robbers on all our extensive staging roads in the* Great West *including the* Black Hills *country. Our messengers having numerous fights with the Indians and Road Agents, they always protected our coaches as they found the* Freund Improved Rifle *was superior to any other of any kind. If it could be manufactured at a reasonably cheaper rate, it would be universally used.*

Respectfully,

Luke Voorhees, Supt.

* * *

Mr. Freund's gun is the best we know of by reputation and trial.

—Wyoming Meat Co.

* * *

I hereby gladly recommend Mr. Freund's rifle, especially as to sights.

—Chas. Witsuk
Steward County Hospital, Laramie County, Wyoming

* * *

I have used one of Freund's guns on my recent trip to Yellowstone National Park, and found it first-class.

—D.H. Wallace
People's Bank, New Castle, Pennsylvania

* * *

Mr. Freund's gun is the best, we know.

—Big Horn Cattle Company (A.J. Winn, D.H. Reynolds)

* * *

Mr. Freund's rifle is the best single barrel rifle I have ever seen.

—J.H. Douglass-Willan
Douglass-Willan, Sartoris & Co.

* * *

So say I.

—W.A. Wyman, M.D.
Wyoming

* * *

(From a letter from Mr. Zimmerman, a gun maker of long and practical experience in Europe and America.)

Dodge City, Kan., March 2, 1880

From the first breech-loading needle gun of Prussian manufacture, to the very last of the new patterns today, your improved Sharps towers above them all, where the points of value are settled. An arm, to be effective these days, must possess the highest grade of shooting qualities, combining therewith these three other imperative qualities, to-wit: rapidity of action, durability, and safety. The improvement of yours on the Sharps covers all three requirements. It combines all these points, which thousands of gunsmiths have tried to harmonize, and failed. The arm pleases every one who sees it, and attracts the undivided attention of stockmen, officers of the army, and hunters.

Maj. Smith, of Fort Dodge, is well pleased with the rifle you sent him. All the officers at the Fort examined the arm, and were gratified with it in every particular. The breech mechanism was, they said, well adapted to hard service. They praised the fine workmanship upon the arm, and expressed themselves astonished that the Sharps Rifle Company did not adopt the improvement at once, instead of making the hammerless gun now turned out by them, which, in everybody's opinion in this section, is as worthless as it is hammerless.

* * *

The Sharps Rifle—Freund's Improvement
What An Old Hunter Says About It.

I am a hunter by profession, if such an occupation can be dignified with such a title. As it requires large and varied experience, coupled with no small degree of skill, acquired at the expense of arduous and perilous labor, to entitle one really to such a distinction, I think I may classify my occupation among the professions.

I have hunted on "the plains" and in the Rocky Mountains for seventeen years. The objects of pursuit have varied with the seasons through all the gradations of game, from the "jackrabbit" to the grizzly bear and buffalo. I hunt not only for sport, but also for the purpose of pecuniary gain; so that I am interested in all the latest improvements in gun manufacture, and endeavor to keep posted in regard to the most important changes. Knowing the requirements of a good weapon, I can easily distinguish between the catch-penny patent (styled an "improvement"), and an alteration or addition which accomplishes a valuable purpose. Of course, I am speaking more particularly with regard to the demands of the Frontiersman and mountaineer. What would answer admirably in "the States" for shooting quail, geese and ducks, would be poorly adapted to an encounter with a mountain lion or a grizzly bear. I have hunted and killed not only these, but also antelope, deer, elk, buffalo and the "bighorn" or mountain sheep, which last are so numerous in some parts of our country that I have ranged over as to give geographical name to an extended and remarkable range of mountains—the Big Horn mountains. I have hunted from Montana to Arizona, through all the intervening territory. Nor do I confine myself at all to Spring and Summer, but am engaged as assiduously in the depth of Winter, killing game and curing large quantities of meat, which I haul to market in the spring.

Many men call themselves hunters who go out only in fine weather, in the Fall, for "sport." Of course, such are easily deceived by the appearance of a gun, not requiring, as I do, to know the points of excellence or weakness. My gun is my best friend, serving as my means of subsistence and being my most reliable dependence in the moment of danger; so that, as another one loves his horse or his dog, I love my gun, and when I shoot a good one I know it as well as the manufacturer. I have tried quite a variety in my experience—the muzzle-loader, the fine English breech-loader, the Winchester, the needle-gun and the Remington—but the best, most complete and reliable gun I ever used is Freund's improvement on the Sharps rifle—in short, it is the "chief" for the Rocky Mountain region, and all hunters should have one at whatever cost. I consider Freund's breech the finishing touch on the Sharps gun, forcing in the cartridge with the greatest facility, and, at the same time, half-cocking the piece. With its double-extractor, and its perfect adjustment and finish, excluding all dirt, there is nothing more to be desired. Being closed in the rear, and operating without any obstruction and with unerring precision, it obviates the necessity for cleaning until an almost unlimited number of discharges have been made. The needle-gun was a vast improvement over the now discarded muzzle-loader, but the weakness and imperfection of the breech mechanism have rendered them unreliable, and even dangerous to the operator. For the purpose of consuming the old stock of muzzle-loading barrels, it was an excellent device, but since heavier charges are demanded, they have become unserviceable.

—Harry Yount

APPENDIX II:
FREUND FAMILY TREE

Frank W. Freund
1837–1910

Clotilda Gasperrini
1854–1941

William F.
1879–1926

Elsie
died age 11

Clotilda
died as infant

Angie
1881–1954

George
died as infant

Joseph
1891–1945

Stella
1894– ?

Jeanette
1897–c. 1960

George Freund
1840–1911

Ida Gasperrini
? – ?

Clarence Jethro
c. 1891– ?

INDEX

A

Amber, John   344
American Field Magazine   230, 231, 237
"American Frontier" Freund-marked Rifles   211, 216, 226, 320, 322, 325, 327, 337, 336, 338, 339, 340, 341, 343, 346
Andrews, A.J.   119
Appel, Morris   324, Appendix I
Arapahoe Indians   28
Ashworth, Richard   344, Appendix I
Atlantic City, Wyoming Territory   27
Axtell, Tom   238

B

Baker, Jim   48
Baker, N.A.   13
Ballard Rifle Company   89
Ballard Rifles   18, 74, 76, 215
Barnum's New York Museum   39
Bear River City, Wyoming Territory   18, 19, 20
Beecher, Lt. Frederick H.   345
Beecher, Reverend Henry Ward   345
Benett, Mr.   249
Benton, Wyoming   16
Big Horn, Wyoming   317, 330
Bixby, Lawrence   328
Black Hills, Dakota Territory   51, 53, 54, 206
Black Hills, Wyoming   16
"Boss Gun"   211, 317, 320, 323, 325, 327, 333
Bradley, P.C.   231, Appendix I
Brigham City, Utah Territory   23
Brooklyn Navy Yard   103
Bronson, Edgar Beecher   345, Appendix I
Bronson, W.N.   69
Brookhard, Senator   244
Brown, Charley   55
Browning, Jonathan   1, 20, 31, 33
Browning, John Moses   20, 31
Bryan City, Wyoming Territory   16
Bryson, T.E.   350
Burdet, Thompson & Law   243
Burnam Company   245
Burnett, Edward   71, 72

C

Campbell, Henry and James   55
Carissa Mine   27
Carlisle Ranch   354
Casement, Dan   11, 12, 18
Casement, General Jack   11, 12, 13, 18
Catalogs, Freund Brothers
    Christmas, 1879 Catalog   250, 253-272
    Colorado Armory Catalog No. 389   287-306
    F.W. Freund Patent Rifle Catalog   273-282
    Hunter's Sights Catalog   283-286
Central Pacific Railroad   9, 20, 25
Chester, E.L.   332
Cheyenne Armory   16, 23, 27, 327
Cheyenne *Sun*   91
Cheyenne *Daily Leader*   13, 14, 53, 57, 86

Cheyenne, Dakota Territory   3, 14, 26
Cheyenne-Deadwood Stage Line   315
Cheyenne, Wyoming Territory   8, 18, 24, 30, 33, 38, 45, 49, 51-99, 113, 208, 309
Clark, Captain William Philo   94, 217, 219
Cody, William F. "Buffalo Bill"   336
Collinson, Frank   72
Colorado Armory   35, 115-120, 122, 124, 125, 224, 240
Colorado Industrial Association   43
Colt Revolvers   16, 40, 57, 236, 309, 350
Cooper, Walter   237, 338, 341
Corinne, Utah Territory   20, 22, 23, 25, 30
Crazy Horse, Chief   94
Crook, General George   74, 87, 330, 331, Appendix I
Custer's 7th Cavalry   215
Custer's Black Hills Expedition   51

D

Daily Rocky Mountain News   38, 43-45, 49
Dale City, Dakota Territory   15, 16, 27
Dale Creek Bridge   15, 16
Dammann, F.A.   56, 74, 76-79, 81, 341
Davis, N.R. & Company   79, 81, 341, 342
Davis, Polk, Wardwell, Gardner and Reed   245
Dawson, Wyoming Territory   347
Delaney, W.   339, Appendix I
Denver, Colorado   20, 33-50, 310, 312
Denver Deutscher Schuetzenverein   43, 205
Dodge, Major-General Grenville   11
Dodson, W.T.   236, Appendix I
Driscoll, J.W.   332
Dupont, E.I. deNemours & Company   10, 13, 22, 24, 26
Durango, Colorado   44, 83, 112-125, 224, 240, 354, 356

E

Easton, Pennsylvania   101
"E.E.G."   340

F

"False-Hammer" Cavalry Pattern Freund Carbine   217, 218, 221, 351
"False-Hammer" Freund Rifle   217, 221, 222, 351, 352, 353
"Famous American Frontier" Freund-marked Rifles   211, 315, 322, 326, 333, 341, 342, 345, 354
Farris, E.W.   333
Fisher, Homer   342
Flayerman, Norm   238
Forester, Baron Esmond de   5
Fort Lewis, Colorado   114, 115
"Freund American Frontier"-marked Rifle   344

Freeman, Leigh   19
Freund and Brother/Freund & Bro.   10, 16, 20, 21, 24-30, 35, 38, 39, 41, 44, 45, 49, 58, 63-66, 68-70, 72, 74, 205, 211, 222, 223, 234, 317, 322, 325, 335
Freund Bros. Manufacturing, Cheyenne   317
Freund Brothers   14, 16, 22, 23, 25, 35, 38, 41, 43, 44, 50, 51, 53, 54, 57, 74, 76, 80, 81, 84, 205, 209, 222, 309
Freund Family
    Freund, Angie   90, 91
    Freund, Carl   2, 4
    Freund, Christian   2
    Freund, Christoph Wilhelm   2
    Freund, Clarence Jethro   116
    Freund, Clotilda Gasperrini   56, 57, 69, 70, 73, 77-87, 90-92, 98, 101, 103, 109, 111, 114, 116, 243, 245, 246
    Freund, Elsie   5, 90
    Freund, Frank William   1, 5-7, 9, 12, 14, 18, 21-23, 31, 32, 34-36, 43-46, 52, 54, 55, 60, 61, 64, 65, 71-74, 77, 79, 80, 81, 83, 85-94, 97, 98, 100-111, 113, 114, 121, 124, 205, 207-209, 211, 213-215, 217, 218, 222, 223, 225-227, 229-234, 236, 237, 239, 241, 243, 245-247, 310, 312, 313, 319
    Freund, Georg C.   2, 3
    Freund, George   5, 12, 18, 23, 35, 36, 43, 44, 52, 61, 74, 80, 81, 84, 85, 90, 112-115, 117-119, 121-125, 205, 207, 224, 225, 236, 240, 241, 310, 354, 355
    Freund, Ida Gasperrini   85, 116, 117
    Freund, Jeanette   5, 90, 103, 106, 109, 111, 243, 245, 246
    Freund, Jean Louis   2
    Freund, John Christoph Wilhelm   2
    Freund, Johannes   2
    Freund, Louise   5
    Freund, Ludwig   2
    Freund, Maria Sitze   2, 5
    Freund, Stella   103
    Freund, Wilhelm   2, 5
    Freund, William   5, 76, 91, 109, 111
Freund Model 1879 Rifle   332
Freund's Castle, Wyoming   71, 72, 111
"Freund's Inventions" marking   325
Freund's Patent   339, 346, 351
Freunt, Ulrich   1
Frontier Index   19
Furstenau, Germany   2

G

Gallatin, E.L.   223
"General" Freund   36, 37, 52, 70
Gerdel, Henry   317
Goodman & Franks   28, 29
Gove, Carlos & Company   49
Greener Firearms   39
Green River, Wyoming Territory   16
Grinnell, George Bird   47, 206

H

Hamilton City, Wyoming Territory 27
Hanna, Oliver 71, 72
Hanson, Margaret Brock 71
Hanson, L.P. 107
Harris, E.S. 41
Hartley & Graham 89, 99, 220, 221, Appendix I
Hayes, President Rutherford B. 330
Heidelberg, Germany 2, 113
Henry Repeating Rifle 16
Hull, A.C. 17
Hull, F.A. Jr. 68
Hull and Belden Co. 68
Hunting at High Altitudes 206
Hyde, Frank 342
Hygienic Institute 45

I

"Improved Old Reliable" Sharps Rifle 332
Inter Ocean Hotel 70

J

Jackson, William Henry 15, 23, 25
Julesburg, Colorado Territory 12-14

K

Keegan, Angie Freund 111
Keegan, John 90
Kelton, General John C. 244
Kirkland, C.D. 78
King, Clarence 348, 349, Appendix I
King, Nelson 329
Kittredge, Benjamin and Company 41, 320
Kupfer, John 23

L

Laramie City, Wyoming Territory 15-18, 24, 30, 234
Lawrence, R.S. Sight 239, 328
Lee Fire Arms Company 22
Little Big Horn, Montana, Battle of 215
Little Wolf 94
Lorillard, Pierre 212, 341, Appendix I
Lovell, J.P. 41, 321
Lower, John P. 49, 50, 316, 317, 322, 355, 356, Appendix I

M

Mackey's *Masonic Ritualist* 22
Manifest Destiny 9
Martin, Colonel Luke 14
Masonic Order 22, 28, 31, 32, 106, 108, 109, 123
May, Boone 315, Appendix I
Mayer, Frank H. 114
Maynard Rifles 47, 207
McKenzie, Lynton 344
McLaughlin, (hunter) 55
Meigs, Montgomery C. 319
Meline, J.A. 55, 214, Appendix I
Miller Brothers 343
Milner, Moses "California Joe" 47, 206
Miner's Delight Mine 27
Miner's Pocket Knife 122
Missouri River 9, 10
Mitchell, J.S. 85
Moore, R.L. Jr., M.D. 43, 78, 83, 238
More-Light Sight, Freund's Patent 81, 83, 114, 120-125, 224, 227, 228, 240, 241, 316, 318, 321-323, 325, 326, 335, 337, 342, 344, 346, 348, 356
Munsell, H.M. 97, 220
Murphy, Wm. G. 12

N

National Armory, Springfield, Mass. 243, 244
Nebraska City, Nebraska Territory 9-11
Nebraska City *News* 9, 10
Nebraska Land Company 328
Nebraska *Statesman* 9
Neinsh Company 248, 249
North Platte, Nebraska Territory 11, 12, 14

O

Ogden, Utah Territory 20, 22,
"Old Reliable" Sharps Rifles 71
Oregon Trail 27

P

Patents, Freund
153432 44, 126, 128, 129
160762 45, 126, 130-183
160763 45, 126, 134, 135
160819 45, 126, 136, 137, 225, 226
162224 45, 126, 138-142
162373 45, 126, 143-145
162374 45, 126, 146, 147
168834 126, 148, 149, 243
180567 54, 126, 150-152, 214, 328
183389 54, 126, 153-156, 222
184202 54, 126, 157-164, 215
184203 54, 126, 165-167, 215
184854 54, 126, 168-171
185911 54, 126, 172-175, 214, 246, 328
189721 54, 126, 176, 177, 225, 227
211728 76, 126, 178-160
216084 54, 126, 181-183, 214
229245 81, 121, 184-186, 227
268090 90, 126, 187, 188
273156 126, 127, 189, 190
289768 126, 127, 191, 192
292642 126, 127, 193-195
297375 126, 127, 196, 197
313414 126, 127, 198, 199
496051 93, 126, 200, 201, 227
D22406 121, 126, 202, 203
Pershing, General John 70
Picard, Mr. 10
Pioneer Gun Store 21
Pond, C.H. 247
Post, M.E. 93, 94, 218, 219, Appendix I
Powder River country 71
Promontory Point, Utah Territory 9, 11, 22, 23, 25

R

Ramsey, C.R. "Bull" 333
"Red Jacket" Whiskey 23
Reed, General S. 12
Remington Arms Company 7, 39, 46, 47
"Remington Cast Steel" Barrel Marking 340
Remington, Eliphalet 7, Appendix I
Remington Percussion Revolvers 16
Remington Rolling Block Rifles 45, 46, 226, 311, 312
Remington Rifles 63, 205, 207, 206
Reynolds, Charley 53
Richards, Barlett 328
Richards & Co. 95, 219
Rifle, The 237
Rocky Mountain Directory and Colorado Gazetteer 35
Roosevelt, Theodore 102, 105, 204, 222, 223, 237, 325, 352, 353, Appendix I
Russell, Andrew J. 11, 13, 19
Russell, Majors & Waddell 11

S

Salt Lake City, Utah Territory 20, 21, 22, 24, 30, 234
Sargent, Frank P. 46
Schafer, Harold 222
Schenk, Th. 22
Scheutzen Rifle Matches 310
Schmidt, John V. 7
Schuyler, Hartley & Graham 41
Schuyler, Phillip W. 346, Appendix I
Seymour, George D. 247-249
Sharps Percussion Carbines 313, 314
Sharps Rifles 18, 45, 47, 54, 55, 57, 71, 72, 74, 75, 246
Sharps Rifle Manufacturing Company 35, 38, 39, 41, 43, 48, 58, 60-66, 68, 72, 85, 89, 96, 111, 204, 207-209, 211, 213, 217, 246, 247, 313, 329, 331
Shaw, W.E. 332
Sheridan, General Philip 87, Appendix I
Sherman, Wyoming 16
Simonds, W.E. 246, 247
"Skull-and-Crossbones" George Freund Trademark 355, 356
Smelter City Spoon 120
Smith, Duane A. 115
Smith & Wesson Pistols 57
Smithsonian Institution 206
South Pass City, Wyoming Territory 27, 28, 30-33
South Pass *News* 27, 28
Sovitz, Germany 113
Spade Ranch 328
Spencer Repeating Rifles 16
Spies, Kissam & Co. 41
Sportmans Depot 38, 50
Spring Gulch, Wyoming Territory 27
Springfield Rifles 12, 47, 48, 206, 215, 217, 244
Squires, H.C. 323
Stahl(e), Ed F. 338, Appendix I
Stanley, G.W. 342, Appendix I
Stetson, Jennings, and Russell 245
Stevens "Bicycle Rifle" 236, 238
Stowe, Harriet Beecher 345
Strong, General W.E. 88, Appendix I
Sweetwater Armory 16, 27, 28, 30
Sweetwater County, Wyoming Territory 27
Sweetwater Mines, The 16, 27

U

Union Metallic Cartridge Company 84, 97
Union Pacific Railroad 9, 11-18, 20, 24, 25, 49, 111, 205, 234
U.S. Military Railroad Construction Corps 11
Utah Central Rail Road 20

V

Vaughn, J.A. "Tennessee" 332
Virginia City, Montana Territory 31, 32

W

Warren, Francis E. 70, Appendix I
Webber, Ed 344
Westcott, E.G. 72-74, 247
Whell, Mr. 116
White, Jim 71, 72
Whitney Rifles 47
Williams, Rodger D. 47, 48, 206
Wilson, R.L. 350
Winchester Patent Repeating Rifles and Carbines 14, 16, 22, 57, 236
Winchester Repeating Arms Company, New Haven, Connecticut 14, 24, 26, 39
Wyoming Armory 8, 14, 51, 52, 54, 56, 57, 78, 83, 88, 92, 96, 111, 114, 211, 350, 351, 354
Wyoming Saddle Gun 215, 216, 348
Wurtembergishes Urkandenbuch 1

Y

Yankton, Dakota Territory 14
Young, Brigham 20

Z

Zimmerman, F.C. 60, 61

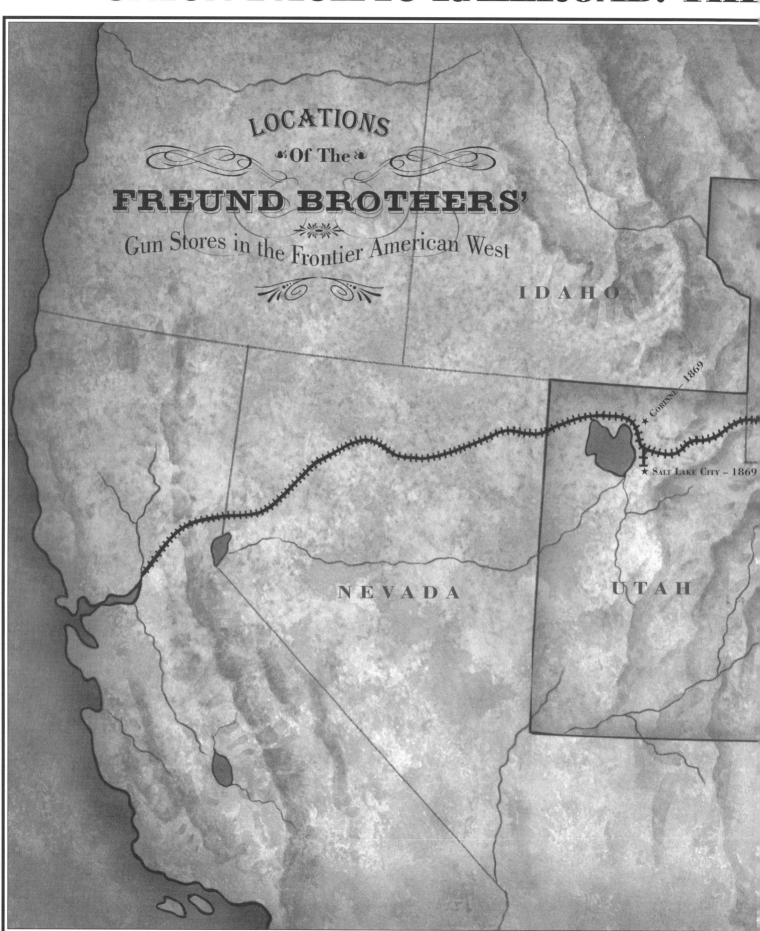

LOCATIONS
Of The
FREUND BROTHERS'
Gun Stores in the Frontier American West

IDAHO

Corinne — 1869

★ Salt Lake City — 1869

NEVADA

UTAH